Compendium of Plant Genomes

Series Editor

Chittaranjan Kole, Raja Ramanna Fellow, Government of India,
ICAR-National Research Center on Plant Biotechnology, Pusa,
New Delhi, India

Whole-genome sequencing is at the cutting edge of life sciences in the new millennium. Since the first genome sequencing of the model plant Arabidopsis thaliana in 2000, whole genomes of about 100 plant species have been sequenced and genome sequences of several other plants are in the pipeline. Research publications on these genome initiatives are scattered on dedicated web sites and in journals with all too brief descriptions. The individual volumes elucidate the background history of the national and international genome initiatives; public and private partners involved; strategies and genomic resources and tools utilized; enumeration on the sequences and their assembly; repetitive sequences; gene annotation and genome duplication. In addition, synteny with other sequences, comparison of gene families and most importantly potential of the genome sequence information for gene pool characterization and genetic improvement of crop plants are described.

Interested in editing a volume on a crop or model plant?
Please contact Prof. C. Kole, Series Editor, at ckoleorg@gmail.com

More information about this series at http://www.springer.com/series/11805

M. K. Rajesh • S. V. Ramesh •
Lalith Perera • Chittaranjan Kole
Editors

The Coconut Genome

Springer

Editors
M. K. Rajesh
ICAR-Central Plantation Crops
Research Institute
Kasaragod, Kerala, India

S. V. Ramesh
ICAR-Central Plantation Crops
Research Institute
Kasaragod, Kerala, India

Lalith Perera
Coconut Research Institute
Lunuwila, Sri Lanka

Chittaranjan Kole
ICAR-National Institute for Plant
Biotechnology
New Delhi, India

ISSN 2199-4781 ISSN 2199-479X (electronic)
Compendium of Plant Genomes
ISBN 978-3-030-76648-1 ISBN 978-3-030-76649-8 (eBook)
https://doi.org/10.1007/978-3-030-76649-8

This Springer imprint is published by the registered company Springer Nature Switzerland AG
The registered company address is: Gewerbestrasse 11, 6330 Cham, Switzerland

This book series is dedicated to my wife Phullara and our children Sourav and Devleena

Chittaranjan Kole

Foreword

Coconut (*Cocos nucifera* L.) is often referred to as the "tree of life" owing to its multitude of uses such as a source of vegetable oil and raw materials for a plethora of industries, housing materials, cosmetics, detergents, among other products, in addition to improving the aesthetics of landscapes. The palm is inextricably linked to the people's history and lives in the tropical and subtropical regions in the general and fragile island and coastal ecosystems. The crop is progressively recognized as an invaluable functional food and treasure trove of nutraceuticals worldwide due to the reputation of products such as virgin coconut oil, tender coconut water, and inflorescence sap.

Despite the fact that the rapid strides of R & D led improvements in crop genomics, it is rational to acknowledge the fact that perennials like coconut, compared to the annuals or seasonals, warrant altogether different crop management strategies. Further, coconut has been the source of revenue of 20 million farmers and their dependents worldwide, and hence unforeseen crop failures pose serious threats to the livelihood security of these mostly small and marginal growers. Genomics-assisted breeding has accorded a giant leap in the crop improvement programs of cereals, pulses, and other annuals. In contrast, the application of genomic science in enhancing perennials such as coconut palm has been severely lagging for want of helpful and practical palm genomic resources.

In view of this, the book entitled "The Coconut Genome" under the series "Compendium of Plant Genomes" is a well-timed introduction to the growing body of the scientific literature on palm-based biotechnology and genomics. The editors have done a tremendous job of bringing in experts who have vast experience in coconut genomics and biotechnology to author the chapters.

I express my sincere wishes to the editors for taking untiring efforts to bring out this compilation with the state-of-the-art information on coconut genomics. I am hopeful that this compilation would serve the need of young researchers and students, who wish to have a glimpse of the accomplishments in coconut genomics and enlighten them in visualizing the way forward. Undoubtedly, this book would serve and be of utmost value to all the biologists involved in palm breeding, biotechnology, genetics and cytogenetics, and other related disciplines. I wish that this book would lay a strong foundation for stimulating further discussions in palm genomics and contribute to the development of coconut varieties with desired traits to meet the ever growing demand for coconut-based products.

Jelfina C. Alouw
Executive Director
International Coconut Community (ICC)
Jakarta, Indonesia

Preface to the Series

Genome sequencing has emerged as the leading discipline in the plant sciences coinciding with the start of the new century. For much of the twentieth century, plant geneticists were only successful in delineating putative chromosomal location, function, and changes in genes indirectly through the use of a number of "markers" physically linked to them. These included visible or morphological, cytological, protein, and molecular or DNA markers. Among them, the first DNA marker, the RFLPs, introduced a revolutionary change in plant genetics and breeding in the mid-1980s, mainly because of their infinite number and thus potential to cover maximum chromosomal regions, phenotypic neutrality, absence of epistasis, and codominant nature. An array of other hybridization-based markers, PCR-based markers, and markers based on both have facilitated constructions of genetic linkage maps, mapping of genes controlling simply inherited traits, and even gene clusters (QTLs) controlling polygenic traits in a large number of model and crop plants. During this period, a number of new mapping populations beyond F_2 were utilized and a number of computer programs were developed for map construction, mapping of genes, and mapping of polygenic clusters or QTLs. Molecular markers were also used in the studies of evolution and phylogenetic relationship, genetic diversity, DNA fingerprinting, and map-based cloning. Markers tightly linked to the genes were used in crop improvement employing the so-called marker-assisted selection. These strategies of molecular genetic mapping and molecular breeding made a spectacular impact during the last one and a half decades of the twentieth century. But still they remained "indirect" approaches for elucidation and utilization of plant genomes since much of the chromosomes remained unknown and the complete chemical depiction of them was yet to be unraveled.

Physical mapping of genomes was the obvious consequence that facilitated the development of the "genomic resources" including BAC and YAC libraries to develop physical maps in some plant genomes. Subsequently, integrated genetic–physical maps were also developed in many plants. This led to the concept of structural genomics. Later on, emphasis was laid on EST and transcriptome analysis to decipher the function of the active gene sequences leading to another concept defined as functional genomics. The advent of techniques of bacteriophage gene and DNA sequencing in the 1970s was extended to facilitate sequencing of these genomic resources in the last decade of the twentieth century.

As expected, sequencing of chromosomal regions have led to generation of too much data difficult to store, characterize, and utilize with the then available computer softwares. Neverthelss, the development of information technology made the life of biologists easier by offering a swift and sweet marriage of biology and informatics, and a new subject—bioinformatics——was born.

Thus, the evolution of the concepts, strategies, and tools of sequencing and bioinformatics reinforced the subject of genomics—structural and functional. Today, genome sequencing has traversed much beyond biology and involves biophysics, biochemistry, and bioinformatics!

Thanks to the efforts of both public and private agencies, genome sequencing strategies are evolving very fast, leading to cheaper, quicker, and automated techniques right from clone-by-clone and whole-genome shotgun approaches to a succession of second-generation sequencing methods. The development of software of different generations facilitated this genome sequencing. At the same time, newer concepts and strategies were emerging to handle sequencing of the complex genomes, particularly the polyploids.

It became a reality to chemically—and so directly—define plant genomes, popularly called whole-genome sequencing or simply genome sequencing.

The history of plant genome sequencing will always cite the sequencing of the genome of the model plant *Arabidopsis thaliana* in 2000 that was followed by sequencing the genome of the crop and model plant rice in 2002. Since then, the number of sequenced genomes of higher plants has been increasing exponentially, mainly due to the development of cheaper and quicker genomic techniques and, most importantly, the development of collaborative platforms such as national and international consortia involving partners from public and/or private agencies.

As I write this preface for the first volume of the new series "Compendium of Plant Genomes," a net search tells me that complete or nearly complete whole-genome sequencing of 45 crop plants, eight crop and model plants, eight model plants, 15 crop progenitors and relatives, and three basal plants is accomplished, the majority of which are in the public domain. This means that we nowadays know many of our model and crop plants chemically, i.e., directly, and we may depict them and utilize them precisely better than ever. Genome sequencing has covered all groups of crop plants. Hence, information on the precise depiction of plant genomes and the scope of their utilization are growing rapidly every day. However, the information is scattered in research articles and review papers in journals and dedicated Web pages of the consortia and databases. There is no compilation of plant genomes and the opportunity of using the information in sequence-assisted breeding or further genomic studies. This is the underlying rationale for starting this book series, with each volume dedicated to a particular plant.

Plant genome science has emerged as an important subject in academia, and the present compendium of plant genomes will be highly useful to both students and teaching faculties. Most importantly, research scientists involved in genomics research will have access to systematic deliberations on the plant genomes of their interest. Elucidation of plant genomes is of interest not only for the geneticists and breeders, but also for practitioners of an array of plant science disciplines, such as taxonomy, evolution, cytology, physiology, pathology, entomology, nematology, crop production, biochemistry, and obviously bioinformatics. It must be mentioned that information regarding each plant genome is ever-growing. The contents of the volumes of this compendium are, therefore, focusing on the basic aspects of the genomes and their utility. They include information on the academic and/or economic importance of the plants, description of their genomes from a molecular genetic and cytogenetic point of view, and the genomic resources developed. Detailed deliberations focus on the background history of the national and international genome initiatives, public and private partners involved, strategies and genomic resources and tools utilized, enumeration on the sequences and their assembly, repetitive sequences, gene annotation, and genome duplication. In addition, synteny with other sequences, comparison of gene families, and, most importantly, the potential of the genome sequence information for gene pool characterization through genotyping by sequencing (GBS) and genetic improvement of crop plants have been described. As expected, there is a lot of variation of these topics in the volumes based on the information available on the crop, model, or reference plants.

I must confess that as the series editor, it has been a daunting task for me to work on such a huge and broad knowledge base that spans so many diverse plant species. However, pioneering scientists with lifetime experience and expertise on the particular crops did excellent jobs editing the respective volumes. I myself have been a small science worker on plant genomes since the mid-1980s, and that provided me the opportunity to personally know several stalwarts of plant genomics from all over the globe. Most, if not all, of the volume editors are my longtime friends and colleagues. It has been highly comfortable and enriching for me to work with them on this book series. To be honest, while working on this series I have been and will remain a student first, a science worker second, and a series editor last. And I must express my gratitude to the volume editors and the chapter authors for providing me the opportunity to work with them on this compendium.

I also wish to mention here my thanks and gratitude to the Springer staff, particularly Dr. Christina Eckey and Dr. Jutta Lindenborn, for the earlier set of volumes, and presently Ing. Zuzana Bernhart, for all timely help and support.

I always had to set aside additional hours to edit books besides my professional and personal commitments—hours I could and should have given to my wife, Phullara, and our kids, Sourav and Devleena. I must mention that they not only allowed me the freedom to take away those hours from them but also offered their support in the editing job itself. I am really not sure whether my dedication of this compendium to them will suffice to do justice to their sacrifices for the interest of science and the science community.

New Delhi, India Chittaranjan Kole

Preface

The monotypic genus *Cocos* belongs to the family Arecaceae (palm family), and *Cocos nucifera* L., the coconut palm, is an economically important palm species cultivated in the coastal environments of humid and sub-humid tropical regions. Botanically, coconut is an arborescent monocot that generally grows to a height of 4–25 m, bearing economic produce, its fruit—a drupe. Nonetheless, all the palm parts are utilized as food, fuel, folk medicine, cosmetics, building materials, manure, etc.Although coconut forms an integral component of folk medicine since time immemorial, the advancements in phytochemistry have lent further credence to the pharmacological and nutraceutical effects of the crop.

In the realm of science-based crop improvement programs, till recently, the palm has largely witnessed the application of conventional breeding and cytogenetics approaches with little advancement in the *in vitro* propagation techniques. However, considering the perennial nature, outcrossing-induced breeding behavior, and the consequent heterozygosity, this economically significant palm warrants exploitation of modern genomics and biotechnological tools. The advent of the genomics revolution in model plants has greatly influenced the researchers dealing with other less privileged crops to tread the path of genome-based crop improvement programs. It is an exciting time to discuss the genomics science of coconut since a handful of relatively good quality genome sequences (of both the tall and dwarf morphoforms representing diverse geographic origins) is publicly available. The genetic control of *in vitro* recalcitrance, pest and pathogen resistance, oil biosynthesis metabolic pathways, and abiotic stress tolerance to develop climate-smart coconut are some of the principal outcomes expected of genomic research in coconut. Further accelerated and targeted development of SSR or SNP molecular markers, genome sequence-based genotyping efforts, and a wide array of investigations, including genetic diversity assessment, fine-mapping, gene flow characteristics, gene–environment interactions, and exploration of novel genomic selection approaches, are possible in coconut.

Notwithstanding the rapid growth of the original scientific literature on coconut genomics, a comprehensive compilation on palm genomics is conspicuously lacking. For this reason, this volume entitled "The Coconut Genome" would fulfill the requirements of scholars involved in coconut crop improvement programs. To set the stage, the introductory chapters are dedicated to emphasizing the world economic importance of the crop, botanical description and advancements in cytology, repertoire of genetic resources, and conservation strategies being adopted. These are followed by a chapter on breeding strategies currently being followed utilizing the extant resources. In the absence of whole-genome sequences, various molecular markers, viz., RAPD, AFLP, RFLP, and SSRs, have been utilized to characterize the coconut genetic diversity, and their applicability to coconut breeding programs is discussed. Major efforts undertaken to unleash the QTLomics potential of coconut to improve agronomic traits of importance are presented in the following chapter. The crux of this volume, nuclear and organellar genomes of coconut, and efforts made in other multi-omics approaches such as metabolomics, proteomics, and ionomics are deliberated comprehensively. These chapters are followed by the chapters dedicated to exploring key traits: oil biosynthesis, aroma trait, and resistant gene analogues. To complete the volume, a chapter has been included to present the prospectives in coconut genomics, outlining the future line of work while treading genomics-aided crop improvement research.

Further, it has been a great privilege to collaborate with researchers in coconut genomics and biotechnology to bring out this volume. Editors are greatly indebted to all the authors who have contributed their expertise and time in shaping up this book volume.

We hope this compilation would serve as a reference for those novices who join the palm research community in general and coconut genomics, particularly those who have already significantly contributed to the research in coconut genomics.

Kasaragod, India M. K. Rajesh
Kasaragod, India S. V. Ramesh
Lunuwila, Sri Lanka Lalith Perera
New Delhi, India Chittaranjan Kole

Contents

Editors and Contributors

About the Editors

Dr. M. K. Rajesh is serving as Principal Scientist (Biotechnology) in the Division of Crop Improvement, ICAR-Central Plantation Crops Research Institute (ICAR-CPCRI), Kasaragod, Kerala, India. He has obtained his Bachelor's degree in Agriculture, and Master's and Doctoral degrees in Biotechnology from Tamil Nadu Agricultural University (TNAU), Coimbatore, India. He has over 23 years of experience in the use of molecular markers for assessment of genetic diversity and tagging of important traits, *in vitro* culture and *in vitro* conservation of economically important palms, especially coconut and arecanut. His current research focuses on genome, transcriptome, proteome, and metabolome analyses of coconut, with special focus on somatic embryogenesis, oil biosynthesis and host-pathogen interactions. He has published more than 100 research papers in peer-reviewed journals, written over 30 chapters, and edited five books. He is currently serving as Member of Editorial Board of BMC Plant Biology and is the Editor of Journal of Plantation Crops, published by the Indian Society of Plantation Crops (ISPC).

Dr. S. V. Ramesh is currently serving as Senior Scientist (Biotechnology) at ICAR-Central Plantation Crops Research Institute (ICAR-CPCRI), Kasaragod, Kerala, India. His field of specialization is plant molecular biology and biotechnology having obtained his master's and doctorate degrees in plant biochemistry. He is a recipient of DBT-CTEP Travel Award (2013) and AABSc-sponsored Young Plant Biotechnologist Award (2015). He did his postdoctoral research work in the field of small non-coding RNAs at Washington State University, USA, under DBT-CREST Fellowship (2013–14). He has authored many scholarly articles—spread over diverse fields, viz., plantation biology, plant–virus interactions, and nutrition biochemistry—in the journals of international and national repute and presented his research findings in several international scientific fora. Currently, his research areas of focus include transcriptomics (non-coding RNAs) of abiotic stressors and lipid biochemistry of plantation crops. He also features in the Editorial Board of international journals PLOS ONE and Measurement: Food.

Dr. Lalith Perera represents Coconut Research Institute (CRI) of Sri Lanka since 1990 and currently serves as Institute's Additional Director. He was Former Deputy Director (Research) and Former Head of the Genetics and Plant Breeding Division of the CRI.He was Former Chairman of the Steering Committee of the Coconut Genetic Resources Network (COGENT) and is currently Member of the Technical Working Group of the International Coconut Community (ICC). His research work has earned him numerous awards including the Presidential Research Award (Sri Lanka), Endeavor Research Fellowship (Australia), Scholarship from Commonwealth Universities to pursue his Ph.D. in UK, and Visiting Scientist Fellowship (FAO). He has authored and co-authored close to 100 publications, chapters, scientific communications, and conference/workshop proceedings.

Prof. Chittaranjan Kole is an internationally acclaimed scientist with an illustrious professional career of spanning over thirty-six years and original contributions in the fields of plant genomics, biotechnology, and molecular breeding leading to the publication of more than 150 quality research articles and reviews. He has edited over 130 books, and his contributions and editing acumen have been appreciated by seven Nobel Laureates including Profs. Norman Borlaug, Arthur Kornberg, Werner Arber, Phillip Sharp, Günter Blobel, Lee Hartwell, and Roger Kornberg. He has been honored with a number of fellowships, and national and international awards including the Outstanding Crop Scientist Award conferred by the International Crop Science Society. He has served in the academia as Vice-Chancellor, BC Agricultural University; Project Coordinator of Indo-Russian Center of Biotechnology in India; Director of Research of Institute of Nutraceutical Research of Clemson University, USA; and Visiting Professor in the Pennsylvania State University and Clemson University, USA. In the recent past, he was awarded with the Raja Ramanna Fellow by the Department of Energy, Government of India. He holds the Presidentship of International Climate-Resilient Crop Genomics Consortium and International Phytomedomics and Nutriomics Consortium.

Contributors

Jorge Gil C. Angeles Philippine Genome Center—Program for Agriculture, Livestock, Forestry and Fisheries, Office of the Vice Chancellor for Research and Extension, University of the Philippines Los Baños, Los Baños, Laguna, Philippines

Ginny Antony Central University of Kerala, Kasaragod, Kerala, India

V. Arunachalam ICAR-Central Coastal Agricultural Research Institute, Ela Old Goa, Goa, India

B. Kalyana Babu ICAR-Indian Institute Of Oil Palm Research, Pedavegi, Andhra Pradesh, India

K. P. Chandran ICAR-Central Plantation Crops Research Institute, Kasaragod, Kerala, India

P. Chowdappa ICAR-Central Plantation Crops Research Institute, Kasaragod, Kerala, India

Iván Córdova-Lara Centro de Investigación Científica de Yucatán, Mérida, México

K. B. Hebbar ICAR-Central Plantation Crops Research Institute, Kasaragod, Kerala, India

S. Jayasekhar ICAR-Central Plantation Crops Research Institute, Kasaragod, Kerala, India

B. A. Jerard ICAR-Central Island Agricultural Research Institute, Port Blair, Andaman and Nicobar Islands, India

Anitha Karun ICAR-Central Plantation Crops Research Institute, Kasaragod, Kerala, India

Chittaranjan Kole Raja Ramanna Fellow, Government of India, ICAR-National Institute for Plant Biotechnology, Pusa, New Delhi, India

Jickerson P. Lado Genetics and Molecular Biology Division, Institute of Biological Sciences, College of Arts and Sciences, University of the Philippines Los Baños, Los Baños, Laguna, Philippines

Rita P. Laude Genetics and Molecular Biology Division, Institute of Biological Sciences, College of Arts and Sciences, University of the Philippines Los Baños, Los Baños, Laguna, Philippines

Antonio C. Laurena Philippine Genome Center—Program for Agriculture, Livestock, Forestry and Fisheries, Office of the Vice Chancellor for Research and Extension, University of the Philippines Los Baños, Los Baños, Laguna, Philippines

K. A. Lineesha Central University of Kerala, Kasaragod, Kerala, India

A. Manickavelu Central University of Kerala, Kasaragod, Kerala, India

R. Manimekalai ICAR-Sugarcane Breeding Institute, Coimbatore, Tamil Nadu, India

K. S. Muralikrishna ICAR-Central Plantation Crops Research Institute, Kasaragod, Kerala, India

V. Niral ICAR-Central Plantation Crops Research Institute, Kasaragod, Kerala, India

Carlos Oropeza-Salín Centro de Investigación Científica de Yucatán, Mérida, México

Evangeline D. Pascual Genetics and Molecular Biology Division, Institute of Biological Sciences, College of Arts and Sciences, University of the Philippines Los Baños, Los Baños, Laguna, Philippines

S. Paulraj ICAR-Central Plantation Crops Research Institute, Kasaragod, Kerala, India

Santy Peraza-Echeverría Centro de Investigación Científica de Yucatán, Mérida, México

Lalith Perera Coconut Research Institute, Lunuwila, Sri Lanka

Carlos Puch-Hau Departamento de Recursos del Mar, Centro de Investigación Y de Estudios Avanzados del Instituto Politécnico Nacional, Unidad Mérida, Mérida, Mexico

M. K. Rajesh ICAR-Central Plantation Crops Research Institute, Kasaragod, Kerala, India

S. V. Ramesh ICAR-Central Plantation Crops Research Institute, Kasaragod, Kerala, India

Luis Sáenz-Carbonell Centro de Investigación Científica de Yucatán, Mérida, México

K. Samsudeen ICAR-Central Plantation Crops Research Institute, Kasaragod, Kerala, India

Ajeet Singh ICAR-Central Plantation Crops Research Institute, Kasaragod, Kerala, India

R. Sudha ICAR-Central Plantation Crops Research Institute, Kasaragod, Kerala, India

Abbreviations

2D-GE	Two-dimensional gel electrophoresis
5mC	5-methylcytosine
ADB	Asian Development Bank
AFLP	Amplified fragment length polymorphism
AMADH2	Amino aldehyde dehydrogenase 2
AMT	Accurate mass and time tag
AT	Atlantic Tall
AYD	Andaman Yellow Dwarf
BADH2	Betaine aldehyde dehydrogenase homologue 2
BAT	Bali Tall
BLAST	Basic local alignment search tool
bp	Base pairs
BYD	Bali Yellow Dwarf
CACP	Commission for Agricultural Costs and Prices
CAGR	Compound annual growth rate
CAPS	Cleaved amplified polymorphic sequence
CAS	CRISPR associated
CDS	Coding sequence
CGD	Chowghat Green Dwarf
CGRD	Coconut Genetic Resources Database
CICY	Centro de Investigacion Cientifica de Yucatan, Merida
CIRAD	Centre de Cooperation Internationale en Recherche Agronomique pour le Developpement
cM	Centimorgan
COD	Chowghat Orange Dwarf
COGENT	International Coconut Genetic Resources Network
CPCRI	Central Plantation Crops Research Institute
CRD	Cameroon Red Dwarf
CRI	Coconut Research Institute
CRISPR	Clustered regularly interspaced short palindromic repeat
CATD	Catigan Green Dwarf
ddRAD	Double-digest restriction site-associated DNA

DDRT-PCR	Differential-display reverse transcription-PCR
DEG	Differentially expressed gene
DRM2	Domains rearranged methyltransferase 2
EST	Expressed sequence tag
FAOSTAT	Food and Agriculture Organization Statistics
FIA-MS	Flow injection analysis–tandem mass spectrometry
FSSAI	Food Safety and Standards Authority of India
GA	Gibberellic acid
Gb	Giga base
GBS	Genotyping by sequencing
GO	Gene ontology
GS	Genomic selection
GWAS	Genome-wide association studies
HPLC	High-performance liquid chromatography
HT	Hainan Tall
ICC	International Coconut Community
ICP-MS	Inductively coupled plasma mass spectrometry
IFAD	International Fund for Agricultural Development
InDel	Insertion–deletion
IPGRI	International Plant Genetic Resources Institute
ISSR	Inter simple sequence repeat
ISTR	Inverse sequence-tagged repeat
iTRAQ	Isobaric tag for relative and absolute quantitation
KASP	Kompetitive allele-specific PCR
Kb	Kilobase
KEGG	Kyoto Encyclopedia of Genes and Genomes
KO	KEGG orthology
KYD	Kulasekharam Yellow Dwarf
LC-MS	Liquid chromatography–mass spectrometry
LCT	Laccadive Ordinary Tall
LD	Linkage disequilibrium
LIBS	Laser-induced breakdown spectrometry
LIDAR	Light detection and ranging
LINE	Long interspersed nuclear element
LTRs	Long terminal repeats
MALDI-TOF	Matrix-assisted laser desorption/ionization–time of flight
MAS	Marker-assisted selection
MAWA	Malayan Yellow Dwarf x West African Tall
Mb	Mega base
METI	Methyltransferase 1
MGD	Malayan Green Dwarf

miRNA	microRNA
MS	Mass spectrometry
MSP	Minimum support price
MYD	Malayan Yellow Dwarf
N50	Minimum contig length
NAA	Neutron activation analysis
NCBI	National Center for Biotechnology Information
ncRNA	Noncoding RNA
NGS	Next-generation sequencing
NMR-MS	Nuclear magnetic resonance–mass spectrometry
nt	Nucleotide
NYD	Nias Yellow Dwarf
ORF	Open reading frame
OTC	Open top chamber
PCA	Philippine Coconut Authority
PCR	Polymerase chain reaction
QTL	Quantitative trait locus
QTLs	Quantitative trait loci
QTOF-MS	Quadrupole time of flight–mass spectrometry
RAD-seq	Restriction site-associated DNA sequencing
RAPD	Random amplified polymorphic DNA
RCA	Revealed comparative analysis
RdDM	RNA-directed DNA methylation
RFLP	Restriction fragment length polymorphism
RGC	Resistance genes candidates
RIT	Rennell Island Tall
RNAi	RNA interference
RNA-seq	RNA sequencing
RWD	Root (wilt) disease
SCAR	Sequence characterized amplified region
SCoT	Start codon-targeted polymorphism
SGD	Sweet Green Dwarf
SIMS	Secondary ion mass spectrometry
SNP	Single-nucleotide polymorphism
SSCP	Single-strand conformation polymorphism
SSR	Simple sequence repeat
STANTECH	Standardized research techniques in coconut breeding
STRs	Short tandem repeats
TALEN	Transcription activator-like effector nuclease
TE	Transposable element
TF	Transcription factor

WAT	West African Tall
WCT	West Coast Tall
WTA	World Trade Agreement
XRF	X-ray fluorescence spectroscopy
Y2H	Yeast two-hybrid

World Economic Importance

1

S. Jayasekhar and K. P. Chandran

Abstract

The chapter illustrates the global scenario of the coconut sector by narrating the production, trade, competitiveness, and price aspects. Multiple data sources such as UN Comtrade Database, Statistical Year Book of International Coconut Community, and WTO data sets were used for the analysis. Market share analyses revealed comparative advantage analysis, ratio method-price instability measure were the major analytical tools employed for the study. Stagnancy and instability observed in the global production of coconuts in the last decade is a worrisome fact. Productivity has also registered a negative growth rate in the same period. It was found that the share of coconut oil in global production and trade of edible oils is meagre. The concentration in production, exports, and imports of the coconut sector implies the vulnerability of the sector even for the slightest supply and market shocks. The high-value coconut products such as desiccated coconut are widely consumed in high-income countries across the world. The Philippines revealed to be the most competent exporter of the coconut and coconut products in the world. Though coconut oil is directly competing with major edible oils such as soybean and palm oil, in the quality front, coconut oil stands as a premium product. The high international prices of coconut oil in comparison with the competing edible oils weakened the competitiveness of the coconut oil in the international market. Moreover, the price volatility of coconut oil in the international markets has been found to be comparatively high.

S. Jayasekhar (✉) · K. P. Chandran
ICAR-Central Plantation Crops Research Institute, Kasaragod, Kerala, India
e-mail: jayasekhar.s@icar.gov.in

K. P. Chandran
e-mail: chandran.kp@icar.gov.in

1.1 Introduction

Coconut is considered as one of the most important crops for the Asia and Pacific region, providing food, nutrition, and livelihood to millions of coconut farmers in the region. Despite the economic importance of the palm, coconut production continued to show stagnancy in production, productivity, and trade in the recent decade (Sairam and Jayasekhar 2018). The coconut sector in the past has been dominated by products, copra and coconut oil, and the international coconut trade used to be driven by the demand for coconut oil. However, demand for coconut oil has witnessed a sharp decline during the last decades due to increased competition from other edible oils, such as palm oil and soybean. Therefore, the

M. K. Rajesh et al. (eds.), *The Coconut Genome*, Compendium of Plant Genomes,
https://doi.org/10.1007/978-3-030-76649-8_1

price of coconut oil is influenced by the supply and demand of competing edible oils (Sportel 2013). The coconut sector has been inextricably linked to the coconut oil, the most dominant product of the palm, from time immemorial. Such a strong dependency on a single product had indubitably made the sector vulnerable for supply and price shocks. It is also noteworthy that the issues of trade and market prices are increasingly playing a key role in sustaining the livelihood of those who are dependent on this sector. Hence, the role of product diversification and value addition assumes greater importance than ever before (Jayasekhar et al. 2016). A robust and well-strengthened value chain plays a key role in ensuring the sustainable livelihood of all the stakeholders' of the coconut sector. In this regard, it is crucial to assure the judicious distribution of revenue share along the chain from producer to the consumer (Muralidharan et al. 2019). In recent times, moving beyond the increase in unilateral productivity, the management of surplus production at domestic and international levels assumes a critical role (Sportel and Véron 2016). Obviously, in recent times, there is a changing trend wherein coconut is increasingly being processed into diversified value-added products. This emerging trend has influenced the current production, processing, and trading system in coconut. To take advantage of these emerging opportunities in the world market, coconut growers must look into the trade patterns, performance, and global competitiveness of coconut and coconut products. This chapter analyses the world coconut production scenario, trade aspects of major coconut products, the comparative advantage of major trading countries, price trends and also provides suggestions for a sustainable coconut economy at the international level.

1.2 Analytical Frame

The aim of this chapter is to provide a heuristic view of the global coconut scenario. The data on global aspects of coconut and coconut products were collected from the International Coconut Community (ICC)[1] statistical yearbook (APCC 2019) and UN Comtrade database (UN Comtrade 2020). Revealed comparative analysis (RCA) tool was used for analysing the comparative advantage of major coconut and coconut products exporting countries. For the analysis, we have relied on the classic work by Balassa (Balassa 1965, 1977) which defines the comparative performance of a country in a particular commodity as the ratio of the share of that particular commodity in the country's total exports to the world and the world share of that commodity in the total exports of the world.

The RCA index for a particular commodity (x) and country (y) is calculated as

$$\begin{aligned}&(\text{Revealed Comparative Advantage})_{xy}\\&= \left(P_{xy}/P_{wx}\right)/\left(P_y/P_w\right)\end{aligned}$$

where

P_{xy} export of commodity 'x' by country 'y'
P_{wx} total world export of commodity 'x'
P_y country 'y's total export
P_w world export (total).

The instability of major edible oil prices was calculated by measuring the standard deviation of the log (P_t/P_{t-1}) over a period, where P_t is the price in period 't', and P_{t-1} is the price in period $t-1$. This is, in other words, the same as the standard deviation of the growth rates (ratio method) (Sekhar 2004).

1.3 Production Scenario

According to the latest production statistics brought out by the ICC, the coconut production in the world is estimated at 68,833 million nuts from an area of 12.08 million ha (ICC 2019). The world productivity of coconut stands at 5,777 nuts ha^{-1} (Table 1.1). It is noteworthy that the area and production of the coconuts in the world are by and large skewed, wherein 70% of the total area and production is concentrated in India, Indonesia, and

[1]Formerly Asian and Pacific Coconut Community (APCC)

the Philippines. India is the largest producer of coconuts with a share of 31% of the total production. In productivity, India is much ahead of the major coconut producers' with an average yield of 9,815 nuts ha^{-1}. The huge gap in average world productivity of coconuts and the potential production, which has been demonstrated by countries such as India and Brazil (11,923 nuts ha^{-1}), warrants a detailed analysis. The skewness in area and production is a matter of concern; if any one of the major producing countries' experiences a production shock, the effect will be detrimental to the world supply and translating into a price shock across the international trade sphere (Jayasekhar et al. 2019).

The major concerns of stagnancy and instability in the production of coconut observed especially from the year 2008 onwards are depicted in Figs. 1.1 and 1.2. On the other hand, in the initial years of the decade 2000, an impressive growth rate was experienced in the world coconut production with a compound annual growth rate (CAGR) of 1.83% during 2000–06. A decline in the growth rate of production was documented in the subsequent years with CAGR 0.98 in the period 2006–12 and a negative growth rate (−0.10) during 2012–18.

In the case of world productivity, a sharp dip with negative growth rates has been observed (Table 1.2) during the periods 2006–12 (−0.16%) and 2012–18 (−0.51%). It is of utmost importance to further comprehend the production economics to formulate suitable revival strategies for worldwide coconut production and yield.

1.4 Edible Oil Scenario

It is an undeniable fact that coconut oil is still the major value-added product of coconut, which is widely traded across the world in different quality grades. Nevertheless, it is imperative to understand the comparative standing of coconut oil vis-à-vis the major edible oils. The comparison of this aspect is illustrated in Table 1.3.

It is striking that the soybean oil and palm oil together account for more than 60% of the production and consumption of the total edible oils in the world. Among all the edible oils, palm oil is the single largest internationally exported edible oil accounting for a share of 49% of the total trade in edible oils. Coconut oil is a minor player with a share of 1.8% in total edible oil production and consumption. To claim better market access by competitively taking up the share of other major edible oils, it is imperative to produce more coconut oil.

1.5 Country-Wise Contribution in the National Trade

The contribution of coconut export sector in the national exports of each country reflects the importance of the sector to the economy of that country. The depiction in this aspect is provided in Table 1.4.

It is observed that the Pacific Island countries such as Kiribati and Vanuatu are very much dependent on the coconut sector wherein coconut exports account for more than 50% of the total export revenue. For other countries' including the major coconut producers, the importance of the crop is sectoral, which warrants proactive steps to strengthen the coconut economy, including trade in those countries. On the other hand, Pacific Island countries should be cautious enough to ensure sustainable international trade in the coconut sector.

1.6 Demand, Supply, Consumption, and Trade Aspects

1.6.1 Coconut Oil

A glance at the supply and demand balance of the coconut oil in the world (Fig. 1.3) reveals that in the recent years (averages of 2017 and 2018), the utilization of the oil has been equally divided between the sectors, viz., food, industry and trade (around 30% each), and the ending stock is about 9%. The ending stock is an important indicator of the demand (or lack of that) of coconut oil in a particular year in the world market, which can cause the price fluctuation in the ensuing year.

Table 1.1 Area, production, and productivity of coconut in the world

Country	Area ('000 ha)	% share	Production (million nuts)	% share	Productivity (nuts ha^{-1})
Indonesia	3,544	29.3	14,356	20.9	4,530
The Philippines	3,612	29.9	14,049	20.4	4,196
India	2,178	18.0	21,384	31.1	9,815
Sri Lanka	440	3.6	2,450	3.6	6,623
Brazil	216	1.8	2,343	3.4	11,923
Papua New Guinea	221	1.8	1,483	2.2	6,709
Thailand	179	1.5	666	1.0	4,859
Others	1,690	14.0	12,102	17.6	5,662
Total	12,080	100.0	68,833	100.0	5,777

Source ICC (2019)

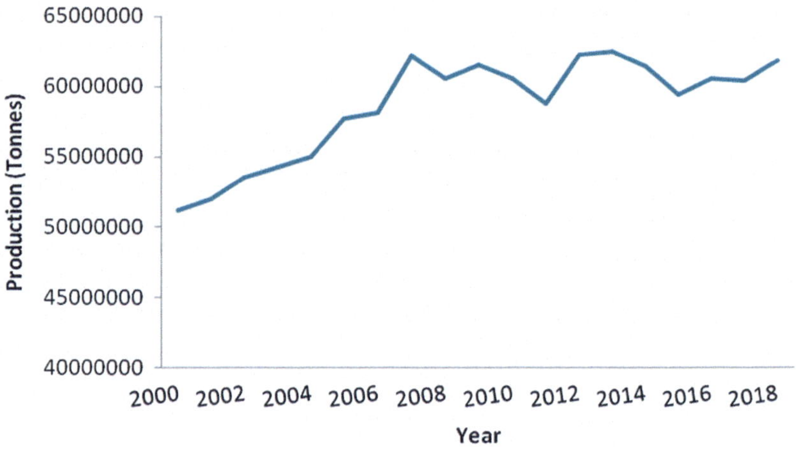

Fig. 1.1 World coconut production (2000–2018)

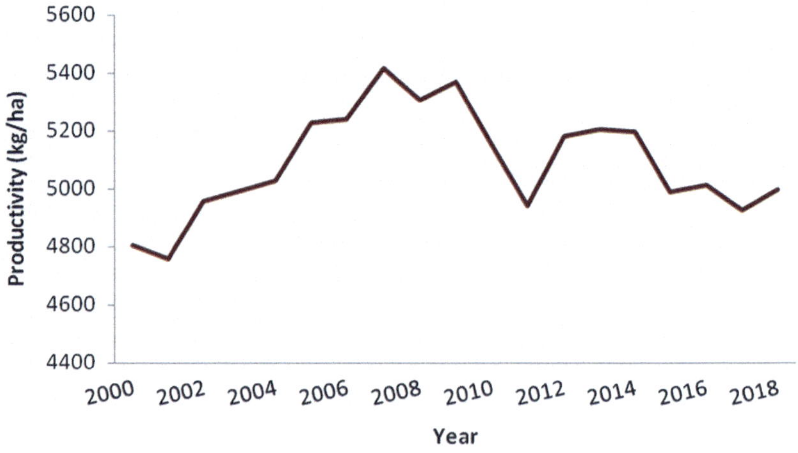

Fig. 1.2 World coconut productivity (2000–2018)

Table 1.2 Compound growth rates in world coconut production and yield

Period	CAGR (%) production	CAGR (%) yield
2000–06	1.83	1.24
2006–12	0.98	−0.16
2012–18	−0.10	−0.51

Table 1.3 World production and utilization of major edible oils (2018–19)

Item	Production	% share	Exports	% share	Consumption	% share
Soybean	55.17	27.8	10.51	13.6	54.63	28.4
Palm oil	70.61	35.5	48.53	62.7	66.31	34.5
Sunflower	18.51	9.3	9.73	12.6	17.15	8.9
Rapeseed	28.1	14.1	4.6	5.9	28.96	15.1
Cottonseed	5.18	2.6	0.1	0.1	5.12	2.7
Peanut	5.95	3.0	0.26	0.3	5.96	3.1
Coconut	3.66	1.8	1.73	2.2	3.39	1.8
Olive	3.26	1.6	1.02	1.3	2.87	1.5
Palm Kernel	8.34	4.2	0.88	1.1	7.82	4.1
Total	198.78	100	77.36	100	192.21	100

Source ICC (2019), UN Comtrade (2020)

Table 1.4 Country-wise coconut export value contribution in national exports

Country	Coconut export value (million US$)	% share in national export
Kiribati	7.92	56.6
Vanuatu	23.32	51.74
Sri Lanka	598.19	5.26
Samoa	2.04	4.63
Solomon Islands	22.07	4.42
The Philippines	2269	3.60
Marshall Islands	1.21	2.05
Indonesia	1487	0.88
Papua New Guinea	58.01	0.70
Fiji	5.00	0.55
F.S. Micronesia	0.12	0.26
India	648	0.23
Thailand	60.38	0.23
Kenya	0.59	0.12
Malaysia	258	0.12
Vietnam	169	0.08

Source Computed from ICC (2019)

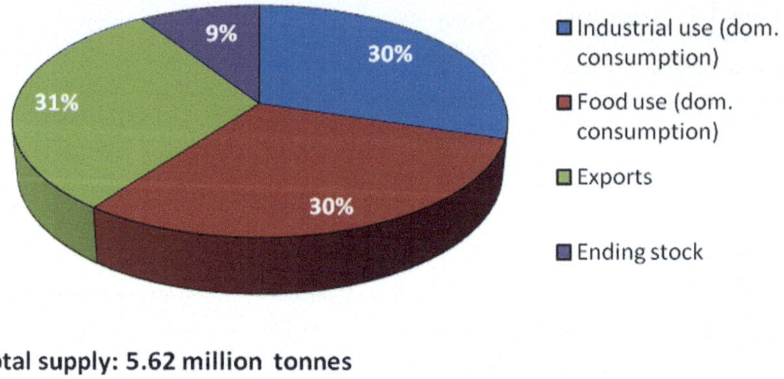

Fig. 1.3 Supply and demand balance of coconut oil

The Philippines exports about 47% of the total world exports of the coconut oil and categorically exhibited its dominance in the sector followed by Indonesia (with a share of 25%). However, considering the period 2012–19, a stagnant pattern for coconut oil exports was experienced in the Philippines. Surprisingly, even if not a major producer of coconut oil, the Netherlands showed a positive growth rate during the same period, with a compound annual growth rate of 1.2% and has emerged as the world's fastest-growing coconut oil exporter. On the other hand, the downward trend over the same period was observed in Malaysia and Indonesia (Research and Markets 2020a) (Table 1.5).

Average annual domestic consumption of coconut oil was found to be 3.375 million tonnes (Fig. 1.4), and the Philippines, European Union, India, USA, and Indonesia are the major coconut oil consumers accounting for more than 75% of the total coconut oil consumption (for domestic use) in the world.

1.6.2 Copra Meal

According to the research, there has been a surging demand for animal-sourced products such as meat, milk, and eggs across the world, and the trend would continue (Headey and Ecker 2013). The importance of copra meal as an excellent animal feed needs to be emphasized in this context, and the projected growth of copra meal market for the next five years is around 3.8% (Research and Markets 2020b). The fact that copra meal is a high-quality cheap source of food for the animals had caused the impressive projection on the market growth of the copra meal. The Philippines and Indonesia are the major exporters of copra meal in the world, and account for 94% of the total world exports (Fig. 1.5).

In the case of domestic consumption of copra meal (Fig. 1.6), India and the Philippines together account for about 60% of the total world consumption (which stands at 1.85 million tonnes). It is striking that for a low-value by-product such as copra meal, there is no huge consumers from high-income countries except South Korea.

1.6.3 Desiccated Coconuts

The Philippines and Indonesia are the major exporters of desiccated coconuts in the world, wherein both the countries together export 52.3% of the total world exports in the year 2018–19 (Fig. 1.7). The major export destinations of this product are the European Union.

The high value-added coconut products such as desiccated coconuts are mostly consumed by the high-income countries of America and European Union (Fig. 1.8), accounting for about 65% of the total world consumption of desiccated coconuts.

However, there exists ample scope for raising the consumption of such products in the Asian countries, especially because of the huge domestic markets in these countries. Proactive

Table 1.5 Exports of coconut oil

Country	Export (M.T.)	% share
The Philippines	943,131	47.0
Indonesia	510,441	25.5
Netherlands	251,550	12.5
Malaysia	102,735	5.1
Others	199,168	9.9
Total	**2,007,025**	**100.0**

Source: ICC (2019)

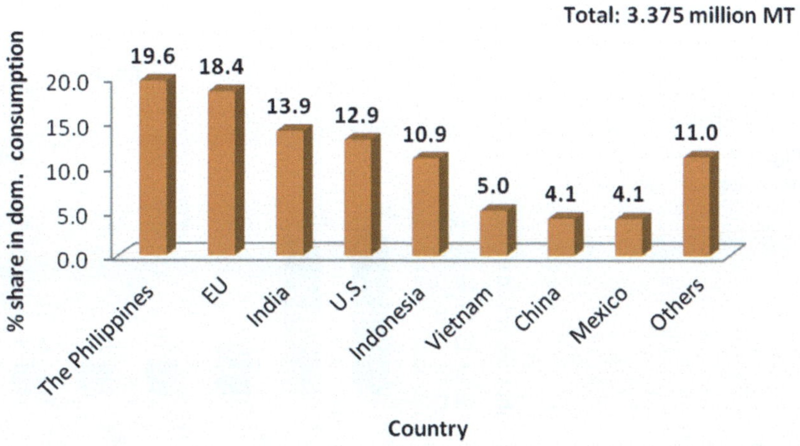

Fig. 1.4 Country-wise share in domestic consumption of coconut oil

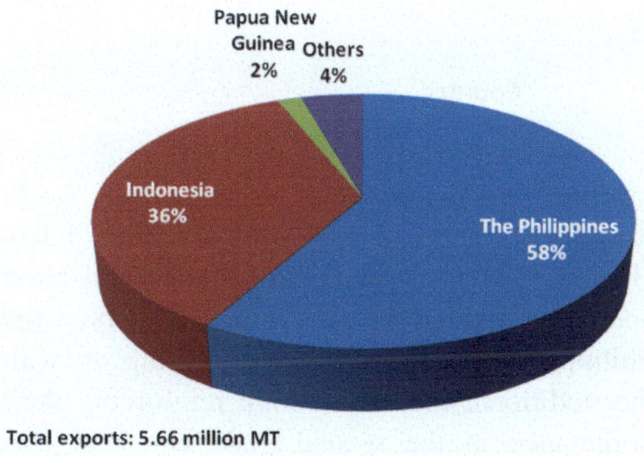

Fig. 1.5 Exports of copra meal: world

consumer awareness and marketing strategy are warranted in this regard.

1.6.4 Copra

In the case of copra trade, Papua New Guinea is the major exporter with a share of 26.2% of the total world exports. There is fair competition among Pacific Island countries and Asian countries for the copra exports, and thereby, the exports shares are diversified among seven major coconut growing countries in these regions (Table 1.6).

Surprisingly, the Philippines is the major importer of copra in the world with a whopping 67% share of the total import of 160,229 MT (Fig. 1.9). This has to be correlated with the fact that the Philippines is the largest exporter of coconut oil and copra meal in the world as both these products are the derivatives of copra.

1.6.5 Coir and Coir Products

In the case of coir and coir products, Sri Lanka and India are the major traditional producers and exporters. In both these countries, the husk processing and coir production are age-old traditional practice involving a large number

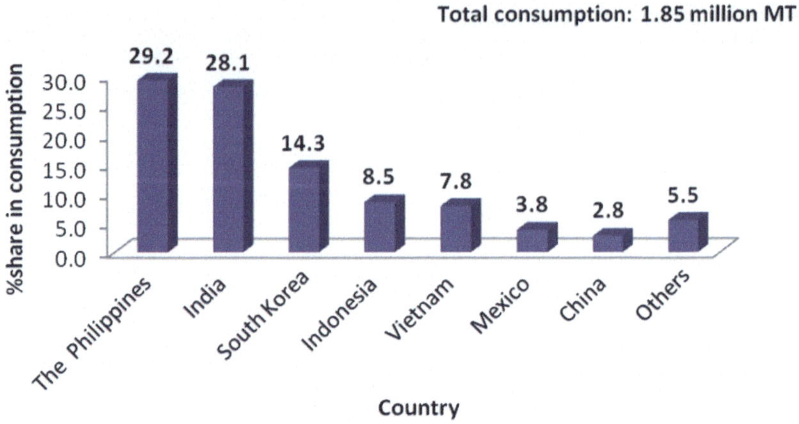

Fig. 1.6 Country-wise share in domestic consumption of copra meal

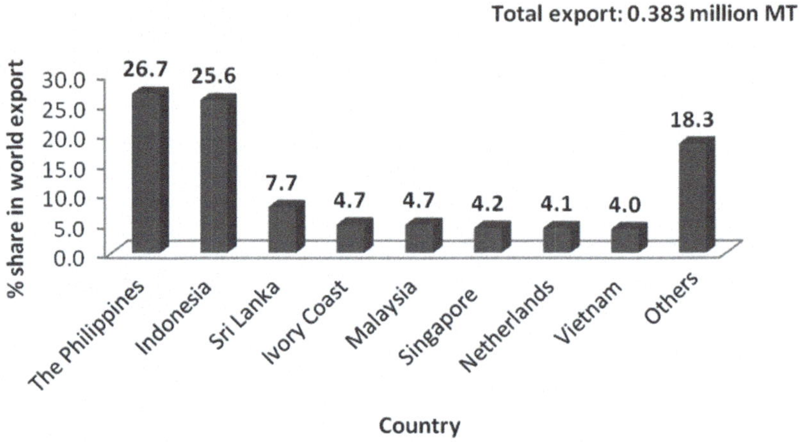

Fig. 1.7 Exports of desiccated coconut (world)

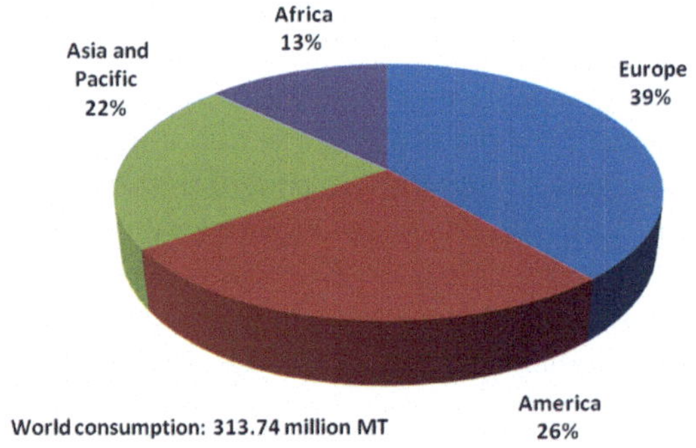

Fig. 1.8 Estimated consumption of desiccated coconut (world)

of skilled manual labourers (men and women). The total export share of Sri Lanka and India together sum up to 71% of the annual world exports of coir and coir products (Fig. 1.10). However, of late, owing to fast mechanization and modernization of coir industry, the Philippines and Indonesia are catching up with the traditional players through reinforcing their dominance in the second-grade coir and coir products, in a cost-effective manner (Bello et al. 2020).

In the case of imports of coir products, more than 80% share comes from developed countries implying the market demand is from the high-value markets for coir products (Fig. 1.11). Major import share was accounted for by Europe (43%) followed by USA (24%) and South Korea (14%).

Table 1.6 Exports of copra (world)

Country	Export (M.T.)	% share
Papua New Guinea	46,600	26.2
Thailand	34,233	19.2
Indonesia	24,425	13.7
Vanuatu	20,394	11.5
Solomon Islands	18,985	10.8
India	18,227	10.2
Kiribati	9967	5.6
Others	5026	2.8
Total	**177,857**	**100**

Source Computed from ICC (2019)

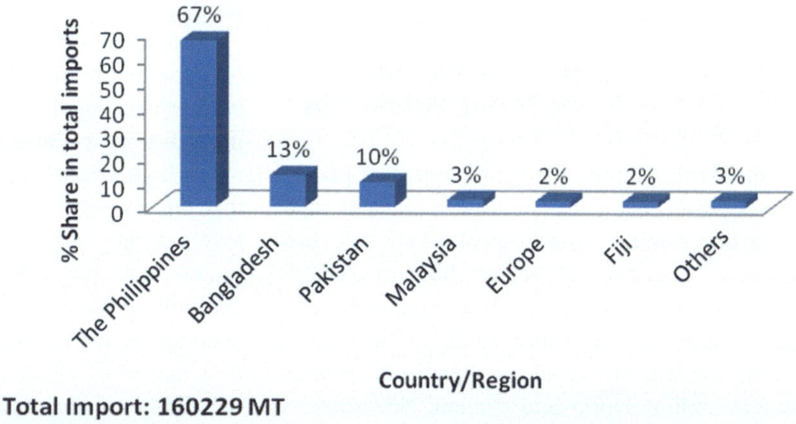

Fig. 1.9 Imports of copra (world)

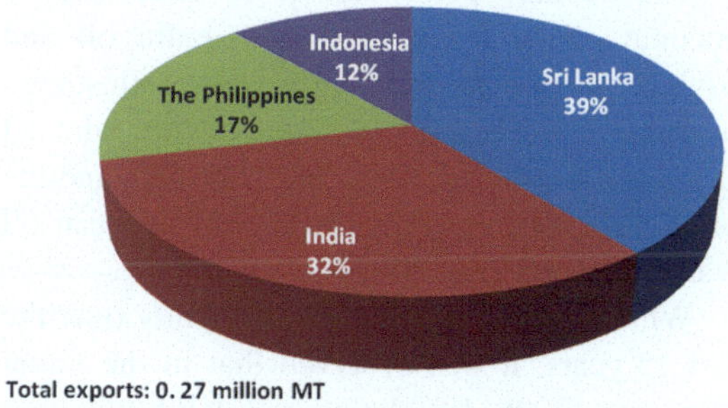

Fig. 1.10 Exports of coir and coir products

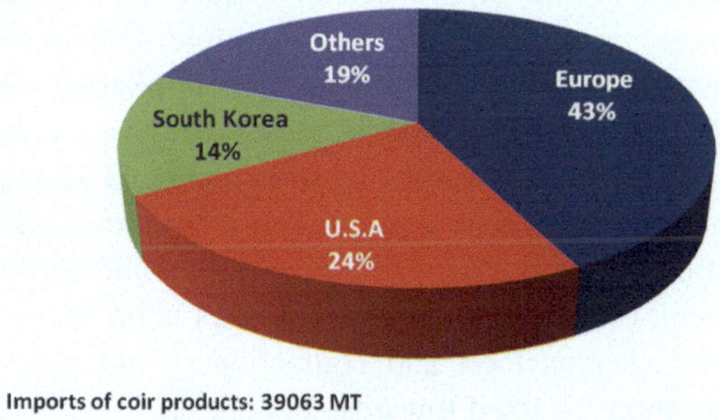

Fig. 1.11 Imports of coir products

1.7 The Comparative Advantage of Coconut Products Exports

The country-wise computed indices on the revealed comparative advantage of major coconut products (exports) are illustrated in Table 1.7. The Philippines categorically has the highest comparative advantage in all the coconut products except coir and coir products, and Indonesia closely follows the Philippines in the products such as coconut oil and copra meal. India, though the largest producer of the coconuts in the world, lacks competitiveness in major coconut value-added products except for the coir and coir products. The comparative advantage

Table 1.7 Country-wise revealed comparative advantage on the export of coconut products

Country	Desiccated Coconut	Coconut oil	Copra meal	Coir products
The Philippines	28.9	32.6	26.1	14.4
Malaysia	4.1	11.9	8.5	na
Indonesia	5.1	21.2	16.1	11.6
Thailand	5.6	1.60	2.1	na
India	1.7	2.10	0.8	24.1

Note Figures are revealed comparative advantage (RCA) indices
na not available

indices provide a valuable indicator for planning trade activities and policy formulations.

1.8 Price Analysis of Coconut Oil Vis A' Vis Major Edible Oils

In the world-edible oil sector, the major players are soybean, palm oil, sunflower oil, and rapeseed oil, wherein with comparatively low production, coconut oil sector had to directly compete with these edible oils in the international trade facet. On the other hand, in the quality front, coconut oil stands as a premium product due to its high lauric acid content. Nevertheless, the international coconut oil prices are very much linked to the supply–demand equations of the other major edible oils and therefore subject to instability and price fluctuations. Palm kernel oil is the close substitute and thereby a close competitor of coconut oil due to comparable levels of the lauric acid contents in both these oils (Boateng et al. 2016). However, in the industrial front, oleochemical industry prefers palm kernel oil, whereas food and confectionery industries are more inclined towards coconut oil.

Adulteration of coconut oil with cheaper oils such as palm oil due to large price differential is a serious issue that affects coconut farmers and also human health. For instance, the Food Safety and Standards Authority of India (FSSAI) has taken some steps and banned production, procurement, and distribution of 72 brands of adulterated coconut oil in 2019 (CACP 2019). It is crucial that appropriate steps need to be taken to check adulteration and stop manufacturing, sale, and distribution of adulterated coconut oil to protect the interests of both consumers and producers.

The price movements of coconut oil, soybean oil, and palm oil for the period 2008–20 are depicted in Fig. 1.12. It is striking that, excluding a few years, mostly coconut oil prices were much higher than the other two edible oils, and this price wedge was especially ruled at highest levels during 2013–18. The higher international prices than the substitutable edible oils will certainly debilitate the competitiveness of the coconut oil in the international market, and this is a matter of grave concern as far as the sustainability of the coconut sector in long-term perspective.

The average percentage price difference of coconut oil in comparison with palm oil and soybean oil for the different period is illustrated in Table 1.8. The price difference with palm oil stood at 44.60% for the period 2008–20, and the price difference in comparison with soybean oil was computed to be 23.39%.

While examining the price instability over the last 15 years, it was observed that in the initial five years (2006–10), the price instability indices of coconut oil (0.035), palm oil (0.040), and soybean oil (0.031) have not shown much of difference. However, in the subsequent period, coconut oil prices at the international level were comparatively volatile than the other two major edible oils (Table 1.9).

It is also noteworthy that the prices almost tend to yield stability in the cases of palm oil and soybean oil. In contrast, coconut oil has shown a tendency to increasing price instability.

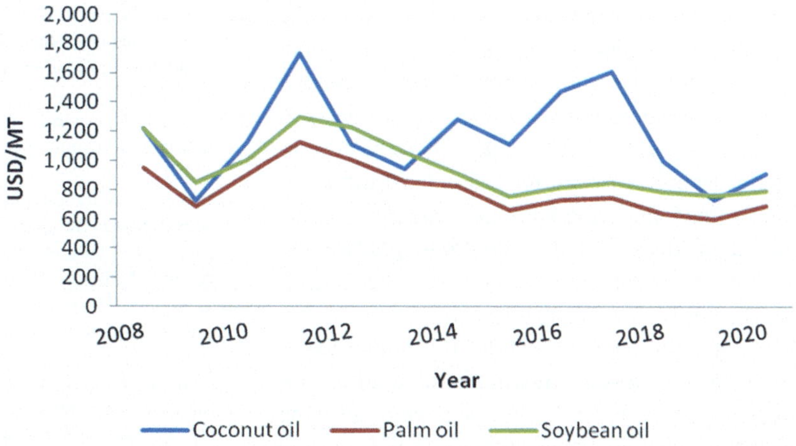

Fig. 1.12 Price movements of major edible oils (2008–20)

Table 1.8 Average price difference (%) with coconut oil

Period	Palm oil	Soybean oil
2008–12	24.86	4.15
2012–16	45.12	24.22
2016–20	70.68	45.37
2008–20	44.60	23.39

Source Computed from data portal: https://www.indexmundi.com/

Table 1.9 Price instability of major edible oils

Period	Soybean oil	Palm oil	Coconut oil
2006–10	0.031	0.040	0.035
2011–15	0.017	0.023	0.037
2016–20	0.016	0.025	0.033

Computed by the author

1.9 Conclusion

Stagnant and fluctuating coconut production across the world is a matter of concern given the growth of coconut industry worldwide. Productivity has recorded negative growth rates in recent times. Therefore, the coconut production sphere in the world warrants a critical analysis, so that a pragmatic revamping strategy could be envisioned and implemented. From the trade point of view, the market access of coconut oil has to be enhanced further. To materialize this, the concerted effort of all the coconut growing countries' in the world is imperative. Several small Pacific Island countries such as Samoa, wherein major share of the gross domestic product is accounted for by the coconut sector. Such countries are vulnerable to the tough trade competition internationally, and therefore, there should be provisions to safeguard the trade interests of such small countries. The Philippines and Indonesia have categorically exhibited their dominance in the world exports of value-added coconut products. The evolving stringency in the international food safety standards can pose a serious threat to the market access of the products from these countries, as it is evident from the current levels of consignment rejections registered. This reinforces the importance of enforcing high-level food safety standards along the global value chain of the coconut sector. High-value coconut products such as desiccated coconuts and coconut milk powder find its way to the markets of developed countries in Europe and America. Still, the potential low-value by-products such as copra meal are not properly positioned internationally. Hence, a strategic framework to popularize such products in the high-end markets is to be formulated. The policy issues, including tariff measures concerning the international coconut trade, should be discussed in a common platform to formulate international policy on the sector, which will be beneficial for each of the coconut growing countries in the world. It is suggested to establish and develop international markets for coconut products in major coconut growing countries, wherein a high profile infallible system of the export supply chain can be developed, monitored, and evaluated.

References

Balassa B (1965) Trade liberalization and "revealed" comparative advantage. Manchester Sch 33(2):99–123

Balassa B (1977) 'Revealed' comparative advantage revisited: an analysis of relative export shares of the industrial countries, 1953-1971. Manchester Sch 45 (4):327–344

Bello RT, Pantoja BR, Tan MFO, Banalo RA, Alvarez JV, Rañeses FP (2020) A study on skills for trade and economic diversification (STED) in the non-traditional coconut export sectors of The Philippines (No. 995072489702676). International Labour Organization

Boateng L, Ansong R, Owusu W, Steiner-Asiedu M (2016) Coconut oil and palm oil's role in nutrition, health and national development: a review. Ghana Med J 50(3):189–196

CACP (2019) Price policy for copra 2020 season. Commission for Agricultural Costs and Prices, Department of Agriculture, Cooperation & Farmers Welfare, Ministry of Agriculture & Farmers Welfare Government of India, New Delhi, p 110

Headey D, Ecker O (2013) Rethinking the measurement of food security: from first principles to best practice. Food Sec 5:327–343. https://doi.org/10.1007/s12571-013-0253-0

ICC (2019) Coconut Statistical Year Book-2017. International Coconut Community, Jakarta, Indonesia, p 351

Jayasekhar S, Chandran KP, Thamban C, Jaganathan D, Muralidharan K (2016) Analyzing the trade competitiveness of Indian coconut sector in the liberalization regime. J Plantn Crops 44(3):147–152

Jayasekhar S, Thamban C, Chandran KP, Muralidharan K (2019) Coconut sector in India experiencing a new regime of trade and policy environment: A critical analysis. J Plantation Crops 47(1):48–54

Muralidharan K, Subramanian P, Mathew AC, Thamban C, Jayasekhar S, Krishnakumar V, Madhavan K (2019) Upgrading a coconut value chain: empirical evidence from North Kerala. Int J Innovative Hortic 8 (1):72–80

Research and markets (2020a) World coconut (copra) Oil —Market analysis, forecast, size, trends and insights, Report ID-4701104, October 2020. Retrieved from https://www.researchandmarkets.com/reports/4701104/

Research and markets (2020b) Global copra meal market (2020 to 2025)—Growth, trends, and forecast, Report ID-4987189, October 2020. Retrieved from https://www.researchandmarkets.com/reports/4987189/

Sairam CV, Jayasekhar S (2018) World coconut economy: sectoral issues, markets and trade. In: The coconut palm (*Cocos nucifera* L.)-research and development perspectives. Springer, Singapore, pp 801–820

Sekhar CSC (2004) Agricultural price volatility in international and Indian markets. Econ Political Wkly 39(43):4729–473

Sportel T, Véron R (2016) Coconut crisis in Kerala? Mainstream narrative and alternative perspectives. Dev Change 47(5):1051–1077

Sportel T (2013) Labour, livelihoods and political narratives: a study of social structures, globalisation and development in the coconut economy of Kerala (Doctoral dissertation)

UN Comtrade (2020) UN comtrade database, United Nations, https://comtrade.un.org/

2 Botanical Study and Cytology

R. Sudha, V. Niral, and K. Samsudeen

Abstract

Coconut palm (*Cocos nucifera* L.) is a monospecific, monoecious and protandrous, tropical perennial tree crop. It belongs to the palm family Arecaceae which contains about 190 genera and 2800 species. It is an unbranched tree with light gray to brown colored trunk of 9–18 m height and occasionally up to 30 m height; dwarf morphoforms are also present. It has pinnate leaves with a length of 4–7 m and width of 1–1.5 m at its broadest portion. Female flowers are spherical and few while male flowers are small and abundant. The fruit is a drupe, ovoid of length up to 30 cm and of width 20 cm. The fruit consists of a thin hard exocarp, fibrous mesocarp, hard endocarp, white thick albuminous endosperm and a cavity filled with liquid endosperm. In young nuts, the coconut water/milk is abundant, which shows a gradual reduction in volume towards maturity. This chapter provides an overview of botanical and cytological aspects of coconut.

R. Sudha (✉) · V. Niral · K. Samsudeen
ICAR-Central Plantation Crops Research Institute, Kasaragod, Kerala, India
e-mail: R.Sudha@icar.gov.in

V. Niral
e-mail: niral.v@icar.gov.in

K. Samsudeen
e-mail: samsudeen.k@icar.gov.in

2.1 Introduction

Coconut (*Cocos nucifera* L.) is a monocotyledonous palm that belongs to Phanerogams division and Angiosperms subdivision, Monocotyledons class, Calycinae series, Arecales order, Arecaceae family, Arecoideae subfamily, Cocosease tribe and Buttinae sub-tribe. Tribe *Cocoseae* encompasses many economically and agriculturally important palm species including the peach palm (*Bactris gasipaes*) and the African oil palm (*Elaeis guineensis*). Coconut has been included under the monotypic genus *Cocos* with *nucifera* being the sole species. About 60 species were placed in the genus *Cocos* earlier; however, those species were redistributed either into *Butia* or *Syagrus* (Beccari 1916).

Coconut palms are normally unbranched, with an upright, cylindrical, pillar-like stem having a height of 25–30 m. The trunk is stout and flexuous, encircled with scars due to fallen old leaves. It forms only one vegetative bud at the tip of the stem that gives rise to all the leaves. The vegetative bud, generally called 'cabbage', comprises several folded embryonic leaves protected by older leaves' sheaths. The palm dies when the bud is removed or is extensively injured.

2.2 The Root

Being a monocotyledon, the coconut palm has a fibrous root system. Uniformly thick primary roots are produced from the base of the stem.

M. K. Rajesh et al. (eds.), *The Coconut Genome*, Compendium of Plant Genomes,
https://doi.org/10.1007/978-3-030-76649-8_2

Rootlets are produced from these primary roots, which spread in all directions. On the primary roots, some pneumatophores or breathing roots are also found. The average length of the main root is about 6 m, having uniform thickness and about 8 mm in diameter. The main and branch roots are long-lived, whereas the rootlets, tertiaries and branches are short-lived.

As the palm ages, the total number of primary roots increases. The number of roots is also influenced by the genotype, health of the palm, fertility of the soil and cultural practices. An adult tree has as many as 2000–3000 main living roots (Menon and Pandalai 1958). The color of the root varies from white at the point of growing to dark brown at the oldest part. The roots are devoid of root hairs. Water and nutrients are absorbed by the white zone, which is about 2–5 cm behind the root tip. The root tip is covered with a root cap. Anatomically, there is no cambium in roots. Its centre has a stele surrounded by a single-celled pericycle sheath. From the rootlets and aerenchymatous (respiratory exchange) protuberances or pneumatophores arise.

The root habit in coconut varies based on the type and depth of the soil. Based on their functions, roots can be classified as follows: (i) vertical or the water roots, (ii) the horizontal roots, (iii) the oblique or prop roots, (iv) the rootlets or the secondary roots and (v) the pneumatophores or the respiratory roots.

Louis and Balasubramanium (1983) investigated the root pattern of 32 palms grown under different agronomic conditions. In an adult palm producing 5570 roots, 41.81% of the roots pass through horizontally at 90° angle in all directions and are placed 30 cm below the surface soil. These roots, at optimum conditions, absorb 30–35 mL of water per day and hence an average of 5000 roots would uptake $\sim$150 L of water. The growth rate of root is 4.2 m per year, and it is highly influenced by the type of soil and soil moisture. Davis et al. (1954) studied the water sucking power of the primary roots and reported that the cut ends had higher power than the normal tips. Sucking power also varied at different lengths on the root.

The pneumatophores appear as white projections throughout the length of the root. They are characterized with epidermis as an outer layer, hypodermis/exodermis just beneath and endodermis as an inner layer. The mesoderm enclosing the air chamber consists of large and loose parenchymatous cells. The inner core of the root is formed by empty cells or laden with tannin. The inner core of the cortex cells help the roots to contact the atmosphere outside during water stagnation.

The roots possess several xylems, phloem vessels located on the periphery of the stele. It has been reported that each root possesses 20 or more xylem and phloem bundles (Louis 2002).

2.3 The Crown

The leaves on the crown are arranged one above the other in whorls of five each. The tall forms of coconut normally bear about 30–36 leaves. Every leaf axil bears inflorescence or fruit bunch except the senile leaves. The shapes of the crown vary and are classified as: (i) circular, (ii) semicircular, (iii) angular and (iv) X-shaped (Ratnambal et al. 1995). The leaves in the angular shaped crown are held upwards in an obtuse angle. These palms possess a long pre-flowering period. The high yielding hybrid palms mostly possess X-shaped crowns. In tall forms, the crowns segregate largely as circular and semicircular shapes (Fig. 2.1) and rarely as X and angular shapes.

2.4 The Stem

During germination, the embryo develops into a conical-shaped plumule and produces a pro-stem from which the growing point develops into a small projection. It consists of meristematic cells, dense cytoplasm and large nuclei. Below this is a layer that possesses numerous procambial strands. The stem is characterized with no secondary thickening since it lacks cambium cells. The procambial strands are dispersed all over the stem and are rich in sugar and starch granules. The cells of the strands multiply by undergoing repeated divisions before differentiating into xylem and phloem vessels. The initially formed protoxylem vessels are smaller than metaxylem,

Fig. 2.1 Tall palm with spherical shaped crown

Fig. 2.2 Leaf scars in the stem

which are formed later. The xylem and phloem vessels are enclosed within a thin-walled cell layer, which thickens subsequently and forms the fibrous sheath. Compared to periphery bundles, vascular bundles are widely spaced. The periderm or the cortex is absent in both the stem and the root. However, a storied type of cork cells, called rhytidome, is formed on the periphery.

The stem is generally unbranched, light gray in color, either upright or slightly curved. The stem is initiated from terminal bud. Foundation of the stem attains its complete development after four years in talls and 2–3 years in dwarfs. The diameter of the base of the trunk, in tall palms, is around 0.8 m, and it tapers to about 0.4 m (Child 1974). The stem generally does not exhibit any changes in width once formed. The stem does not grow laterally since it has no outer cambium. A pattern of triangular-shaped leaf scars are consipicously present on the stem where former leaves had been attached (Fig. 2.2). Internodes are the unscarred areas between the leaf scars. The diameter of the cylindrical stem measures 80–85 cm, and in dwarfs, it is about 75 cm. The base of the stem in the tall type measures 150 cm while it is smaller in dwarfs. The stem elongates and grows to a height of 30–40 m during its life period of 60–80 years.

2.5 The Leaf

Coconut seedlings have pinnate leaves which remain fused for the first year. After the formation of 8–10 such leaves, subsequent leaves tend to split into leaflets (Fig. 2.3). It has a rosette of leaves at the apex of the trunk comprising of opened leaves and those in the bud in various developmental stages. Based on the genotype, growing conditions and the age of the palm, the length of a leaf varies. About 30 months before emerging as an unopened spear leaf, leaf primordia differentiation gets completed (Patel 1938). The life span of the emerged leaf is 2–3 years; however, under favourable conditions, the emerged leaves of good bearing palms remain for 3 to 3 and 1/2 years (Patel 1938). Based on the genotype and growing conditions, the total number of unfolded leaves usually vary from 22 to 35. In tall palms, the average annual leaf production varies from 14 to 16 while in dwarf palms it is about 21. The youngest leaves are folded together and form the palm's 'heart' from where the leaves gradually open out and enlarge. Leaves are arranged in spirals, running both clockwise and anticlockwise. The phyllotaxy of the leaves

Fig. 2.3 Pinnate leaf of coconut with leaflets

is alternate, and the consecutive leaves are placed at an angle ranging from 137 to 140°; hence, each leaf could harvest maximum quantum of sunlight.

The leaf length varies from 4.5 to 6.0 m (Davis 1954). The petiole length is about one-quarter of the total leaf length, but the ratio differs depending on the genotype. The petiole continues as the leaf mid-rib. The petiole is attached to the stem with the help of a sheath firmly holding the stem with its wings almost around it. The leaflets from both sides of the mid-rib are different in length. The leaflets from the base of the leaf are short, and the length of the leaflet increases gradually until reaching the maximum length of 130 cm and about one-third of the mid-rib, again becoming smaller towards the apex of the leaf. The smallest leaflets in the tip may be 25 cm long. In dwarf palms, the maximum length of the leaflet may not be longer than 80–110 cm. According to the genotype, the length and width of leaflets vary. The total number of leaflets of a mature palm leaf varies from 200 to 300. Leaf blades are covered by a thick cuticle. Upper epidermis is thicker than the lower epidermis, and the stomata are positioned in the lower epidermis (Satyabalan 1993).

On an average, a leaflet consists of 170–220 stomata per mm^2, and in general, dwarf genotypes have higher stomatal density than talls (Satyabalan 1993; ICAR-CPCRI 2015). For instance, stomatal densities of the Ranthambli and Sri Lanka Green Dwarf were 189.87 mm^2 and 215.18 mm^2, respectively. Solangi et al. (2010) studied stomata of six coconut varieties and observed parallel rows arrangement of the stomata on the abaxial epidermis, and these stomata are sunken below the epidermis. The stomata is surrounded by two guard cells and four subsidiary cells, two of which are roundish and the other two are lateral in position to the guard cells. The average length of the stomata ranged from 31.66 to 39.06 μm, and the average width ranged from 9.46 to 12.8 μm.

2.6 The Inflorescence

The coconut palm is monoecious and protandrous. Male and female flowers are borne on the same inflorescence, and the inflorescence is enfolded in a double sheath or spathe. The whole structure is called 'spadix' which forms singly in each leaf axil (Fig. 2.4). The male flowers are more abundant than the female flowers. The male flowers are situated on the top portion of spikelets, which are attached to a main axis or peduncle. The female flowers are borne at the base of the spikelets (Fig. 2.5). The inflorescence primordium can be noticed four months after the differentiation of first leaf primordium, and the fully grown spathe opens one year later. Maturation of the fruit takes 12 months from flowering. Coconut palm generally produces 10–14 inflorescences per year under normal conditions in the tropics. However, all the leaf axils do not produce fruits because of premature abortion of inflorescence due to various reasons (Child 1953; Menon and Pandalai 1958; Dransfield et al. 2008). The length of the spadix ranges from 1.0 to 1.5 m, and the length of the inflorescence varies from 0.8 to 1.0 m. The inflorescence is pale yellow to creamy in color. The main rachis consists of 10–30 rachillae/spikelets, with each rachilla/spikelet bearing 50–200 or more male

Fig. 2.4 An unopened inflorescence

Fig. 2.5 Spikelets with male and female flowers

flowers, and 0–3 female flowers at their base. Male and female flowers cannot be differentiated easily at early stages; however, their morphology and size are distinguishable in the later stages. Male and female flowers turn into brownish-orange color when mature (Perera et al. 2010).

2.7 Male Flower

Matured male flowers are 7–9 mm long, sessile, scented and perianth lobes are six in number (Fig. 2.6). The three smaller perianth lobes at the base form the calyx and the other three, the sepals. The sepals touch each other without overlapping. There are six stamens, each held by succulent filaments. The anthers are linear and burst along longitudinal slits. The ovary is small and rudimentary in male flower with three hair-like erect structures called the pistils. Pollen grains are spherical, and it becomes ellipsoidal when dried with a longitudinal groove at the centre. Davis (1985) observed two grooves in some exceptional cases. The pollen normally measures a length of 0.063 mm and a breadth of 0.020 mm. It has been reported that each male flower produces 80–100 thousand pollen grains. Pollen remain viable for 3–5 days at room temperature, up to 16 days in refrigerated conditions and several months under deep freeze conditions. Though the pollen sterility varies from 3 to 33% (Aldaba 1921), sterility of 25% was reported for most of the coconut varieties, and it was least in dwarf (6.4%) genotypes. Genic and chromosomal determinant of male sterility has been reported in coconut.

2.8 Female Flower

Female flowers are 2.5–4.0 cm long, nearly spherical and sessile (Fig. 2.7). Six perianth structures enclosing the ovary are round, concave

Fig. 2.6 Male flowers

and fleshy, the lower three form the calyx and the upper three forms the corolla. The aestivation is imbricate, and the margin of each petal overlaps the neighbour. Two small bracteoles were positioned at the base of the female flower. The ovary is tricarpic, and the stigma is small, fleshy and short. Viscous nectar is secreted from the ductless glands located at the base of the fleshy stigma at the time of maturity. A few days before the maturity, the stigmatic lobes turn to white color and emerge outside the perianth lobe. The stigma remains receptive for 73 h. The period between the receptivity of the first female flower and the last female flower is called as female phase, and it generally lasts for 5–7 days.

2.9 Pollination

The palm is protandrous and male flowers open immediately after splitting of the spadix. Male flowers open in the early morning from 6 to 10 h from the distal to the near end of the rachilla. Typically, all the male flowers of a spadix complete their flowering within 19–25 days. The female flowers open characteristically 2–4 days after the male phase. Female flowers complete their flowering in 5–7 days, which ensures the predominance of cross-pollination. However, in certain circumstances, for instance, when the number of leaves produced per year exceeds 12, it results in an increase in the number of spadices produced. The spadix subtended by the succeeding leaf may open when the female flowers of the preceding inflorescence are still open, and the stigma is receptive which aids self pollination.

Fig. 2.7 Female flowers

Consequently, the pollens from the succeeding spadix could be pollinating the female flowers of the preceding spadix facilitating self-fertilization. In dwarf palms, the female phase occurs within the male phase, generally towards its end period. The male phase prevails for 14–18 days and the female phase for 3–6 days in dwarf palms ensuring self-pollination. Early botanists (Muller 1883; Kunth 1909; Jack and Sands 1922) had believed that coconut was anemophilous (wind-pollinated) because the coconut possesses several botanical features, such as non-synchronized male and female flowering cycle, more pollen production, the absence of any attractive flower color and strong odor, the position of male and female flowers in the inflorescence, and occurrence of protandry, which favour anemophily. Furtado (1924) proposed that coconut is insect-pollinated, and a third group proposed that it is both insect and wind-pollinated (Marechal 1928; Patel 1938). Later, Henderson (1986) had reported that the coconut is both insect and wind-pollinated, but mainly insect-pollinated. However, Kevan and Blades (1989) reported that in the Maldives, the wind was the main pollinating agent as honeybees are reported to be absent in the islands. Ashburner et al. (2001) reported that in the Gazelle Peninsula (Papua New Guinea), 96.3% palms were insect-pollinated, and only 3.7%, wind-pollinated. Meléndez-Ramírez et al. (2004) executed a detailed study in Yucatan peninsula, Gulf of Mexico. They reported that self-pollination (geitonogamy) caused 19% fruit set, xenogamy (cross-fertilization) caused 30% fruit set, cross-fertilization by wind (anaemophily) resulted in 10% fruit set, while 40% fruit setting was attributed to insects (entomophily). Nayar (2017) reported that coconut possesses a mixed mating system, and it may be considered an adaptive mechanism to ensure its survival under varied physiographic conditions.

2.10 Development of Fruit

The development of the fruit had been investigated by Sampson (1923), Tammes (1940) and John (1953). Sampson (1923) had divided the developmental stages into four periods: (i) when the nut is about 168 days old, it is fist-sized, and the endosperm (kernel) has just commenced to form; (ii) when the nut is about 224 days old, the shell has commenced to harden, and the endosperm has formed all around the inside of the shell, the germ is visible; (iii) in a 308 days old nut, the endosperm is fully formed, but the shell has not completely hardened; and (iv) in a 364 days old nut, the shell has fully hardened, and the endosperm is ripe.

However, John (1953) had described the development of the fruit as follows:

1. Button stage: The fertilized flowers just after one month of fertilization are known as buttons. It weighs about 15 g and is covered with perianth lobes. At this stage, no cavity is observed inside the button (Fig. 2.8).
2. Immature nuts: This stage spreads from 2nd to 4th month after fertilization. During second month, the growth of tissues is rapid and is accompanied by the formation of the cavity to accommodate water, meat and shell. Weight at this stage is 35 g. During the third month, the nut weighs 119 g and contains 3–4 g of water. The weight increases rapidly and attains 364 g during the fourth month.
3. Tender nuts: This period is from 5th to 8th month following fertilization. During the fifth month, the nut weight increases to 812 g. Formation of meat and fiber differentiation occurs since the sixth month. The weight of the coconut meat progressively increases from 35 g in the sixth month, through 70 g in the seventh month and 91 g in the eighth month. The volume of nut water continuously increases, and in the seventh-month, the water content is around 224 mL. The nut water at this stage is sweet and suitable for tender nut purpose.
4. Matured nuts: This stage spreads from 8th month to 12th month (Fig. 2.9). At this stage, there are no marked growth, but a slow process of desiccation can be observed. The nut water volume is around 100 mL, and the matured 12-month-old nut weighs about 1.4 kg. The kernel weight increases up to 224 g. The embryo formation is also completed at this stage. Hence, the period since the initiation of female flower development until the formation of matured nut is around three and a half years.

Fig. 2.8 Fertilized female flowers

Fig. 2.9 Nuts at different maturity stages

2.11 Fruit

The fruit is a fibrous drupe which contains an endosperm with an embryo embedded in it and protected outside by a thick pericarp. The pericarp has three different layers: the exocarp/epicarp (outermost tough fibrous layer), mesocarp (husk portion) and endocarp (shell). The husk is fleshy in the tender fruit which becomes fibrous in the matured fruit. The thickness of mesocarp ranges from 2 to 15 cm depending on the genotypes. The shell has three germination pores (eyes/stoma) on its basal side, representing the three carpels of the ovary. One of the eyes is soft, whereas the other two eyes are quite hard. The shell protects the embryo, and is present under the soft eye. The endosperm (kernel) thickness ranges from 0.8 to 2.0 cm, depending on the genotype. A thin layer of brown color testa/seed coat lies between the endocarp and the endosperm and adheres to the endosperm. There is a cavity in the middle of the endosperm, and it is filled with sweet water, also called as liquid endosperm. The central cavity is filled with water in the immature fruits. However, the quantity of water reduces gradually during maturation and on storage for a few months after harvest. The fruits are ready for harvest from 10 to 13 months after pollination. The volume of water in the cavity decreases considerably towards the end of maturation, which may be due to the absorption by the endosperm tissue or evaporation. The fruit loses its germination ability when the nut water is completely exhausted (Menon and Pandalai 1958). Coconut yield is usually assessed in terms of the number of nuts produced per palm or unit area and equivalent copra weight. Kartha and Narayanan (1956) investigated the development of oil in the ripened coconut. Sethi and Kartha (1956) studied the oil distribution in different parts of endosperm and a decreasing trend of oil was observed from the outermost layer to the innermost layer.

2.12 The Embryo

The shell of the coconut has three 'eyes' representing the carpels. Only one embryo develops, and hence, two 'eyes' remain non-functional. The embryo is embedded in the endosperm found beneath the functional 'eye' or the soft 'eye'. The embryo is a minute peg-shaped body, of few millimetres length, with an average weight of 0.119 g, 76.69% moisture and 3.45% of N, 0.68% of P, 2.14% of K, 0.052% of Ca and 0.209% of Mg on a dry weight basis (Nathanael 1958). Corner (1966) reported that the embryo continuously grows since the inception and germination starts about two months after the nuts are sown in the nursery when provided with suitable moisture and temperature conditions. The embryo differentiates into the cotyledon and the stem apex. The surface of the cotyledon secretes enzymes which act on the endosperm tissue to mobilize nutrients for the growing seedling. The embryo grows in two directions, the plumule or the shoot or upper part, grows towards the soft eye of the shell and the basal part of the embryo develops into haustorium which is an absorbent spongy tissue. Haustorium expands and fills the entire cavity within 20–24 weeks after germination (Child 1974; Balachandran and Arumugam 1995). Haustorium has two different parts, the oil-rich yellow outer part and carbohydrate-rich white inner part. Coconut haustorium consists of sugars, minerals, proteins, polyphenols, alkaloids and growth-promoting substances. Balasubramaniam et al. (1973) reported that the total starch content of haustorium increases linearly during maturation, whereas soluble and reducing sugar content increase rapidly and remains at a steady state thereafter. The surplus carbohydrates mobilized from the endosperm are stored as starch in the haustorium. The haustorium consists of thin-walled cells loosely connected with large interspaces and long branching strands running through it to form a vascular system carrying soluble nutrients to the plant's growing parts.

2.13 Cytology

Santos (1929) was the first to report that the chromosome number of coconut ($n = 16$). Later, several Indian researchers namely Janaki Ammal (1945), Venkatasubban (1945) and Sharma and Sarkar (1956) confirmed the chromosome number of coconut (as $n = 16$). It has been reported that both tall and dwarf types are diploids with $2n = 2x = 32$ chromosomes (Sisunandar et al. 2010) with minor differences in the karyotype (Raveendranath and Ninan 1973; Sisunandar and Adkins 2007).

Ninan et al. (1960), Abraham et al. (1961) and Raveendranath and Ninan (1973) performed karyomorphological investigations of the coconut chromosomes. The size and structure of the chromosomes in tall and dwarf genotypes varied significantly. Secondary constrictions in the long arm of Chromosome VI in tall variety and Chromosome III in dwarf were observed by Raveendranath and Ninan (1973). Satellites on the long arm of chromosome VI and the short arm of chromosome IX were observed by Nambiar and Swaminathan (1960). However, Thankamma Pillai et al. (1983) reported satellites in the arm of chromosome XII. Raveendranath and Ninan (1973) also reported the presence of satellites in chromosome VI. The presence of sub-median centromeres in most of the chromosomes was documented by Nambiar and Swaminathan (1960). Four chromosomes in tall, three in dwarf and one in dwarf green possessed sub-median centromeres (Raveendranath and Ninan 1973). It was also reported that talls showed more symmetry in chromosomes than dwarf types.

Perera et al. (2008) studied the cytogenetic differences of ordinary tall and spicata coconut and observed two types of cells, viz., diploid and aneuploid cells in spicata. The chromosome complement of $2n = 32$ was observed in diploid cells and chromosome complement varying from 18 to 24 in an aneuploid cell. However, most of the cells showed diploid chromosome complement and about 35% cells exhibited aneuploid chromosome complement. An earlier report by Raveendranath and Ninan (1973) showed that in spicata, the chromosome complement is $2n = 32$. Also, the relative lengths of chromosome categories were alike for diploid cells of both ordinary tall coconut and spicata. The study also proposed that the spicata trait is incompletely dominant or pleiotropic. However, Hunger (1920) reported that spicata trait is due to mutation and is heritable. Cytological studies by Ninan et al. (1960) and Ninan and Satyabalan (1963) on spicata palms revealed abnormalities such as irregular meiosis with translocations and inversions. Spicata palms are mainly cross-pollinated in nature, hence believed to have originated from tall forms of coconut through mutation.

Forni-Martins and da Cruz (1996) reported that dwarf coconuts have a low frequency of chiasma per bivalent (<1.5 per bivalent). Pereira et al. (2017) characterized the Brazilian Green Dwarf coconut by conventional and differential karyotyping techniques [chromomycin A3 (CMA3)/4′, 6-diamidino-2-phenylindole (DAPI) and telomeric fluorescent in situ hybridization (FISH)]. Various abnormalities were observed, such as anaphase with converging fibers and asynchronous division. The estimated recombination rate was 17.04, equivalent to an average of 1.06 chiasmata per bivalent. It is cytologically unstable since the meiotic index was 79%. The length of chromosomes varied from 5.57 to 2.13 μm. Eleven metacentric and five submetacentric chromosomes pairs were observed and hence, the karyotype was considered asymmetric. Terminal blocks in two chromosome pairs (metacentric and submetacentric) were revealed by the CMA3/DAPI banding, which was coincident with the nucleolus organizing region (NOR) and allowed the satellite characterization (1.36 μm in chromosome IV and 0.85 μm in chromosome VII). There are no evidences for structural rearrangement of chromosomes during its evolution based on the telomeric FISH, and the telomeres of the coconut chromosomes are of *Arabidopsis* type (TTTAGGG).

Lindquist (1960) reported that the tall variety exhibited stable and regular meiotic division, while a study of dwarfs and inbred lines revealed differences in the meiotic division. Several researchers concluded that microsporogenesis is regular in open-pollinated parents than in inbreds. Many meiotic irregularities were observed in dwarfs than talls (Nambiar and Swaminathan 1960). Meiotic behaviour of nine hybrids and cultivars revealed that abnormalities and pollen sterility are common in Green Dwarf and Orange Dwarf. At the same time, these occurrences are very low in T x D and D x T hybrids (Pillai et al. 1983). The high degree of self-pollination occurring in dwarfs may be attributed to chromosomal aberrations and pollen sterility.

Raveendranath et al. (1975) studied the meiotic behaviour of mutants like spicata and bulbiferous palms and found no appreciable karyomorphological differences and reported that point mutation might be responsible for bulbil formation. Ninan et al. (1960) reported meiotic abnormalities such as aneuploidy ($2n - 1$) in some cells of androgena palms.

Tudor and Laude (1996), analysing the meiosis in Dwarf and Tall coconuts, found cytological features such as univalents, laggard chromosomes and premature segregation of chromosomes. The occurrence of rod and ring pairings was observed in dwarf coconut during crossing-over. The rod pairing is characterized with only one chiasma, while the ring-type exhibits two chiasmata (Senda et al. 2005).

2.14 Cytology of Endosperm and Embryo

Abraham and Mathew (1963) observed that the nuclei size varied significantly in the developing endosperm. C-mitotic division and different ploidy levels, viz., triploid ($3n = 48$), hexaploid ($6n = 96$) and dodecaploid ($12n = 192$) in the endosperm cells, were also observed. Lesser oil content was observed in cells with higher chromosome number (Abraham et al. 1965). The cytological study in the makapuno palms (from the Philippines) revealed that the presence of a mosaic type of endosperm layers, and the number of chromosomes ranged from $3n = 48$ to 96. When the embryos were rescued, germinability was retained while most of them aborted.

Ninan and Raveendranath (1965) observed the incidence of a haploid embryo in West Coast Tall (WCT) palm. Karyomorphological characters of somatic chromosomes obtained from 6-months-old embryos were studied by Raveendranath and Ninan (1973) and reported a uniformity in the relative length of the chromosome obtained from the root tip and embryo cells of WCT palms.

Sisunandar et al. (2010) performed karyotype analysis of cryopreserved embryos and revealed that the same chromosome number ($2n = 32$) was observed in all the cultivars investigated. No significant differences were observed regarding chromosome type, length of the long and short arms, arm length ratio and centromeric index between control and cryopreserved material. However, an idiogram analysis observed a larger number of black banding on chromosomes isolated from cryopreserved material.

2.15 Gene Action

The genetic studies are limited in coconut. Sukumaran and Bavappa (1982) reported that the height of the seedling, number of leaves and girth at the base and width of the fifth leaf were controlled by more than a pair of genes. In contrast, traits such as time taken for germination, leaves produced per year, the number of spadices produced, and husk thickness are controlled by a dominant gene. Polygenes control the number of leaflets per leaf, number of nuts per year, thickness of the kernel and weight of the kernel. Ninan (1960) reported that the spikes in coconut are controlled by single gene *NN*, and spicata form with the branchless, single spike is controlled by heterozygous form *Nn*. Cedo et al. (1984) reported that spicata is a mutant form in coconut and open-pollinated progenies of spicata segregated in 1:1 ratio of normal and spicata palms. Hence, it is believed that spicata character to be in (*Ss*) heterozygous state and dominant over normal (*ss*) inflorescence (Ninan and Satyabalan 1963). Makapuno is also a mutant coconut form, which is controlled by a single recessive gene (*mm*), associated with lethality (Cedo et al. 1984).

Dengune and Remireys (1971) reported that the normal thick kernel in tall variety is controlled by single pair of gene *MM*, while in Makapuno, where endosperm is curd or jelly-like, is controlled by heterozygous form *Mm*. The recessive gene *m* is lethal in action, and it is being carried in heterozygous form *Mm*. The embryo bearing *mm* genotype aborts and nuts with the particular genotype never germinates. Louis and Chidambaram (1976) observed the segregating population to study the genetic basis of the right and left twist of the coconut leaves and concluded that this phenomenon is not controlled genetically. Davis (1969) studied the aestivation of perianth in coconut and reported that the twist of the perianth structure in the male and female flower is pre-determined. Hence, this phenomenon may be controlled by extranuclear DNA in the cytoplasm. Davis (1985) reported complementary gene action for bisexual flowers in the monoecious floral bunch. Also, he reported the pleiotropic effect of the gene controlling the bisexualism for production of barren nuts. Louis and Chidambaram (1976) undertook inheritance studies on the phyllotaxy of coconut and reported segregation of 1:1 ratio of right and left in F_1.

Color of the petiole has been an important character used to identify hybrid seedlings. Bourdeix (1999) reported that two genes (*R-G*: Brown, *rrG*: Green, *R-gg:* Orange and *rrgg:* yellow) control the color of petiole in coconut.

2.16 Conclusion

Palms exhibit many distinguishing structural features, such as fibrous roots, waterproof leaves, extremely lignified wood, starch storing stems, and oil storing seeds (Tomlinson 1990). Research on morphology, vegetative and reproductive growth aspects of coconut palm had begun decades back. Many insights have been obtained in the field of coconut botany; however, there are numerous lacunae with respect to our knowledge of the ontogeny of the anthers and ovules, fertilization, apomixis, genetic control of different economically important traits, and their underlying molecular phenomenon, etc. Planning and executing research on some of these areas would immensely aid future crop improvement programs, in general, and genomics-assisted breeding, in particular.

References

Abraham A, Mathew PM (1963) Cytology of coconut endosperm. Ann Bot 27(3):505–512

Abraham A, Mathew PM, Ninan CA (1961) Cytology of *Cocos nucifera* L. Cytologia 26:327–332

Abraham A, Ninan CA, Gopinath P (1965) Cytology of development of abnormal endosperm in Philippine Makapuno coconuts. Caryologia 18(3):395–408

Aldaba VC (1921) The pollination of coconut. Philipp Agric 10:195–207

Ashburner GR, Faure MG, James EA, Thompson WK, Halloran GM (2001) Pollination and breeding system of a population of tall coconut palms *Cocos nucifera* L. (Arecaceae) on the Gazelle Peninsula of Papua New Guinea. Pacific Conservation Biol 6:344–351

Balachandran C, Arumugam C (1995) Triglyceride deposition in tissues of germinating coconut (*Cocos nucifera* Linn). J Oil Fat Ind 72(6):647–651

Balasubramaniam H, Atukorala TMS, Wijesundera S, Hoover AA, De Silva MAT (1973) Biochemical changes during germination of coconut (*Cocos nucifera*). Ann Bot 37:439–445

Beccari O (1916) *Cocos nucifera* Linn *forma palmyrensis* Becc. In: Rock JF (ed) Palmyra Island with a description of its flora. College of Hawaii Bull Honolulu, Honolulu

Bourdeix (1999) Genetic determination in dwarf coconut germ color. Oleagineux 43:371–374

Cedo LO, de GEV, Rimando TJ (1984) Controlled pollination of embryo-cultured makapuno coconut (*Cocos nucifera* L.). Philippine Agric 67:100–104

Child R (1953) Coconut. New biology, vol 15. Penguin Books Ltd., London

Child R (1974) Coconuts, 2nd edn. Longman, London

Corner EJH (1966) The natural history of palms., London

Davis TA (1954) A leaf sampling knife suited for coconut and similar palms. Indian Coconut J 8:26–29

Davis TA (1969) Aestivation of coconut flowers. Ceylon Cocon Q 20:123–130

Davis TA (1985) Nature and extent of utilization of genetic resources in the improvement of coconut. Proc Int Symp SEA Plant Gen Res 127–135

Davis TA, Anandan AP, Menon KPV (1954) Hermaphroditism in *Cocos nucifera* L. Indian Cocon J 7:133–142
Dengune GP, Remireys TA (1971) Philippines April Feb March 9&10. Philip University
Dransfield J, Uhl NM, Asmussen CB, Baker WJ, Harley MM, Lewis CE (2008) Genera palmarum: the evolution and classification of palms. Royal Botanic Gardens, Kew
Forni-Martins ER, da Cruz ND (1996) Recombination indices in species of *Erythrina* L. (Leguminosaea, Papilionoideae). Bot J Linn Soc 122:163–170
Furtado CX (1924) A study of the coconut flower and its relation to fruit production. Gard Bull 3:261–274
Henderson A (1986) A review of pollination studies in the Palmae. Bot Rev 52(3):221–225
Hunger TWF (1920) Manual of plant breeding for the tropics
ICAR-CPCRI (2015) Annual Report 2014–15, ICAR-Central Plantation Crops Research Institute, Kasaragod, 671124, Kerala, India, p 126
Jack HW, Sands WN (1922) The dwarf coconut in Malaya. Malayan Agri J 10:4–12
Janaki Ammal EK (1945) In: Darlington CD, Janaki Ammal EK (eds) Chromosome atlas of cultivated plants. George Allen & Unwin Ltd, London
John CM (1953) The story of the coconut palm. Bull (Indian Central Cocon Ctte) 7:9–18
Kartha ARS, Narayanan R (1956) Development of oil in the ripening coconut. Indian J Agric Sci 26:319–327
Kevan PG, Blades DCA (1989) Insect visitors to flowers of coconut in the Maldive Islands. Unpublished report (TCP/MDV/ 8851, A). FAO, Rome
Kunth P (1909) Hand book of flower pollination
Lindquist K (1960) Cytogenetic studies in the seriola group of *Lactuca*. Hereditas 46:75–157
Louis H (2002) Coconut—The wonder palm. Hi-tech coconut corporation, Nagercoil
Louis H, Balasubramanium K (1983) Effect of cultural practices, irrigation and manuring on the development of root systems in the coconut palm. In: Nayar NM (ed) Proceedings of the International Symposium on Coconut Research and Development. CPCRI, Kasaragod
Louis H, Chidambaram (1976) Inheritance study on the phyllotaxy of coconut palm. Ceylon Cocon Q 27:22–24
Marechal H (1928) Observations and preliminary experiments on the coconuts with a view to developing improved seed nuts. Fiji Agric J 1:16–45
Meléndez-Ramírez V, Parra-Tabla V, Kevan PG, Ramírez-Morillo I, Harries H, Fernández-Barrera M, Zizumbo-Villareal D (2004) Mixed mating strategies and pollination by insects and wind in coconut palm (*Cocos nucifera* L. (Arecaceae)): importance in production and selection. Agr For Entomol 6(2):155–163
Menon KPV, Pandalai KM (1958) The coconut palm A monograph. Indian Central coconut committee, Ernakulam
Muller (1883) Dritter Nachtrag. Katalog der herpetologischen Sammlung des Basler Museums. J.G. Bauer, Switzerland
Nambiar MR, Swaminathan MS (1960) Chromosome morphology, microsporogenesis and pollen fertility in some varieties of coconut. Indian J Gen 20:200–211
Nathanael WRN (1958) Ceylon Coconut Research Institute Annual Report 1958
Nayar NM (2018) Opportunistic flowering and pollinating system as a survival strategy of the coconut (*Cocos nucifera* L.), a monotypic species of the Arecaceae family. Genet Resour Crop Evol 65:333–342. https://doi.org/10.1007/s10722-017-0537-x
Ninan CA (1960) Promising varieties and hybrids in coconuts. Coconut Bull 14:289–295
Ninan CA, Raveendranath TG (1965) A naturally occurring haploid in the coconut palm (*Cocos nucifera* L). Caryologia 18:619–623
Ninan CA, Satyabalan K (1963) Cytogenetic studies in the genus *Cocos*. II. Some observations on the spicata character in coconuts (*Cocos nucifera* L). Indian Cocon J 16:109–114
Ninan CA, Pillai RV, Joseph J (1960) Cytogenetic studies in the genus *Cocos*. 1. Chromosome number in *C. australis* and *C. nucifera* L. varieties. Spicata and androgena. Indian Cocon J 13:129–134
Patel JS (1938) The coconut: a monograph. Government Press, Madras
Pereira TNS, Neto MF, de Souza MM, da Costa Geronimo IG, da Costa Geronimo IG, de Melo CAF, Pereira MG (2017) Cytological characterization of Brazilian Green Dwarf coconut (*Cocos nucifera* L.) via Meiosis and conventional and differential karyotyping. Cytologia 82(2):167–174
Perera PIP, Hocher V, Verdeil JL, Bandupriya HDD, Yakandawala DMD, Weerakoon LK (2008) Androgenic potential in coconut (*Cocos nucifera* L.). Plant Cell Tissue Organ Culture 92(3):293–302
Perera PIP, Hocher V, Weerakoon LK, Yakandawala DMD, Fernando SC, Verdeil JL (2010) Early inflorescence and floral development in *Cocos nucifera* L. (Arecaceae: Arecoideae). South African J Bot 76 (3):482–492
Pillai NG, Kamaladevi CB, Wahid PA, Nambiar CKB (1983) Coconut Research and Development. In: Nayar NM (ed) Proceedings of the 1st International Symposium, CPCRI, Kasaragod, Dec 27–31, Wiley Eastern Pub, New Delhi, pp 144–151
Ratnambal MJ, Nair MK, Muralidharan K, Kumaran PM, Rao EB, Pillai RV (1995) Coconut descriptors, Part. I. CPCRI, Kasaragod, Kerala, India
Raveendranath TG, Ninan CA (1973) A study of somatic chromosome complements of tall and dwarf coconut (*Cocos nucifera* L.) and its bearing on intervarietal variation and evolution in coconuts. J Plantn Crops 1 (1):17–22
Raveendranath TG, Nair PN, Ninan CA (1975) Chromozomes of bulfiferous palm. Cocon Bull 6:2
Sampson HC (1923) The coconut palm. John Bale, Sons and Danielsson Ltd, London
Santos J (1929) Cytological study of *Cocos nucifera* Lin. Philippine J Sci 37:417–437

Satyabalan K (1993) The coconut palm-botany and breeding. Asian Pacific Coconut Community, Jakarta

Senda T, Hiraoka Y, Tominaga T (2005) Cytological affinities and interfertilities between *Lolium temulentum* and *L. persicum* (Poaceae) accessions. Hereditas 142:45–50

Sethi AS, Kartha ARS (1956) Oil distribution in different parts of the coconut kernel. J Sci Industr Res 15b:105–106

Sharma AK, Sarkar SK (1956) Cytology of different species of palms and its bearing on the solution of problems of phylogeny and speciation. Genetica 28:361–488

Sisunandar, Adkins SW (2007) Karyotype of some Indonesian coconut (*Cocos nucifera* L.) cultivars. Chromosome Res 15:73

Sisunandar PAS, Yohannes MS, Rival SA, Adkins SW (2010) Dehydration improves cryopreservation of coconut (*Cocos nucifera* L). Cryobiol 61:289–296

Solangi AH, Anwar Arain M, Muhammad S (2010) Stomatal studies of coconut (*Cocos nucifera* L.) varieties at coastal area of Pakistan. Pakistan J Bot 42(5):3015–3021

Sukumaran CK (1982) Assessment of specific and general combining abilities of nine coconut cultivars In: Proceedings of PLACROSYM-V on Plantation Crops, CPCRI, Kasaragod, 12–15 Dec 1982

Tammes PML (1940) Over ontwikkeling van de vrucht van den klapper. Landbouw 16:385–395

Thankamma Pillai PK, Vijayakumar G, Ravindran PN (1983) Cytology of coconut. In: Nayar NM (ed) Coconut research and development. Wiley Eastern Ltd., New Delhi

Tomlinson PB (1990) The structural biology of palms. Oxford University Press

Tudor MA, Laude RP (1996) Cytogenetics of coconut (*Cocos nucifera* L.) varieties Catigan Green Dwarf (CAT) and Laguna Tall (Lag) and their F1 hybrid. Philipp J Crop Sci 21:8

Venkatasubban KR (1945) Cytological studies in Palmae: chromosome numbers in a few species of palms of British India and Ceylon. Proc Indian Acad Sci B 22:193–207

3 Germplasm Resources: Diversity and Conservation

V. Niral, B. A. Jerard, and M. K. Rajesh

Abstract

Coconut, *Cocos nucifera* L., is a monospecific palm in the genus *Cocos* under the subtribe Attaleinae, is widely distributed in the coastal tropics, and is regarded as a semi-domesticated species. In the absence of truly wild forms and no known wild relatives, the origin of the coconut palm has been postulated based on historical accounts and fossil records, encompassing a vast region ranging from Northern Indian Ocean, Malesia, Melanesia, and Micronesia. However, domestication of coconut cultivation is believed to have occurred independently in the Pacific and Indian Ocean regions, with substantial genetic differentiation between the coconut populations of these two regions. Considerable intra-specific diversity and heterozygosity are observed in coconut and systematic classification of the genetic diversity has been attempted by many workers taking into consideration the palm stature, features of the leaf/inflorescence/fruit, and also pollination type. The germplasm resources are mainly conserved in *ex situ* field gene banks in most coconut-growing countries and also in the five designated International Coconut Gene banks, while *in situ* conservation and also complementary conservation strategies are also recommended. The diversity in forms observed in the crop, common varieties, the varietal classification systems advocated by various workers, nomenclature used, as well as the status of germplasm conservation and cataloging are described in this chapter.

3.1 Introduction

Coconut, *Cocos nucifera* L., an ancient semi-domesticated species, is marked by an absence of truly wild forms or wild relatives. Being under cultivation for long, coconuts have been used by humans as a source of potable water, food, and oil, since time immemorial and are presently projected as a nutritive health food. The genus *Cocos* is monospecific, following the transfer by Beccari (1916) of nearly 90 other *Cocos* species to other genera, predominantly to *Butia* and *Syagrus*. *Cocos nucifera* L. is placed in the family Arecaceae Brecht. and J. Presl. (formerly Palmae) and the subfamily Arecoideae (one of the five sub families). It belongs to the tribe Cocoseae (one of the 14 tribes) and subtribe Attaleinae (one of the three subtribes). The

V. Niral (✉) · M. K. Rajesh
ICAR-Central Plantation Crops Research Institute, Kasaragod, Kerala, India
e-mail: niral.v@icar.gov.in

M. K. Rajesh
e-mail: rajesh.mk@icar.gov.in

B. A. Jerard
ICAR-Central Island Agricultural Research Institute, Port Blair, Andaman and Nicobar Islands, India
e-mail: augustinejerard.b@icar.gov.in

M. K. Rajesh et al. (eds.), *The Coconut Genome*, Compendium of Plant Genomes,
https://doi.org/10.1007/978-3-030-76649-8_3

monotypic genus *Cocos,* is one of the 11 genera under the subtribe Attaleinae (Moore 1973; Dransfield et al. 2008). The taxonomic classification of the species is detailed below.

Kingdom - Plantae

Sub kingdom - Tracheobionta

Super division - Spermatophyta

Division - Magnoliophyta

Class - Liliopsida

Sub class - Arecidae

Order - Arecales

Family - Arecaceae

Sub family - Arecoideae

Tribe - Cocoseae

Sub tribe - Attaleinae

Genus - Cocos

Species - *nucifera* L.

The origin of coconut has been conjectured by several workers taking into consideration historical accounts, fossil records, and present status of dissemination and distribution of the palm. The present-day pan-tropical distribution of this palm, coupled with the absence of wild relatives, and truly wild coconut populations, adds to the complexity. Most authorities claim coconut to be native to Asia (de Candolle 1886; Beccari 1917; Mayuranathan 1938), while a few researchers claim its origin in West Coast of Central America (Martius 1850; Cook 1901) and on the adjacent islands of the Pacific (Child 1974; Purseglove 1968, 1985).

The tribe Cocoseae (with 18 genera arranged in three subtribes) to which *Cocos* belongs, is distributed predominantly in the New World, with only a few members in the Old World. Interestingly, among the 11 genera under the sub tribe Allaleinae (to which *Cocos* belongs), with the exception of coconut, one genera is endemic to South Africa and another two to Madagascar, while the remaining genera are distributed in South America. Accounting for the present-day distribution, the likely origin for the whole Cococeae tribe in western Gondwanaland appears more convincing and hence it is hypothesized that the tribe differentiated before the break up of that super-continent, with members radiating to different continents: becoming very diverse in the Americas; some drifting to the African and Madagascar Plates, where they survive to the present day; others rafting on the Indian plate, where they are now extinct. Dispersal of coconut, owing to their capability to float and drift through ocean currents, has become independent of plate tectonics (Harries 1990, 1992, 1995).

The progenitor of present-day coconut and/or its wild relatives are postulated to have become extinct, probably due to adverse geological and climatic events in the past. The plausible origin of the coconut palm, on the basis of phylogenetic studies, while ruling out an American origin, encompasses a vast region ranging from the Indian archipelago, northern Indian Ocean, Southeast Asia, Philippines, New Guinea, and the west Pacific, covering northern Indian Ocean, Malesia, Melanesia, and Micronesia. However, based on phylogenetic analysis of Cocoseae, Gunn (2004) postulated a South American origin of the lineage that eventually gave rise to the coconut. Phylogenetic studies on palms conducted in the past two decades have shown affinities of *Cocos* with the genera *Attalea, Syagrus, Lytocaryum, Parajubaea, Jubaeopsis, Butia* and *Allogoptera* (Gunn 2004; Baker et al. 2009, 2011; Noblick 2013; Meerow et al. 2009, 2014). Subsequently, Noblick and Meerow (2015) proposed the merger of *Lytocaryum* with *Syagrus*. Hence, closest affinity of *Cocos* has been reported with *Attalea* and *Syagrus,* both of which are predominantly distributed in South America (Noblick 2017).

Within *Cocos nucifera*, phylogenetic studies indicate genetic differentiation between the coconut populations of the Pacific and Indian Ocean basins, implying independent origins of coconut cultivation in these two regions and absence of substantial genetic admixture between these subpopulations, in spite of movement of coconuts by humans between the two oceanic basins and the absence of reproductive barriers (Gunn et al. 2011). Nayar (2017) has reviewed the available literature and elucidated the coconut timeline, considering coconut phylogeny with reference to plate tectonics/continental drift. The enigmatic coconut palm, regardless of its origin, is found distributed across much of the tropical coast, its dispersion plausibly aided by sea currents and sea farers. Therefore, the coconut palm is considered as a semi-domesticated species, with location-specific populations having evolved with varying degrees of dependence on humans.

3.2 Diversity in Forms

The coconut palm, despite the evolutionary bottlenecks, exhibits considerable variability in their morphological features, with several distinct populations and ecotypes. The available variability in crop genetic resources is an important pre-requisite for crop improvement research, to face emerging challenges to crop productivity and development of improved varieties.

The most visible diversity in case of the coconut palm is in terms of features of the fruit (Cook 1901; Patel 1938) and also plant growth habits. On the basis of growth habit, coconut palms are categorized into two broad classes, namely dwarfs and talls (Narayana and John 1949). The color of the fruit in coconut varies from shades of yellow, green, orange (red), and brown. Further, differences in shape and size of the fruit are also observed. The fruits, on the basis of size, are categorized as large, medium, small, and also very small or micro. Coconut fruits, on the basis of their shape are categorized as round, oval, ovoid, angled, triangular, oblong, egg-shaped, pear-shaped, or elliptic. Further, based on the equatorial view, the coconut fruits can be classified as angled, round, or flat based on the curvature of the fruit and the presence of ridges. The shape of the husked fruit (after removal of husk) varies from round, oval, oblong, pointed, ovoid, almost round, or oblate (IPGRI 1995; Ratnambal et al. 1995, 1999). In addition, differences are observed for almost all features of the coconut palm, starting from root, stem, leaf, inflorescence, and the fruits (Jerard et al. 2017a; Niral and Jerard 2017, 2018; Karun and Niral 2019).

3.2.1 Variability in Stem

Coconut palms normally have a single, unbranched stem. However, in rare instances variability such as polyembryony, branching, and suckering has been reported by various coconut researchers and from different countries. These are basically considered as freaks, as they are not transmitted across generations.

Branching in coconut is quite rare, but has been reported from various regions of the stem/stages of growth (Ridley 1907; Burkill 1923–25; Furtado 1923–25; Quisumbing 1926–27; Jacob 1935; Davis 1950, Indiresh 1992;

Mao and Lai 1993). Branching is most likely induced as a result of injury to the growing bud of the coconut palm, due to lightning, storm, pest attack, or some physical injury (Davis 1968; Balaga 1975).

Suckering, synonymous with branching but restricted to the underground portion of the stem (bole), is another rare phenomenon in coconut. Patel (1938) observed as many as 40 suckers arising from the base of a 25-year-old unflowered coconut palm in India. Davis (1956a) and Chatterjee (1959) also reported instances of suckering in coconut palms. More recently, suckering has been observed in three seedlings belonging to three different varieties at the ICAR-CPCRI in India. Suckering offers the possibility for vegetative propagation of the palms through stimulation of the possibly suppressed, rare basal buds in the bole region. The possibility of inducing suckering by causing mechanical injury has been demonstrated in coconut with very limited success, unviable for commercial propagation (Davis 1960; Michael and Verghese 1963).

Polyembrony, a phenomenon wherein more than one seedling is produced from one coconut, has been reported sporadically from different countries. As many as five seedlings have been reported from one coconut, as a consequence of the development of multiple embryos from one ovule (Furtado 1927; Davis 1979).

In addition to these novelties, another important variability observed in the stem, is pertaining to the differential annual rate of stem elongation, on the basis of which coconut palms are classified into two categories viz., talls and dwarfs (Fig. 3.1). This is a heritable trait, with dwarf/tall plant habit being transmitted from generation to generation. The general features of tall and dwarf coconuts are elucidated in Table 3.1.

Fig. 3.1 Tall coconut palm, with dwarf palms in the background

3.2.2 Variability in Leaf

The leaves of the coconut palm do not show much variability, except for variation in size, with dwarf palms, in general, having shorter leaves as compared to talls. However, in rare instances, palms with fused leaflets, wherein the leaflets do not separate as in ordinary palms, are observed in nature (Fig. 3.2), and are attributed to genetic and environmental factors (Patel 1938). While Hunger (1928) referred to these as sacred 'fetish palms', Zuniga et al. (1970) advocated the term 'plicata'. Plicata palms, show reduced leaf length, are late to flower and in general are poor yielders (Sugimura et al. 1994; Arunachalam et al. 2001). Though palms with boron deficiency also exhibit fused leaflets, these respond to borax application, unlike the plicata palms. Other rarities pertaining to leaves are forked leaves, twin leaves, multi-leaf in coconut palms and these are not heritable variants (Davis 1956b).

3.2.3 Variability in Inflorescence

Unlike the leaf, much more variations and abnormalities are reported in the coconut inflorescence (Niral and Jerard 2017, 2018). The coconut inflorescence referred to as a 'spadix' is borne on leaf axils and is enclosed by a spathe. The length of the coconut inflorescence ranges

Table 3.1 Characters associated with tall and dwarf coconut types

Trait	Tall types	Dwarf types
Length of stem at particular age	Relatively long	Relatively short
Girth of stem	Sturdier, with a prominent bole	Thinner, without prominent bole
Spacing of leaf scars on stem	Widely spaced	Closely spaced
Leaf	Longer, with a broader leaf base	Shorter, with narrow leaf base
Leaflets	Long and broader	Shorter and slender
Flowering initiation	Late (5–10 years)	Early (2–4 years)
Pollination type	Often cross-pollinated	Often self-pollinated
Fruit color within a population	Variable; mixture of greens and browns	Uniform; green, yellow, orange (red-yellow), or brown
Size of fruits	Large to small/micro	Medium to small
Copra (dry endosperm) quality	Good	Poor
Root distribution	Dense and plentiful	Less dense and few
Life span	Longer (60–80 years)	Shorter (30–40 years)

Fig. 3.2 A coconut palm showing fused leaflets

from 60 to 200 cm depending on the variety, cultural conditions, season, and age of bearing (Patel 1938; Menon and Pandalai 1958; Niral et al. 2008). The inflorescence has a central axis/rachis (peduncle), which bears about 30–35 branches, the spikelets, on which the male and female flowers are borne. The male flowers are numerous, while the few female flowers (0–5) are borne near the base of the spikelets. Sporadically, palms with unbranched inflorescences (a central axis, occasionally with one or two branches), referred to as Spicata (Fig. 3.3), are reported from all coconut-growing regions (Beccari 1916; Jacob 1941; Davis 1980). The inflorescences of the Spicata palm bear a large number of female flowers (125–130) directly on the central spike and have very few male flowers (50) borne towards the apical end and in between the female flowers, in stark contrast to the normal inflorescences. The Spicata palms cross freely with other varieties. The spicata trait is heritable but associated with meiotic abnormalities and on selfing the spicata palms, only 50 percent of the progenies breed true for this trait (John and Narayana 1949; Ninan and Raveendranath 1972).

Bulbil producing palms, wherein the flower bunches are transformed into bulbils, are another rarity, observed in nature (Ridley 1907; Venkataraman 1928; Davis 1948a, b). In such exceptional palms, the spadix is modified into a large bulbil shoot, or the various spikes transform into smaller bulbils, or in some cases, the female and/or male flowers transform into tiny vegetative shoots resembling bulbils and such clonally propagated bulbil shoots produce only vegetative shoots in place of normal inflorescences. However, such features are not associated with any chromosomal aberration (Raveendranath et al. 1975). Jerard et al. (2014a) reported a bulbiferous twin palm among West Coast Tall

Fig. 3.3 Coconut inflorescence: normal (left) and spicata (right) type

population, producing only bulbil shoots and confirmed through microsatellite analysis the genetic similarity of the twin mother palms and their bulbil progenies. Hence, bulbiferous palms offer scope for clonal propagation, provided their seed-fertility can be restored, and need to be conserved as a rare genetic resource for utilization in breeding programs.

In addition, some non-heritable rarities like double spadix (double inflorescence), multiple spathes, secondary branching of spikelets, partial suppression of spikelets, twin spikelets, fasciated spikes, terminal inflorescence (midget palm), are also reported in coconut (Niral and Jerard 2017, 2018). While in a normal coconut inflorescence, distinct male and female flowers are observed, in some instances a few hermaphrodite flowers are observed in small proportions (Gopal Rao 1948). Hermaphrodite flowers, resemble the female flower, but are much smaller and show development of the staminodal ring into stamens (ICAR-CPCRI 2014). Another type of bisexual flower due to the rare development of pistillodes in the male flowers is also reported (Furtado 1927; Patel 1938). Further, minor variation with respect to number of pistillodes, stamens, perianth of flowers has been reported (Menon and Pandalai 1958). Occasionally palms producing robust inflorescences with only male flowers and referred to as male coconut palm or androgena have been reported (John and Narayana 1942), while palms with only female flowers have not been reported so far. Spicata palms, while showing tendency towards femaleness, includes a much smaller number of male flowers.

3.2.4 Variability in Fruits

The coconut fruit, botanically a drupe, has a smooth exocarp, a fibrous mesocarp (husk), and hard endocarp (shell) encompassing the internal endosperm with embryo embedded in it. A thin layer of testa/seed coat (brown in color) is found to the outer surface of the endosperm, to the endocarp. The endosperm cavity (at the center of the fruit) is filled with sweet water, and is referred to as liquid endosperm. In the immature coconut fruits, the central cavity is completely filled with water (liquid endosperm), and the quantity of water gradually reduces during fruit development and on storage after harvest. A range of variation for fruit color (Fig. 3.4), size and fruit component characters are observed in coconut germplasm (Ratnambal et al. 1995, 1999; Bourdeix et al. 2010; Niral et al. 2013).

Fig. 3.4 Diversity for fruit colors in coconut

While a relatively wide range of variation is observed for weight of nut and endosperm content, very less variation is observed for the traits viz., the endosperm thickness and copra oil percentage per se (Niral et al. 2009; Gayathri et al. 2016). Rare palms producing very miniature-sized fruits of the size of an arecanut was first reported from Lakshadweep in India (ICAR-CPCRI 2002, 2009, 2012; Jerard et al. 2014c, 2017b). The mature fruits of the Mini Micro palm are similar in size to the buttons or immature fruits of a normal tree, but the size of the embryo is similar to that of normal nuts. However, these nuts do not germinate on their own, owing to the limited quantity of endosperm in the nut, and when the embryo of these nuts are transferred to culture media, the embryos germinate, thus enabling its propagation (Jerard et al. 2014c, 2017b).

Variations in fruit development, giving rise to aberrant types, are reported from different coconut-growing countries. A normal coconut fruit is tricarpic with an embryonic ovule, but with only one embryo developing in the fruit, and the three carpels are distinguished by the three ridges running from the stigmatic to the stalk end and three eyes on the endocarp of the coconut fruit. Rare instances of mono, bi, and tetracarpic fruits (but being one-seeded) have also been reported (Davis 1948c), with young fruits having upto 10 carpels also being reported (Menon and Pandalai 1958). Further, very rare instances of complete or partial apocarpic fruits (three carpels fused at the base and free at the apex), have been reported (Gadd 1924). Jacob (1940) reported a two-seeded coconut, with a leathery septum separating the two seeds, inside a common endocarp, while Forbes (1879) reported a fruit with 14 carpels, producing a tree with fourteen stems united at the base.

Horned coconuts, coconut fruits with flat horns ranging from one to six, are another

Fig. 3.5 Horned coconuts-with variable horn-like appendages

interesting but rare phenotype (Fig. 3.5). The development of horn-like appendages is attributed to viz., development of staminodes (Nair and Sadanandan 1976), duplication of segments of gynoecium (Petch 1924) and hypertrophy of perianth (Furtado 1927). Interestingly, this is a heritable trait, with progenies exhibiting the horned trait in varying frequencies (ICAR-CPCRI 2012; Jerard et al. 2014b).

Rare palms with sweet, edible immature mesocarp, in ancient times chewed by sea farers as an antidote to sea sickness, has been reported across countries by distinctive names: Kaithathali (India), Navasi Thembili (Sri Lanka), Preug-wan (Thailand), Cay dua bong (Vietnam), Kalapa Tebu (Indonesia), Klapa Logo (Malaysia), Cayamis, Caumanis, Taban, Tamisan, Mais (Philippines), Nu Wa or Cocos sucre (New Caledonia) and Uta (Rotuma, Fiji Islands).

Coconut fruits normally have a firm endosperm with the central cavity. The endosperm (kernel) of mature coconuts are normally firm, but variants with buttery consistency, referred to as makapuno coconut, have been reported first from Philipiness and Indonesia (Omar 1919; Adriano and Manahan 1931; Copeland 1931; Torres 1937). In India, it is referred to as "thairuthengai", considering that the entire internal cavity of the shell is filled with a jelly-like substance of the consistency of thick curd (Patel 1938; Jerard et al. 2013). Similar buttery endosperm variants are reported from many countries: Kopyor (Indonesia), Klapa Dadeh (Malaysia), Mapharao Khati (Thailand), Sap (Vietnam), Dikri Pol (Sri Lanka), Coco gra (Seychelles), Niu Garuk (Papua New Guinea) and Pia (Polynesia). This trait is heritable, being controlled by recessive genes (Zuniga 1953), and associated with higher deposition of galactomannan in makapuno endosperm due to reduced activity of enzymes involved in galactomannan degradation namely, alpha-D-galactosidase, beta-mannanase and beta-mannosidase (Samonte et al. 1989) and epigenetic effects (Angeles et al. 2018). The sweet kernel coconut (Mohachao Narel) is another important mutant form reported in India, wherein the endosperm (kernel) is sweet with very less fibers, making it extremely palatable for table purpose and use as a salad (Samsudeen et al. 2010, 2013a). Like the makapuno, the sweet endosperm trait is expressed only in a few fruits within a bunch. But unlike the makapuno, the sweet endosperm trait is heritable and expressed in seed propagated progenies of Mohachao Narel. At ICAR-CPCRI, India, in the conserved Mohachao Narel population, the percentage of sweet kernel fruits varied from 4.5% to 60% (ICAR-CPCRI 2017).

Another interesting novelty is the aromatic coconut (Omar 1919), wherein the liquid endosperm (tender nut water) has a pleasant "pandan-like" aroma, similar to the aroma of aromatic rice, due to the presence of 2-acetyl-1-pyrroline (2AP). This trait is heritable and governed by a single recessive gene (Saensuk et al. 2016). Aromatic coconut has been reported from different countries viz., Klapawangi—Malaysia, Aromatic Green Dwarf—Philippines, and Namhom—Thailand (Niral and Jerard 2017, 2018).

Minor variations, like differences in color of testa in coconut fruits from the normal brownish color, are also reported, with rare instances of white testa (Tampake et al. 1982). Variations in husk consistency are also observed, with an easily peelable husk observed in rare palms, unlike the firm adhering husk of normal coconuts, and thin husked fruits (Kelapa Bawang/ Onion Coconut—Indonesia, Lupison—Philippines, Nudam—Fiji). Unlike the normal white husk seen in tender coconut fruits, in rare instances, pink husked fruits are reported (Ran Thembili—Sri Lanka, Guelle Rose—Mauritius. Jerard et al. (2016) reported a coconut palm bearing large fruits with pink colored mesocarp in San Ramon Tall population in India (Fig. 3.6); wherein two types of male flowers were observed in an inflorescence: one with dark pink and another with yellow color anther filaments. Although there were no morphological difference in size of male flowers and anthers, the flowers with pink filaments could be easily identified by the pinkish tinge at the bottom of tepals. High phenolic, flavonoid, and anthocyanin content were recorded in husk tissues of the palm with pink mesocarp, compared to palms with normal husk color (ICAR-CPCRI 2017).

Fig. 3.6 Tender coconut fruit with pink color mesocarp

Further, thick-shelled nuts have been reported across the major coconut-growing countries viz. Sri Lanka (Poropol) and Philippines (Tutapaen), India, China, Indonesia (Niral and Jerard 2017, 2018).

3.3 Varieties

The presence of intra-specific variability in coconut has been reported by researchers across the world. The documentation of the different varieties of coconut has been done by a number of workers, starting with the listing of 13 local varieties of Netherlands Indies in Rumphius' *Herbarium Amboinense* (1741), followed by the listing of 18 varieties with Latin names and descriptions of the coconut populations in Dutch colonies (Miquel 1855). Watt (1889) reported seven varieties from South Asia, followed by many more variants reported from different countries in the region by various authors (Cook 1901; Burkill 1919; Omar 1919; Copeland 1931). There have been similar reports on the variability observed in different countries of South and South East Asia as well as the Neotropics (Patel 1938; John and Narayana 1949; Gangolly et al. 1957; Harries 1971; Zizumbo et al. 2005).

3.3.1 Classification of Varieties

Narayana and John (1949), acknowledging the need for a proper understanding of the variability observed in coconut, categorized coconut palms of Asia (South and South East Asia) into two groups, viz. Talls and Dwarfs. Within these two groups, they further assigned varieties and forms, to account for the variability reported from the major coconut producing countries: Tall group with three varieties viz., var. Spicata (one form—unbranched inflorescence), Androgena (one form—male tree), Typica [nine forms—Ramona, Kappadam, Gigantea, Nova-guineana, Cochin-chinensis, Malayensis, Siamea, Laccadive, Pusilla (Fig. 3.7)] and Dwarf group with two varieties namely, Nana (two forms—Nana, Maldiviana) and Javanica (one

Fig. 3.7 Laccadive Micro Tall, var. typica: a small-fruited cultivar from India

form—intermediate between tall and dwarf). However, since the varieties from the Pacific islands, Africa or America were not included, this classification was not considered comprehensive. Liyanage (1958) adopted this classification with some revisions for Sri Lankan coconut and advocated three groups viz., Typica (with eight tall forms viz., Typica, Kamandala, Bodiri, Navasi, Ran Thembili, Gonthembili, Parapol, Dikripol), Nana (with three dwarf forms viz., pumila, eburnea, regia) and Aurantiaca (with two semi-tall forms namely, King Coconut/thembili, Navasi Thembili). Carlos (1963) reported four botanical varieties in the Philippines viz., Typica, Nana, Javanica, and Spicata.

While the earlier classification systems, were based on palm/fruit morphology, Fremond and co-workers (Fremond et al. 1966) suggested a system of classification based on system of pollination with two groups, viz., autogamous (self-pollinating as in the case of most of the Dwarfs) and allogamous (cross-pollinating as in the Talls). However, this was regarded as cumbersome, as it required studying the pollination behavior of the local populations/varieties. Further, Talls are also capable of self-pollination and cross-pollination has also been reported in Dwarfs when surrounded by Talls (Whitehead 1965; Rognon 1976).

To overcome these limitations, Harries (1978) advocated use of fruit component analysis to

Fig. 3.8 West Coast Tall—representative of variety Typica

enable comparison of varieties across countries/growing environments and introduced the Niu Kafa-Niu Vai concept of coconut variety identification. Niu Kafa represents coconuts evolved through natural selection in uninhabited islands and coral atolls and characterized by long angular, thick-husked fruits, with the capability of late germination, which facilitates its survival under natural conditions. Niu Vai types are large fruited, with higher copra content, and are under commercial cultivation in inland areas. This classification did not include the dwarf varieties as these were considered to have evolved much later and are capable of surviving only under cultivation. The ancestral Niu Kafa and Niu Vai types are observed to maintain their distinctive characteristics in the Central American isthumus, where these populations are geographically isolated (Richardson et al. 1978), while in Asia and Pacific, human intervention has brought these forms together, with introgression of the two forms resulting in the development of many of the cultivated types available today. Harries (1981), classified a few of the present-day coconut populations into Niu Vai and Niu Kafa types: Niu Kafa type—West African Tall and the coconuts along the Atlantic and Caribbean coast of America, such as Jamaica Tall; Niu Vai type—Malayan Tall; Niu Vai-Niu Kafa introgressed population—Tahiti Tall. Further, it was suggested that Niu Vai and Niu Kafa types are only minor variants in most of the Pacific Ocean islands and the coconut populations in these islands are predominantly introgressed forms similar to the Tahiti Tall but sufficiently different to warrant distinct names.

Satyabalan (1997) utilized available information on fruit size and fruit component traits in coconut to categorize known tall and dwarf varieties/cultivars. The tall varieties/cultivars were grouped into five groups: Group 1: Fruits very large, thin husk, thin shell, thin endosperm, more water; copra content >300 g, (<30% of husked fruit weight); Group 2: Fruits large, thin husk, thick shell, thin endosperm, more water; copra content 200–300 g (<30% of husked fruit weight); Group 3: Fruits large, thin husk, thick shell, thick endosperm, less water; copra content 200–300 g (>30% of husked fruit weight); Group 4: Fruits medium-sized, thick husk, thick shell, thick endosperm, less water; copra content <200 g (>30% of husked fruit weight); Group 5: Fruits medium-sized, thick husk, thick shell, thin endosperm, more water; copra content <200 g (<30% of husked fruit weight). The dwarf varieties were categorized under three fruit colors namely, green, yellow, and red/orange with three groups in each category. While the green dwarfs were grouped into three: Group 1: Fruits small; high copra (32–43%) and shell content (26–39% of husked fruit weight); Group 2: Fruits medium-sized; less copra (23%) and slightly less shell content (23–27% of husked fruit weight); Group 3: Fruits medium-sized; slightly higher copra (25–35%) and lesser shell content (20–25% of husked fruit weight). On the other hand, considering the similarity for fruit component traits within the yellow and red/orange groups, three similar groups have been proposed under each of these two color categories: Group 1: Fruits medium small; high copra (24–35%) and shell content (20–29% of husked fruit weight); Group 2: Fruits medium-sized; less copra (21–30%) and shell content

(16–19% of husked fruit weight); Group 3: Fruits medium-sized; slightly higher copra (24–31%) and slightly lesser shell content (21–27% of husked fruit weight). However, this categorization was also not comprehensive due to non-inclusion of the brown dwarfs and the distinctive Niu Leka Dwarf. Hence, classification based on morphology and growth habit continues to be popular among coconut researchers and growers, with varieties being categorized into Tall and Dwarf, with varietal nomenclature prefixed by the name of the country/place of origin and/or fruit color (in case of Dwarfs).

Studies on molecular marker diversity in coconut populations (Gunn et al. 2011), has indicated two major groups: the distinct Indo-Atlantic group (includes Niu Kafa coconut types) and the more diverse Pacific group (includes the domesticated coconut types) with five sub-groups (Southeast Asia, Micronesia, Melanesia, Polynesia, and the Pacific coast of Central and South America). Human intervention through migration and cultivation, however, has brought the divergent forms together and the opportunity for cross-pollination has resulted in development of intermediate introgressed forms. As a result, there is no single method of classification that can comprehensively account for all variability reported in the global coconut populations.

The variations observed in coconut populations have been reported by several authors, from different countries/varied growing regions. While the term variety has different interpretations, in the case of coconut, the term variety is adopted to denote a group of palms having similar characteristics and which can be reproduced from generation to generation (Menon and Pandalai 1958; Ohler 1984). Copeland (1931) opined on the inability to determine whether coconut varieties breed true, while Jack and Sands (1922) reported on low percentage of palms being true to type and hence many coconut workers instead of the term varieties have used the terms ecotypes, biotypes, strains, races, or forms to describe the different forms of coconuts, in addition to agricultural varieties. Nayar (2018), stated that the coconut palm has not undergone significant ecological differentiation (in comparison to rice, for example) and hence the epithet ecotype is inappropriate and suggested the use of the term morphotypes for the native variability in forms observed in the coconut-growing regions. Niral and Jerard (2018), taking into consideration the wisdom of the earlier and present coconut researchers, suggested refinement of nomenclature to describe the different forms of coconut and remove the existing chaos in nomenclature among the coconut workers. In taxonomical terms, variety is ranked just below the level of species, but above subvariety and form. Hence, in coconut, the four broad categories of coconut palms suggested by Bourdeix et al. (2017) viz. *typica, nana, javanica* and *aurantica* can be considered as botanical varieties. The variety *Typica*, representing the predominantly observed cross-pollinating tall type coconut palms, would include the traditional tall types such as Malayan Tall, Indian West Coast Tall (Fig. 3.8), West African Tall, Solomon Tall, Rennell Tall. Variety *Nana* representing self-pollinating palms with dwarf plant habit, would include the typical dwarfs, viz., Chowghat Green Dwarf, Chowghat Orange Dwarf (Fig. 3.9), Malayan Orange Dwarf, Malayan Yellow Dwarf, Rangiroa Red Dwarf. The Niu Leka Dwarf (Fig. 3.10), with dwarf plant habit and cross-pollinating nature, and other compact dwarfs will be classified as variety *Javanica*. Variety *Aurantica* encompassing palms with semi-tall plant habit and with variable reproduction modes (cross/self-pollinating), would include Navasi Thembili and King Coconut and the Dwarf × Tall (Fig. 3.11) and Tall × Dwarf hybrids (Niral and Jerard 2018).

3.4 Germplasm Resources

Plant genetic resources are invaluable and an essential requirement for undertaking crop improvement research. Variability in the gene pool forms the basis for selection, identification of parental lines for hybridization and undertaking improvement for targeted traits, overcoming vulnerability to various biotic/abiotic factors, meeting challenges emerging from climate

Fig. 3.9 Chowghat Orange dwarf—representative of variety Nana

Fig. 3.11 Chandra Sankara (D × T hybrid from India)—representative of variety Aurantica

Fig. 3.10 Niu Leka dwarf—representative of variety Javanica

change threats, grower/consumer preferences, and market-driven product diversification.

3.4.1 Germplasm Conservation

Coconut, being a perennial palm, enables long-term conservation over successive generations. Conservation of coconut genetic resources can be undertaken either ex situ or in situ. Traditionally, conservation of coconut germplasm resources has been undertaken *ex situ* in field gene banks. In the case of coconut, *in situ* conservation of adapted germplasm in the local environment can be beneficial considering the long economic life span of >50 years. Together with *ex situ* field gene banks, on-farm conservation can enable protection of a much larger gene pool (Samsudeen et al. 2013b), though the success of in situ on-farm conservation, would depend on profitability of coconut-based farming systems (Sangalang 2004). Further, although pollen competition may occur (Sangare 1981), the absence of inter or intra varietal sterility in coconut fruit setting, favors easy management and maintenance of populations in gene banks. The year-round production of nuts in coconut palms, facilitates sequential planning of breeding program as well as multiple uses in experiments over a longer period of time.

Considering the importance of coconut in sustaining livelihood security of marginalized sections of the society and resource-poor communities and the fact that about 96 per cent of the coconut farmers worldwide are smallholders and also the falling productivity of coconut in traditional coconut communities, the IPGRI in 1991, created the International Coconut Genetic

Resources Network (COGENT) to promote conservation and use of coconut genetic resources and promote research to ensure the sustainability of the coconut economy. COGENT, beginning with 15 coconut-growing countries as members, currently has 39 member countries that are divided into five sub-networks: Southeast and East Asia (Philippines, Indonesia, Malaysia, Myanmar, Vietnam, Thailand, and China); South Asia and Middle East (India, Sri Lanka, Bangladesh, Pakistan and Sultanate of Oman); South Pacific (Papua New Guinea, Fiji, Solomon Islands, Cook Islands, Vanuatu, Samoa, Kiribati, and Tonga); Africa and the Indian Ocean (Côte d'Ivoire, Benin, Ghana, Madagascar, Kenya, Nigeria, Mozambique, Tanzania, and Seychelles); and Latin America and the Caribbean (Mexico, Brazil, Colombia, Jamaica, Costa Rica, Cuba, Guyana, Trinidad-Tobago, Haiti, and Honduras), and strives to coordinate and strengthen germplasm conservation and related research efforts among the different stakeholders, spread across the different countries. Further, to promote effective access and safe movement of germplasm, COGENT established five multi-site International Coconut Genebanks (ICGs), spread across the five regions: Southeast and East Asia—Indonesia; South Asia and Middle East—India; South Pacific—Papua New Guinea; Africa and the Indian Ocean—Côte d'Ivoire; Latin America and the Caribbean—Brazil (Batugal et al. 2005; Niral et al. 2010). In recent times, considering the rising threats of genetic erosion, including the challenges from climate change vagaries, COGENT, on the basis of interactions with more than 90 key stakeholders encompassing different nationalities/institutions and subject matter experts, have developed a comprehensive strategy document on conservation and use of coconut genetic resources (COGENT 2017).

Acknowledging the importance of coconut germplasm resources for enhancement of productivity and mitigating genetic erosion in native habitats, collection and conservation of coconut germplasm have received considerable attention in almost all major coconut producing nations, with Philippines, India, Indonesia, and Sri Lanka establishing ex situ field gene banks in the early part of the twentieth century. Coconut germplasm collection activities were first made between 1901 and 1910 in Indonesia and Madagascar (Ohler 1984), with Samoan Tall planted in Solomon Islands in 1912 being the oldest accession registered in the COGENT's Coconut Genetic Resources Database (CGRD), encompassing 1416 accessions maintained in the field gene banks, spread across 23 countries (Batugal et al. 2005). Presently, 1837 accessions are conserved in the 24 gene banks (Nampoothiri and Parthasarathy 2018).

A large area, considerable manpower as well as financial resources are required for establishment and management of the gene bank. The COGENT recommends conservation with a population size of 90 palms for Talls and 45 palms for the more homogenous Dwarfs. Further, the planting density adopted for coconut, varies from 80 to 250 palms per hectare, in tune with varieties and cultivation practices (COGENT 2017). The time taken for flowering, after field planting, varies from one year in the most precocious Dwarfs to more than seven years for certain Tall types. The time taken for seed development is also long: requiring about 10 to 12 months from pollination of the female flower to maturity as a viable seed nut, followed by another 6–12 months for obtaining seedlings for planting. Further, the recovery percentage of good seedlings (for field planting) in the nursery is around 65%. The fruits being recalcitrant are not amenable for long-term storage. The bulkiness of the coconut seed nut and seedling makes transportation of planting material expensive for farmers (from seed gardens/nurseries to fields) and gene banks (international movement) alike. On the positive side, the perennial nature of the coconut palm, with fruits being borne all year through, is conducive for long-term conservation in field gene banks, and facilitates undertaking of many experiments over the long life span of the palm.

In addition to the conservation in field gene banks, there is a need for complementary conservation of germplasm. Cryopreservation is the best alternative for *in vitro* long-term conservation of recalcitrant seed species and hence suitable for cost-effective complementary conservation of coconut germplasm. Whitehead

(1966) demonstrated the possibility of freeze-drying coconut pollen and subsequently the simplicity and affordability of pollen cryopreservation technique have been demonstrated by a number of researchers (Karun et al. 2006, 2014). Pollen cryopreservation technique can be applied for conservation of pollen of accessions conserved in the field gene banks and also pollen of original palms collected from farmers gardens for utilization in controlled pollination for regeneration of the accessions. On the other hand, cryopreservation of embryos can be useful in regeneration of germplasm as well as exchange of germplasm, and hence conservation and establishment of cryo-genebank of coconut embryos is important. However, this would require production of a much larger number of seeds of each accession through controlled pollination to ensure the pedigree of the genotype being conserved. Though techniques for cryopreservation of embryos have been developed in different coconut producing countries (Assy-Bah and Engelmann 1992; Sisunandar et al. 2010; Sajini et al. 2011), the protocol still requires refinement. Further, the number of embryos to be conserved per accessions is yet to be finalized, with some researchers proposing conservation of 600 embryos per accession (COGENT 2017). Work on refinement of embryo culture protocol has resulted in the development of another alternative conservation system based on the cryopreservation of embryogenic calli, produced from plumular tissue of the embryos, with potential for utilization in clonal production (Welewanni et al. 2020). However, cryopreservation of embryogenic calli will require controlled pollination for production of seed nuts to be utilized for extraction of sufficient embryos for production of embryogenic callus, followed by conservation of callus from each embryo, with targeted conservation of about 75 sets of embryogenic callus per genotype (COGENT 2017).

3.4.2 Germplasm Cataloging

To enable better utilization of genetic resources it is essential that the accessions are properly cataloged. In the case of coconut, about 1837 accessions are conserved in the 24 gene banks (spread across 23 countries), including 730 accessions in the five international gene banks, much of these germplasm resources remain underutilized due to inadequate funding support for systematic characterization, evaluation, and documentation of the accessions. Considering the need to facilitate, popularize and create public awareness about coconut genetic resources among different stakeholders, COGENT, taking into consideration the standardized descriptors for coconut palms and the methods detailed in the STANTECH manual, developed and released the coconut genetic resources database (CGRD) in 1999 (Hamelin et al. 2005). The CGRD provides information on conserved coconut germplasm under two main parts: passport data and priority characterization, and evaluation data. The accessions referenced in the International Coconut Genetic Resources Database encompasses 90,000 living palms, equivalent to 530 hectares of coconut plantations (COGENT 2017).

India is one of the few countries, where germplasm conservation has received considerable focus, with introduction of coconut cultivars from the major coconut-growing countries viz., Philippines, Malaysia, Indonesia, Sri Lanka, Vietnam, and Fiji way back in 1924. In India, ICAR-CPCRI hosts the National Active Germplasm Site (NAGS) for plantation crops and hosts the largest collection of 455 accessions, representing coconut germplasm from 28 countries, encompassing South and South East Asia, Caribbean Islands, Indian Ocean Islands, Pacific Ocean Islands, and African countries. In addition, India also hosts the International Coconut Genebank for South Asia and Middle East, with 91 accessions, comprising designated Indian germplasm, germplasm from regional member countries viz. Sri Lanka and Bangladesh and accessions collected through prospection from the Indian Ocean Islands of Madagascar, Mauritius, Seychelles, Maldives, Comoros, and Reunion (Niral et al. 2019). The first catalog of coconut germplasm, targeted for coconut researchers, was also published in India (Ratnambal et al. 1995): entitled 'Coconut descriptors', 48 accessions were described using color plates of its different parts along with two pages of textual information following standardized coconut descriptors (IPGRI 1995). Subsequently, a second volume of the publication, describing another 26 accessions was published and distributed in CD-ROM (Ratnambal et al. 2000). These earliest coconut catalogs did not document information on ethnobotany nor historical description of the germplasm resource. Considering this lacunae, the COGENT initiated the development of Catalogue of Conserved Coconut Germplasm and the Catalogue of Farmers' Coconut Varieties, including technical information, general information, interesting antecedents as well as attractive photographs of the material to cater to the needs of a wider stakeholder group, including researchers, farmers and general public (Bourdeix and Batugal 2005). While the catalog of Conserved Coconut Germplasm provides one page each of pictorial description and textual information of important germplasm conserved in genebanks of COGENT member countries, the other describes in a similar manner distinct coconut varieties identified in farmers' fields in different coconut communities of COGENT member nations under the Asian Development Bank (ADB) and the International Fund for Agricultural Development (IFAD)-funded projects (Bourdeix et al. 2010). More recently, considering the importance of the coconut palm in supporting livelihood security especially in resource-poor communities and the importance of genetic resources for sustenance, COGENT with inputs from researchers as well as other stakeholders has developed the Global Strategy for Conservation and use of Coconut Genetic Resources (COGENT 2017).

3.5 Conclusion and Future Prospects

Diversity in coconut forms has been reported across all coconut-growing countries and conservation of the variability has also received support in major coconut producing nations, resulting in the establishment of field gene banks in 23 countries. Further, taking into consideration the importance of coconut in sustaining livelihood of small and marginal farmers and fragile coastal communities and to promote safe conservation of genetic diversity for posterity in the context of genetic erosion emanating from climate change vagaries as well as crop substitution and pressures on native coconut plantations from urbanization, the IPGRI-COGENT has facilitated establishment of five international gene banks spread across the five diverse regions, one each in South East Asia, Pacific, South Asia and Middle East, Africa and Indian Ocean, and South America. These gene banks in addition to conservation of the distinct coconut subpopulations are also preserving some of the mutant types having potential applications in the breeding program.

To enable better utilization of the conserved germplasm a comprehensive documentation of the available germplasm wealth is imperative so as to utilize the available genepool. Characterization of genetic resources is an important prerequisite in crop improvement research for understanding the variability available for different traits and also efficient selection of parents for breeding. The focus of cultivators has always been to obtain higher yields. Since ancient times discerning farmers have practiced the selection as an art, to achieve higher productivity. In the early nineteenth century, to meet the food needs of the burgeoning population, scientific crop improvement programs were developed with focus on development of improved varieties suitable for different agro-ecological zones, leading to the enhancement of production and productivity in all major crop commodities. With coconut being no exception, crop improvement researches have focused primarily on higher yield, in terms of fresh fruit, copra, or oil. Subsequently, growing pressures on productivity and profitability of coconut have resulted in the emergence of novel areas of focus: climate resilience, pest/disease resistance, tender nut water quality, yield and quality of inflorescence sap, and also suitability for product diversification/by product development. Further, there is a need for more focused work on identification of trait-specific donors, as well as genetically diverse lines for better exploitation of heterosis and undertake trait-specific crop improvement program. Comprehensive characterization of genetic resources, through robust phenotyping and genotyping techniques and application of genomic resources will aid in better utilization of invaluable coconut genetic resources for meeting the requirements of the coconut growers and the industry and to ensure the prosperity of coconut communities.

References

Adriano FT, Manahan M (1931) The nutritive value of green, ripe and sport coconut (buko, niyog and makapuno). Philipp Agric 20:195–198

Angeles JGC, Lado JP, Pascual ED, Cueto CA, Laurena CA, Laude PR (2018) Towards the understanding of important coconut endosperm phenotypes: is there an epigenetic control? Agron 8(10):225

Arunachalam V, Jerard BA, Elangovan M, Ratnambal MJ, Dhanapal R, Rizal SK, Damodaran V (2001) Unexploited diversity in coconut palm (*Cocos nucifera* L.). Plant Genetic Resources Newslett 127:39–43

Assy-Bah B, Engelmann F (1992) Cryopreservation of mature embryos of coconut (*Cocos nucifera* L.) and subsequent regeneration of plantlets. Cryo-Lett 13:117–126

Baker WJ, Norup MV, Clarkson JJ, Couvreur TLP, Dowe JL, Lewis CE, Pintaud JC, Savolainen V, Wilmot T, Chase MW (2011) Phylogenetic relationships among arecoid palms (Arecaceae:Arecoideae). Ann Bot 105:1417–1436

Baker WJ, Savolainen V, Asmussen-Lange CB, Chase MW, Dransfield J, Forest F, Harley MM, Uhl NW, Wilkinson M (2009) Complete generic-level phylogenetic analyses of palms (Arecaceae) with comparisons of supertree and supermatrix approaches. Systematic Biol https://doi.org/10.1093/syslro/sypo21

Balaga HY (1975) Induction of branching in coconut. Kalikasan 4:135–140

Batugal P, Ramanatha Rao, V, Oliver J (eds) (2005) Coconut genetic resources. International Plant Genetic Resources Institute–Regional Office for Asia, the Pacific and Oceania (IPGRI-APO), Serdang, Selangor DE, Malaysia

Beccari O (1916) Il genere *Cocos* Linn. e le palme affini. L'Agricoltura Coloniale 10:435–437, 489–532, 585–623

Beccari U (1917) The origin and dispersal of *Cocos nucifera*. Philipp J Sci Series C: Bot 12:27–43

Bourdeix R, Batugal P (2005) Standardized catalogues of coconut germplasm: Catalogue of conserved germplasm and farmers' varieties. In: Batugal P, Ramanatha Rao V, Oliver J (eds) Coconut genetic resources. International Plant Genetic Resources Institute, Serdang, pp 456–462

Bourdeix R, Batugal P, Oliver JT, George MLC (eds) (2010) Catalogue of Conserved Coconut Germplasm. International Coconut Genetic Resources Network (COGENT), Bioversity International, Regional Office for Asia, the Pacific and Oceania. Serdang, Selangor Darul Ehsan, Malaysia, p 388

Bourdeix R, Perera L, Rivera RL, Saena-Tuia V, Masumbuko L (2017) Global Coconut communities—status and strategies in *in situ* diversity management and utilization. In: Chowdappa P, Niral V, Jerard BA, Samsudeen K (eds) Coconut. Daya Publishing House, A Division of Astral International Pvt. Ltd. New Delhi, India, pp 407–441

Burkill IH (1919) Some notes on the pollination of flowers in the Botanical Garden, Singapore and other parts of Malay Peninsula. Gardens' Bulletin Singapore, 2:165–176

Burkill IH (1923–25) The fertility of branched coconut palms. Garden's Bull. Straits Settlement 3:1–2

Carlos T (1963) Agronomic studies in some cultivars of coconut in Philippines. M.S Thesis, UPLB College. Laguna, Philippines

Chatterjee D (1959) On multiple seedling of coconut. Sci Culture 25:370

Child R (1974) Coconuts, 2nd edn. Longmans, Green & Co., London

COGENT (2017) A global strategy for the conservation and use of coconut genetic resources, 2017–2027. (Bourdeix R, Prades A: compilers). Bioversity International, Montpellier, France

Cook OF (1901) Origin and distribution of cocoa palm. Contr U.s Nat Herb 7:57–97

Copeland EB (1931) The Coconut, 3rd edn. Macmillan & Co., London

Davis TA (1948a) Proliferation in two coconut palms (*Cocos nucifera* L.). Ind Coconut J 2: 11–17

Davis TA (1948b) Abnormal palms of Travancore III. A bulbiferous coconut palm. J Bombay Nat Hist Soc 47:527–529

Davis TA (1948c) Abnormal palms of Travancore IV. Polycarpy in a coconut (*Cocos nucifera* L.). J Bombay Nat Hist Soc 48:704–706

Davis TA (1950) Branching in some Indian palms. Ind Coconut J 3:135–145

Davis TA (1956a) Freaks of nature. I Suckering in Coconut Palms. Coconut Bull 9:179–180

Davis TA (1956b) Forked and fused leaves of the coconut palm. Coconut Bull 10:51–53

Davis TA (1960) Clonal propagation in coconuts. Curr Sci 29:273

Davis TA (1968) Difficulties in the genetic improvement of the coconut: a promising alternative method. Ind J Genet 28A:154–164

Davis TA (1979) Some unusual formations in palms. Principes 23:80–83

Davis TA (1980) Super abundant female flower production in a coconut: a report. Philippines J Coconut Studs 5:23–30

de Candolle A (1886) Origin of cultivated plants, 2nd edn. Hafner Publ. Co., New York, p 468
Dransfield J, Uhl NW, Asmussen CB, Baker WJ, Harley MM, Lewis CE (2008) *Genera palmarum*. The evolution and classification of palms. Royal Botanic Gardens, Kew, p 732
Forbes HO (1879) Notes on *Cocos Nucifera*. J Bot 17:193–194
Fremond Y, Ziller R, de Nuce de Lamothe M (1966) The Coconut Palm. International Potash Institute, Berne/Switzerland
Furtado CX (1923–1925) Branched coconut palms and their fertility. Gdns' Bull. Singapore 3: 274–279
Furtado CX (1927) Teratological notes. A. Abnormalities in coconut palms. Gdns' Bull Singapore 4:78–84
Gadd CH (1924) An abnormal inflorescence from a nut. Year Book, Dept Agric, Ceylon, 21–23
Gangolly SR, Satyabalan K, Pandalai KM (1957) Varieties of the coconut. Ind Coconut J 10:3–28
Gayathri UK, Niral V, Jerard BA, Samsudeen K (2016) Evaluation of fruit component traits in conserved coconut accession of Pacific Ocean region. In: Chowdappa P, Muralidharan K, Rajesh MK, Ramesh SV (eds) Abstracts: 22nd Plantation Crops Symposium (PLACROSYM 22), 15–17 December, 2016, ICAR-CPCRI, Kasaragod, pp 86
Gopal Rao T (1948) A note on the occurrence of a hermaphrodite flower in coconut *(Cocos nucifera* L.). J Indian Bot Soc 27:208–211
Gunn B (2004) The phylogeny of the Cocoeae (Arecaceae) with emphasis on *Cocos nucifera*. Ann Missouri Bot Gard 91:505–522
Gunn BF, Baudouin I, Olsen KM (2011) Independent origins of cultivated coconut (*Cocos nucifera* L.) in the old world tropics. PLoS One 6(6): e21143
Hamelin C, Bourdeix R, Baudouin L (2005) The international coconut genetic resources database. In: Batugal P, Ramanatha Rao V, Oliver J (eds) Coconut genetic resources. International Plant Genetic Resources Institute, Serdang, pp 427–438
Harries HC (1971) Coconut varieties in America. Oleagineux 26:235–242
Harries HC (1978) The evolution, dissemination and classification of *Cocos nucifera*. Bot Rev 44:265–320
Harries HC (1981) Practical identification of coconut varieties. Oleagineux 36:63–72
Harries HC (1990) Malesian origin for a domestic *Cocos nucifera*. In: Baas P et al (eds) The Plant Diversity of Malesia, pp 351–357
Harries HC (1992) Biogeography of the coconut, *Cocos nucifera* L. Principes 36:155–162
Harries HC (1995) Coconut (*Cocos nucifera* L., Palmae). In: Smartt J and Simmonds NW (eds) Evolution of crop plants, 2nd edn. Longman Scientific and Technical, Burnt Mill, Harlow, UK, pp 389–394
Hunger FWT (1928) Over Fetisch-Klapperboom. Indische Culturen 13:4–6
ICAR-CPCRI (2002) Annual Report 2001–02. Central Plantation Crops Research Institute, Kasaragod, Kerala, India, p 120
ICAR-CPCRI (2009) Annual Report 2008–09. Central Plantation Crops Research Institute, Kasaragod, Kerala, India, p 145
ICAR-CPCRI (2012) Annual Report 2011–12. Central Plantation Crops Research Institute, Kasaragod, Kerala, India, p 128
ICAR-CPCRI (2014) Annual Report 2013–14, ICAR-Central Plantation Crops Research Institute, Kasaragod, Kerala, India, p 139
ICAR-CPCRI (2017) Annual Report 2016–17, ICAR-Central Plantation Crops Research Institute, Kasaragod, Kerala, India, p 236
Indiresh KM (1992) Branching in Coconut. Ind Coconut J 22:14–15
IPGRI (1995) Descriptors for coconut (*Cocos nucifera* L). International Plant Genetic Resources Institute, Rome, Italy, p 61
Jack HW, Sands WN (1922) The dwarf coconut in Malaya. Mal Ag J 10:4–12
Jacob KC (1935) Stem fasciation in the coconut palm. J Bombay Nat Hist Soc 37:901
Jacob KC (1940) A bicellular coconut *(Cocos nucifera* L.). J Bombay Nat Hist Soc 41:905–906
Jacob KC (1941) A new variety of coconut palm (*Cocos nucifera* Linn.) var. spicata (Jacob KC). J Bombay Nat Hist Soc 41:906–907
Jerard BA, Niral V, Chowdappa P (2017a) Breeding. In: Chowdappa P, Niral V, Jerard BA, Samsudeen K (eds) Coconut. Daya Publishing House, A Division of Astral International Pvt. Ltd., New Delhi, India, pp 131–173
Jerard BA, Rajesh MK, Thomas RJ, Niral V, Samsudeen K (2017b) Island ecosystems hosts rich diversity in Coconut (*Cocos nucifera*): evidences from Minicoy islands, India. Agr Res 6:214–226
Jerard BA, Damodaran V, Niral V, Samsudeen K, Rajesh MK, Sankaran M (2013) Conservation and utilization of *Thairu Thengai*– soft endosperm coconut. accession from Andaman Islands. J Plantn Crops 41(1):14–21
Jerard BA, Niral V, Samsudeen K, Gayathri UK (2016) Pink husked coconut selection- a trait of promise. In: Chowdappa P, Muralidharan K, Samsudeen K, Rajesh MK (eds) Abstracts—3rd International Symposium on Coconut Research and Development (ISOCRAD-3), 10–12 December, 2016, ICAR-Central Plantation Crops Research Institute, Kasaragod, pp 21
Jerard BA, Rajesh MK, Elain SE, Sajini KK, Rahman S, Fayas TP, Karun A (2014a) Scope of novel and rare bulbiferous coconut palms (*Cocos nucifera* L.). Genet Resour Crop Evol 61:1–6
Jerard BA, Niral V, Dhanapal R, Damodaran V, Arunachalam V, Rajesh MK, Devakumar K, Samsudeen K, Nair RV, Kumaran PM, Thomas GV (2014b) IND 221—Andaman Horned Cocos (IC0598221; INGR13063), a Coconut (*Cocos nucifera*) Germplasm with Distinct Character of Horny Nuts. Indian J Plant Genet Resour 27(1):76
Jerard BA, Niral V, Samsudeen K, Jacob PM, Devakumar K, Rajesh MK, Karun A, Thomas GV (2014c)

IND 331-Laccadive Mini Micro Tall (IC0598222; INGR13064), a Coconut (*Cocos nucifera*) Germplasm with distinct character of extremely small nuts with very low copra content, not found in any other coconut variety. Indian J Plant Genet Resour 27(1):77

John CM, Narayana GV (1942) The male coconut tree. Madras Agric J (India) 31:75–77

John CM, Narayana GV (1949) Varieties and forms of the coconut. Ind Coconut J 3:209–225

Karun A, Niral V (2019) Coconut genetic resources. In: Rajasekharan PE, Ramanatha Rao V (eds) Conservation and utilization of horticultural genetic resources. Springer Nature Singapore Pte Ltd, Singapore, pp 251–282

Karun A, Sajini KK, Nair M, Kumaran PM, Samsudeen K (2006) Cryopreservation of coconut (*Cocos nucifera*. L) pollen. J Plantn Crops 34:568–571

Karun A, Sajini KK, Niral V, Amarnath CH, Remya P, Rajesh MK, Samsudeen K, Jerard B A, Engelmann F (2014) Coconut (*Cocos nucifera* L.) pollen cryopreservation. Cryo-Lett 35(5):407–417

Liyanage DV (1958) Varieties and forms of the coconut palm grown in Ceylon. Ceylon Coconut Q 9:1–10

Mao ZS, Lai YT (1993) The coconut germplasm of Hainan Island, China. Plant Genet Resour Newsletter 91–92:53–57

Martius CFP 1850 (1823–50) Historia Naturalis Palmarum. 3 Vols. T. O. Weigel, Leipzig

Mayuranathan PV (1938) Original home of the coconut. J Bombay Nat His Soc 40:174–182

Meerow AW, Noblick L, Borrone JW, Couvreur TLP, Mauro-Herrera M, Kuhn DN, Nakamura K, Oleas NH, Schnell RJ (2009) Phylogenetic analysis of seven WRKY genes across the palm subtribe Attaleinae (Arecaceae) identifies Syagrus as sister group of the coconut. PLoS ONE 4(10):e7353

Meerow AW, Noblick L, Salas-Leiva DE, Sanchez V, Francisco-Ortega J, Jestrow B, Nakamura K (2014) Phylogeny and historic biogeography of cocosoid palms (Arecaceae, Arecoideae, Cocoseae) inferred from sequences of six WRKY gene family loci. Cladistics 31:509-534. https://doi.org/10.1111/cls.12100

Menon KPV, Pandalai KM (1958) The coconut palm. A monograph. Indian Central Coconut Committee, Ernakulam, India, p 384

Michael KJ, Verghese EJ (1963) Production of suckers in coconut. Ind Coconut J 16:106–108

Miquel FAW (1855) Flora Van Nederlandsch Indie III:64–72

Moore HE (1973) The major groups of palms and their distribution. Gentes Herb 11:27–111

Nair RB, Sadanandan AK (1976) A note on horns in coconut fruits. J Plantn Crops 4(2):81–82

Nampoothiri KUK, Parthasarathy VA (2018) Varietal improvement. In: Nampoothiri KUK, Krishnakumar V, Thampan PK, Achuthan Nair M (eds). The Coconut Palm (*Cocos nucifera* L.)—Research and Development Perspectives, Springer Nature Singapore Pte Ltd, Singapore, pp 113–156

Narayana GV, John CM (1949) Varieties and forms of coconut. Madras Agric J 36:349–366

Nayar NM (2017) The Coconut: phylogeny, origins, and spread. Academic Press, London, p 240

Nayar NM (2018) Opportunistic flowering and pollinating system as a survival strategy of the coconut (*Cocos nucifera* L.), a monotypic species of the Arecaceae family. Genet Resour Crop Evol 65:333–342

Ninan CA, Raveendranath TG (1972) Why spicata palms do not breed true? Chromosome studies provide the answer. Cocon Bull 3:2–4

Niral V, Devakumar K, Umamaheswari TS, Naganeeswaran S, Nair RV, Jerard BA (2013) Morphological and molecular characterization of a large fruited unique coconut accession from Vaibhavwadi, Maharashtra India. Indian J Genet 73(2):220–224

Niral V, Jerard BA (2017) Botany, growth and development. In: Chowdappa P, Niral V, Jerard BA, Samsudeen K (eds) Coconut. Daya Publishing House, A Division of Astral International Pvt Ltd, New Delhi, India, pp 73–129

Niral V, Jerard BA (2018) Botany, Origin and Genetic Resources of Coconut. In: Nampoothiri KUK, Krishnakumar V, Thampan PK, Achuthan Nair M (eds) The Coconut Palm (*Cocos nucifera* L.)—Research and Development Perspectives. Springer Nature Singapore Pte Ltd, Singapore, pp 57–112

Niral V, Jerard BA, Kavitha KV, Samsudeen K, Nair RV (2008) Variability and association among floral traits and pollen recovery in coconut (*Cocos nucifera* L.). J Plantn Crops 36 (3):186–191

Niral V, Jerard BA, Samsudeen K, Patil PD, Ananda KS, Nair RV, Thomas GV (2010) International Coconut Gene Bank for South Asia-Conservation and Characterization. In: Thomas GV, Krishnakumar V, Augustine Jerard BA (eds). Proceedings of the International Conference on Coconut Biodiversity for Prosperity, 25–28 October 2010, Central Plantation Crops Research Institute, Kasaragod, Kerala, India, pp 35–40

Niral V, Nair RV, Jerard BA, Samsudeen K, Ratnambal MJ (2009) Evaluation of coconut germplasm for fruit component traits and oil yield. J Oilseeds Res 26 (Special Issue):668–670

Niral V, Samsudeen K, Sudha R, Ranjini TN (2019) Genetic resource management and improved varieties of coconut. Ind Coconut J 11–14

Noblick LR (2013) Syagrus.: an overview. Palm J 205:4–30

Noblick LR (2017) A revision of the genus *Syagrus* (Arecaceae), Phytotaxa 294 (1). Magnolia Press, Auckland, New Zealand, p 262

Noblick LR, Meerow AW (2015) The transfer of the genus *Lytocaryum* to *Syagrus*. The Palms 59:57–62

Ohler JG (1984) Coconut: tree of life. FAO Plant Production and Protection Paper 57, Food and Agriculture Organization, Rome, Italy, p 446

Omar ABH (1919) Races of coconut palm. Garden's Bull 2:143

Patel JS (1938) The coconut. A monograph. Government Press, Madras, India, p 313

Petch T (1924) Horned Coconut. Year Book, Dept Agric, Ceylon, pp 20–21

Purseglove JW (1968) The origin and distribution of the coconut. Trop Sci 10:191–199

Pursglove JW (1985) Tropical crops: monocotyledons, 5th edn. Longman, London

Quisumbing E (1926–27) Branching in coconut. Philipp Agric 15:3–11

Ratnambal MJ, Nair MK (1999) Genetic resources of coconut (*Cocos nucifera* L.) in India. Plant Genet Resour Newsletter 118:56

Ratnambal MJ, Nair MK, Muralidharan K, Kumaran PM, Rao EVVB, Pillai RV (1995) Coconut Descriptors. Part. I. Central Plantation Crops Research Institute, Kasaragod, Kerala, India, p 198

Ratnambal MJ, Niral V, Krishnan M, Ravikumar N (2000) Coconut Descriptors Part II, CD-ROM, Central Plantation Crops Research Institute, Kasaragod, Kerala, India

Raveendranath TG, Nair PN, Ninan CA (1975) Chromosomes of a bulbiferous palm. Coconut Bull 6:2

Richardson DL, Harries HC, Balsevicius E (1978) Coconut varieties in Costa Rica. Variedades de cocoteros en Costa Rica. Turrialba 28(1):87–90

Ridley HN (1907) Branching in palms. Ann Bot 21:415–422

Rognon F (1976) Floral biology of the coconut. Duration and sequence of male and female phases in different types of coconut. Oleagineux 31:13–18

Saensuk C, Wanchana S, Choowongkomon K, Wongpornchai S, Kraithong T, Imsabai W, Chaichoompu E, Ruanjaichon V, Toojinda T, Vanavichit A, Arikit S (2016) *De novo* transcriptome assembly and identification of the gene conferring a "pandan-like" aroma in coconut (*Cocos nucifera* L.). Plant Sci 252:324–334

Sajini KK, Karun A, Amarnath CH, Engelmann F (2011) Cryopreservation of coconut (*Cocos nucifera* L.) zygotic embryos by vitrification. Cryo-Lett 32 (4):317–328

Samonte JL, Mendoza EMT, Ilag LL, Cruz NBD, Ramirez DA (1989) Galactomanan degrading enzymes in maturing normal and makapuno, and germinating normal coconut endosperms. Phytochem 28:2269–2273

Samsudeen K, Nagwaker DD, Anitha Karun, Niral V, Jerard BA, Ajith Kumar P, Devadas K, Nair RV (2010) Exploration and collection of sweet endosperm coconut' Mohachao Narel' from Maharashtra, India. In: George V Thomas, Krishnakumar V, Jerard BA, Niral V, Josephrajkumar (eds) Abstracts. International Conference on "Coconut Biodiversity for Prosperity", 25–28 October, 2010, CPCRI, Kasaragod. India, p 19

Samsudeen K, Rajesh MK, Nagwaker DD, Reshmi R, Kumar PA, Devadas K, Anitha Karun (2013a) Diversity in Mohachao Narel, a Sweet Endosperm Coconut (*Cocos nucifera* L.) population from Maharashtra, India. National Academy Sci Lett 36(3):319–330

Samsudeen K, Thamban C, Niral V, Jerard BA, Rajesh MK, Manjula C, Devadas K, Karun A (2013b) *In situ* approach for rapid characterization to aid on farm conservation of coconut germplasm—a case study of two ecotypes from West coast of India. J Plantn Crops 41(3):357–363

Sangalang JB (2004) *In situ* conservation of coconut genetic resources through coconut based farming systems. College, Laguna, Philippines, p 28

Sangare A (1981) Compétition pollinique et légitimité des semences produites dans les champs semenciers de cocotiers. Oleagineux 36:424–427

Satyabalan K (1997) Coconut varieties and cultivars—their classification. Asian Pacific Coconut Community, Jakarta, Indonesia. p 105

Sisunandar RA, Turquay P, Samosir Y, Adkins SW (2010) Cryopreservation of coconut (*Cocos nucifera* L.) zygotic embryos does not induce morphological, cytological or molecular changes in recovered seedlings. Planta 232:435–447

Sugimura Y, Rocat DA, Salud CD, Kamata N (1994) Characteristics of coconut palm with plicate leaves. Jpn J Trop Agric 38:119–123

Tampake H, Kuswara T, Davis TA (1982) Coconut germplasm survey of Nusa Tenggara, Timur Province, the initial step towards inducing drought-resistance in coconut strains. Indonesian Agric Res Dev J 4:52–61

Torres JP (1937) Some notes on Makapuno coconut and its inheritance. Phil J Agric 8:27–37

Venkataraman K (1928) Parthenogenesis in coconut. Yearbook, Dept of Agric Madras, pp 29–32

Watt G (1889) Dictionary of economic products of India. In: *Cocos Linn, Cocos nucifera L.*, Vol I, Government Printer Calcutta, pp 415–459

Whitehead RA (1965) Flowering in *Cocos nucifera* L. in Jamaica. Trop Agr (Trinidad) 42:19–29

Whitehead RA (1966) Progress in the freeze drying of coconut pollen. Oléagineux 21:281–284

Welewanni I, Perera C, Jayasekera A, Bandupriya D (2020) Recovery, histological observations and genetic integrity in coconut (*Cocos nucifera* L.) embryogenic calli cryopreserved using encapsulation-dehydration procedure. J Natl Sci Found 48(2): 175–186

Zizumbo VD, Fernández BM, Torres HN, Colunga GP (2005) Morphological variation of fruit in Mexican populations of *Cocos nucifera* L. (Arecaceae) under *in situ* and *ex situ* conditions. Gen Resour Crop Evol 52:419–432

Zuniga LC (1953) The probable inheritance of the Makapuno character of the coconut. Philipp Agric 36:402–413

Zuniga LC, Villegas LG, Penaflorida G (1970) Collection of coconut cultivars in the Philippines. In: Emata RG (ed) Coconut Production in the Philippines, United Coconut Association of the Philippines, pp 9–24

Breeding Strategies

4

B. A. Jerard, V. Niral, and M. K. Rajesh

Abstract

Coconut is a monocotyledonous palm with wide diversity among the tall and dwarf types. The perennial nature with long pre-bearing phase, high heterozygosity, self- and cross-pollination nature, need for artificial crossing requiring climbing of tall palms, absence of viable vegetative propagation methods, and highly demanding time and monetary resources for prolonged field experiments make the breeding efforts a challenging process. Progress in coconut crop improvement has been mainly achieved through the mass selection of desirable types among the diverse coconut populations and hybridization among the selected types of major global coconut populations. India has been pioneering in developing improved selections and hybrids and reported genetic gains through heterosis. As the major problems associated with coconut cultivation vary with different coconut-growing zones, the breeding strategies vary at different places. Nevertheless, the primary breeding objective is an improvement in the yield of fruits. The use of morphological traits associated with yield and identification of desired cultivars is chiefly employed in breeding programs that comprise mother palm selection, seed nut selection, and seedling selection procedures. To realize a stable, high-yielding, and stress-tolerant performance in this multi-utility crop, early bearing, dwarf statured, and climate-resilient parental lines/hybrids need to be developed with a potential for product diversification.

B. A. Jerard (✉)
ICAR-Central Island Agricultural Research Institute, Port Blair, Andaman and Nicobar Islands, India
e-mail: augustinejerard.b@icar.gov.in

V. Niral · M. K. Rajesh
ICAR-Central Plantation Crops Research Institute, Kasaragod, Kerala, India
e-mail: niral.v@icar.gov.in

M. K. Rajesh
e-mail: rajesh.mk@icar.gov.in

4.1 Introduction

Coconut (*Cocos nucifera* L.) is a perennial palm, diploid having 32 chromosomes ($2n = 2x = 32$), and is widely grown in tropical regions for its varied uses. The palm belongs to the family Arecaceae in the Coccoideae subfamily and is the only species under the genus. The coconut palms are found growing as groves in remote places or cultivated as a plantation crop in most tropical and subtropical countries within 25° on either side of the equator. Better fruit yield of palms has been reported within 20° latitude from the equator and the palms are grown commercially up to about 800 m above the mean sea level (MSL) in these regions. Coconut palms are commonly regarded as *Kalpavriksha* in India, meaning 'tree of life' considering its potential to

M. K. Rajesh et al. (eds.), *The Coconut Genome*, Compendium of Plant Genomes,
https://doi.org/10.1007/978-3-030-76649-8_4

provide all life necessities. Different parts of the coconut palm provide nutritious food, namely oil, milk, sweet water, sugar, dietary fiber, and palm cabbage. Also, the palm provides several products such as husk fiber, leaves, hardy shells, strong wood for the industries, and an array of products having traditional medicinal values. Besides, an abundance of coconut palms in fragile island ecosystems provides environmental services by safeguarding the ecosystem. Coconut palms have attained immense significance, in the realm of human nutrition, by providing healthy foods, viz. coconut milk, virgin coconut oil (VCO), tasty tender fruit water, the delicious tender kernel from immature fruits, unfermented sweet sap from unopened inflorescence, and sap-based sugar. Besides, the coir fiber, coir dust, and wood are economically utilized in preparing several domestic and industrial products, including coir products, geo-textiles, potting media for soilless culture of several high-value crops, and furniture. Although coconut palms are used in numerous ways, as detailed above, the breeding efforts have been largely focussed on identifying and developing high-yielding types in terms of fruit yield and copra out turn under specific locations or conditions of disease-affected or drought-affected regions. For these conditions, the natural coconut diversity, existing among different coconut populations in specific locations, was utilized. The *ex situ* field gene banks have greatly contributed to the crop improvement programs around the coconut-growing countries through identification, conservation, and multiplication of better performing selections from those populations.

The long juvenile phase of coconut palms, inherently higher magnitude heterozygosity of individual palms within the cultivars and the high resources requirement in terms of land and human resources for field experimentation, and lack of efficient reproducible vegetative propagation protocols are considered as the major impeding factors for breeding attempts. Though the present coconut improvement endeavors generally involve conventional approaches of mass selection and hybridization, efforts have also been reported in the utilization of the selected individual palms for novel traits within the targeted population. The application of biotechnological tools has also aided in deciphering the extent of diversity among the conserved germplasm, i.e., selecting desirable parental lines, characterization at the molecular level, and examining the hybridity (Rajesh et al. 2012). The details on the botanical description of the coconut palm, varietal classification, and current and future breeding strategies are discussed.

4.2 Botanical Description

Earlier, more than 30 species, distributed mainly in Central and South America, were included under the genus *Cocos*. Later, it was distinguished as a monotypic genus by Beccari (1916), with *nucifera* as the only species, and the remaining species were listed under nine different genera. This classification was later supported by Tomlinson (1961) based on the data from the anatomical examination of *Cocos*.

The coconut palm produces an extensive adventitious root system from the trunk's swollen base, and the roots continue to develop throughout its lifespan (Fig. 4.1). The number of roots in an individual palm varies depending on the palm's age, the girth of the swollen base (bole), edaphic factors at the location, and general crop management conditions. The number of roots in a palm was reported to range from 1500 to 7000 (Sampson 1923; Copeland 1931; Patel 1938), and in rare cases, up to 11,360 roots have been observed (Menon and Pandalai 1958). The main roots produce many secondary roots that branch profusely to form a large network of roots, forming a wide absorptive surface which facilitates the palm in absorbing water and nutrients from the soil. The main roots generally sustain for many years, whereas the rootlets are short-lived and are frequently recouped with new growth. A large number of pneumatophores develop from the main roots and rootlets, which facilitate the gaseous exchange among the roots and atmosphere. The roots' delicate growing tip has root caps that protect the roots from any mechanical injury during growth.

Fig. 4.1 Root production from base of stem

The palm produces a single, straight stem referred to as trunk, greyish with markings of fallen leaves (Fig. 4.2) and a crown of leaves referred to as fronds on the top. The trunk has no bark, no cambium, and is devoid of any secondary growth features. Hence, once formed, the stem never alters in thickness, except for a mild reduction in thickness as the palm ages.

With only a single apical bud at the top growing point, any major damage to the bud results in the palm's death. Rarely, the terminal bud produces branches upon partial damage to the bud or split due to physical or environmental factors (Davis 1969). The terminal bud is well protected by surrounding closely arranged successive unopened leaves. The woody stem is predominantly fibrous, is stiff with good suppleness, and can tolerate severe winds to a considerable extent. The stem thickness varies with the level of vigor, growing conditions and variety. In most tall palms, the bottom of the stem is bulged and referred to as the 'bole'. The growing stem of the young coconut palm is seen only when the bole attains its complete developmental stage. In the early stage of growth, the stem thickens in a phased manner, and after reaching the considerable size, a notable increase or decrease in girth of the stem is not observed with age progression. During the life of the palm, under unfavourable growth conditions, the stem thickness shows a reduction. The stem's height is generally related to the age of palms, cultivar, and cultivation environment. It is well known that trunks of dwarf cultivars are shorter than the tall varieties placed at similar conditions at a given age. Rapid stem elongation is noticed among palms under more shade or which are planted at closer spacing (Fig. 4.3).

Fig. 4.2 Stem with leaf scars

The early leaves in coconut seedlings remain unsplit and appear to be fused. After the emergence of six to eight fused leaves, the new leaves start splitting into leaflets and continue to grow with apparent leaflets. The total number of leaves present in the palm's crown, the length of leaf, number of leaflets per leaf, and leaflet's breadth differ with cultivar and palm environment. Generally, a well-managed adult palm exhibits about 35 opened leaves on the crown, which includes 10–12 older leaves holding the harvestable fruit bunches, 10–14 next younger leaves, carrying the fruit bunches at different stages of development, 10–12 opened leaves with unopened spadices in their axils, and unfurled leaves at spindle stage and in palm cabbage.

The length of matured leaves varies from three to six metres, with a strong petiole bearing the number of leaflets on either side. The longevity of the leaf, from its emergence to abscission, has been reported to be around five years. The leaf growth including the genesis of leaflets (Venkatanarayana 1957) and steps in the development of leaf primordia into adult leaf (Periaswamy 1965) was investigated and reported by

Fig. 4.3 A grown-up coconut plantation

different researchers (Patel 1938; Padmanabhan 1963). The leaf development commences with the development of a leaf primordium about thirty months before the actual opening from the leaf sheath. The emerged leaves have a life of about 24–36 months on the crown. Patel (1938) has recorded the life of leaves over three years on the crown in good bearing palms under favourable conditions. Generally, the lifespan of leaf was observed to be lesser in poor yielding palms than the palms with better yielding capacity, although growing conditions also influence the leaf shedding. It was acknowledged that better yielding palms possess a higher number of leaves on the crown, indicating the prolonged lifespan of leaves and drought tolerant nature (Satyabalan 1993). One-year-old seedlings generally possess around eight leaves at the time of planting, and the number increases to about 30–35 with the progression of palms' age and attainment of maturity.

Coconut palm is monoecious and produces male and female flowers on the same spadix, which are borne on leaf axils (Fig. 4.4). The first flowering occurs about two to seven years after planting, depending on the variety and other crop management conditions.

The male flowers are numerous in a spadix; anthesis starts from the top of the spikelets. The period intervening the first and last male flower anthesis is categorized as the male phase, which extends for about 20 days, again differing with the cultivar and palm management. The female flowers are few and larger, and the ovary is tricarpellary, with only one fertile ovule. Generally, the receptivity of female flowers commences 20–25 days after the opening of the inflorescence, by which time all the male flowers of the spadix would have shed. The receptivity of the individual female flower lasts mostly for about three days. The time duration between the first and last

Fig. 4.4 Coconut inflorescence with male and female flowers

female flower's stigma receptivity is categorized as the female phase, which extends for about 5–7 days.

Pollination in coconut is known to occur through wind or insects, as reported by several workers. Generally, a time gap of about two to three days exists between the completion of the male phase and the beginning of the female phase in an inflorescence (Ratnambal et al. 1995, 2000). Hence, the possibilities for self- and cross-pollination exist in coconut palms depending on the phases. Self-pollination has been observed in most dwarf types (Jack and Sands 1922) due to the overlapping of male and female flower phases (Ratnambal et al. 1995, 2000). It can also occur due to the overlapping of the male and female phases of successive inflorescences in a palm (Petch 1913; Patel 1938; Copeland 1931). The coconut pollen retains its viability for up to a week under room temperature, whereas it can be prolonged by storing them under refrigerated conditions or in the desiccators.

The fruit is a drupe that contains white, edible endosperm referred to as 'kernel', with an embryo embedded in it. The fruit (Fig. 4.5) is protected externally by a thick pericarp referred to as a husk. The husk has three distinct layers, viz. the exocarp/epicarp (the colored surface of the fruit with tough fibers), mesocarp (husk beneath the epicarp), and endocarp (shell covering the endosperm). The shell has three pores (called 'eyes') representing the three carpels of the ovary, out of which one is soft and the other two are hard. The embedded embryo is well protected by the shell and located beneath the soft eye of the shell. The thickness of the kernel is about one to two centimetres. In between the shell and the white kernel, a thin, brown-colored layer of testa/seed coat remains adhering to the kernel. Inside the endosperm, there is a cavity filled with nut water, referred to as the liquid endosperm. The quantity of this liquid endosperm reduces gradually during kernel development as the fruit matures and storage for a few months after harvest due to drying. Generally, when the quantum of nut water is decreased or liquid endosperm dries on storage, the germination of the nut is reduced (Menon and Pandalai 1958).

4.3 Varietal Groups

Coconut cultivars are broadly categorized into two major groups, viz. tall and dwarf (Narayana and John 1949), based on their growth habit, pollination mechanism, fruit characteristics, etc. The tall-type cultivars are mostly cross-pollinated, while the dwarf cultivars are mostly self-pollinated due to the synchronization of male and female flower opening within the inflorescence. In both types, there are exceptions for this pollination behaviour. However, the breeding behaviour in individual palms of any cultivar is variable depending on the growing environment. Also, the degree of selfing or out-crossing depends on the existence of other palms in the close vicinity. The origin of dwarf-type coconuts has been postulated due to the phenomenon of inbreeding among the tall-type cultivars as the latter show limited self-pollination. Thus, the dwarf-type cultivars are suggested as mutant forms derived from the tall types (Purseglove 1968).

Using the differences in the set of morphological traits, the cultivars have been categorized in another system of classification as var. *typica* (comprising tall types), var. *nana* (comprising dwarf types), and var. *javanica* (intermediate between tall and dwarf types), along with two

Fig. 4.5 Cross section of fruit showing fruit components

mutant types, viz. var. *androgena* (male coconut palm with multiple spikelets having predominantly male flowers) and var. *spicata* (unbranched inflorescence with mostly female flowers and few male flowers). In another system of classification, based on the fruit component characters such as fruit cavity volume, amount of husk, and traits associated with germination, various types are *Niu Kafa* (long fruits with a higher proportion of husk and small nuts with lesser cavity volume) and *Niu Vai* (mostly bigger, round fruits with more of nut composition and bigger round nuts with larger cavity volume holding a higher amount of nut water).

The level of genetic diversity among the present-day coconut populations is considered due to process of natural evolution based on the adaptation of palms to different environments along with the human selection of choice cultivars in different coconut-growing regions for many uses. The tall cultivars (Fig. 4.6a) are preferably grown for their higher fruit yield with higher copra output, their ability to withstand adverse conditions in the growing environment and longer economic life of over 80 years. The different dwarf cultivars are postulated to have evolved from tall populations through mutation (Menon and Pandalai 1958) or because of inbreeding among the selected talls (Swaminathan and Nambiar 1961) followed by human selection. Most dwarf-type cultivars (Fig. 4.6b) are preferred over tall cultivars for their earliness, tender fruits having sweet water, ornamental look, the convenience of accommodating these cultivars in home gardens, and ease of climbing due to the formers' relatively less height. Some of the dwarf cultivars are also increasingly used for breeding purposes as they are found to possess resistance and/or tolerance to diseases like lethal yellowing, root (wilt) disease, etc. The distinguishing features and uses of tall and dwarf cultivars are given in Table 4.1.

Selected tall and dwarf cultivars are used to develop hybrids to combine the desirable traits of the parental populations.

4.4 Breeding Objectives

In the major coconut-growing areas, the crop is mainly used for oil-yielding copra production. Hence, the breeding objectives mainly focus on achieving higher fruit yield with more out turn of kernel/copra, which contributes to milk (juice of white kernel)/oil yield from copra. However, targeted breeding efforts for better oil quality having a defined fatty acid composition have not attained a real focus, which may help put the coconut oil in competition with other edible oils.

Fig. 4.6 A typical tall (**a**) group of palms and a dwarf type (**b**)

As the juvenile phase of tall coconut cultivars is longer (over 5 to 9 years), precocity in flowering and fruit yield are considered as the important breeding objectives. Marked improvement has been achieved in early flowering and fruiting by combining the different location-specific dwarf selections with promising tall cultivars.

Product diversification has been in the focus for increasing the profitability of coconut cultivation. Hence, traits governing the tender fruit, including taste, aroma, and amenability for processing, have received importance in improvement programs. Though the dwarf selections are suitable for tender nut purpose, they are mostly low yielders compared to hybrids and tall cultivars and susceptible to adverse biotic and abiotic stress factors. Hence, there is a need to incorporate the superior tender fruit traits of dwarf selections along with dwarfness and early flowering traits into the hybrids involving superior tall selections.

Global warming and climate change effects cause moisture stress, which is a serious problem in many coconut-growing areas. Several plantations have been lost in India in the past two decades, with significant production loss in the surviving palms. Therefore, breeding for moisture stress tolerance has become an important objective. Besides, climate changes have resulted in the transformation of otherwise minor pests into major ones, and emergence of invasive pests in newer areas, which pose major challenges to coconut production. However, need-based management practices are developed to tackle the pests. Breeding coconut types for resistance or tolerance to coconut mite (*Aceria guerreronis*) has become one of the important breeding objectives in India as the pest has invaded most coconut-growing areas during the past two decades. During the recent past, spiralling whitefly has caused a severe production loss, especially in the dwarf cultivars and in hybrids. In contrast, the pest incidence had less influence on the yield of the tall cultivars.

In many coconut populations of the world, the spread of phytoplasma and viroid diseases of coconut like lethal yellowing, foliar decay, cadang-cadang, and Weligama wilt have been reported as major threats, necessitating crop improvement efforts to tackle those diseases. Breeding efforts for resistance against coconut root (wilt) disease have been an important objective that impacts the production potential and productivity levels in the major coconut area in Indian state of Kerala and its adjoining regions.

With the single growing bud at the top, all coconut palms grow taller over the years, irrespective of their morphoforms (tall or dwarf or hybrid cultivars), necessitating skilled human resources for palm management operations, including harvesting. Due to several socio-economic factors such as urbanization and

Table 4.1 Distinguishing features and uses of tall and dwarf cultivars

Particulars	Tall types	Dwarf types
Distribution of root	Root mass is dense and abundant	Root mass is thinner and lesser number of roots
Stem	Sturdy with prominent bole	Mostly thin and cylindrical, bole is absent
Stem growth rate	Annual increase of over 50 cm in the active age	Mostly, the annual increase is less than 50 cm
Appearance of leaf scars	Sparse with more gap	Seen closely arranged
Initial flowering after the vegetative phase	Categorized as late, takes about five to nine years	Categorized as early, takes about three to four years
Breeding behaviour	Most cultivars are out-crossed	Most cultivars are self-pollinated
Synchronization of flowering phases	Definite time gap between male and female flower phases observed, in exception overlapping observed over different inflorescences of the same palm	Complete overlapping of male flower opening and female receptivity observed. In a few cases, both intra- and inter-spadix overlapping occur
Fruit and petiole color	Progeny palms produce shades of greens or brown that differ between individuals	Progenies true to mother palm with green or yellow or red or brown
Fronds and bunches	Produce longer fronds rigidly attached having broader leaflets	Produce shorter fronds with fragile attachment to stem having narrower leaflets
Fruit size	Wide range—generally exceptionally large to exceedingly small	Narrow range—generally small to medium
Alternate bearing habit	Rarely occur	Commonly occur
Phenotypic variation	High within and between different cultivars	Low within cultivars and high between cultivars
Reaction to abiotic stresses (drought, wind, temperature)	More tolerant	Medium to less tolerant or susceptible
Reaction to major pests	More tolerant	Medium to less tolerant or susceptible
Suitability of fruits	Mostly suitable for copra/oil production, sap production, fiber production	Mostly for tender coconut water purpose
Productive lifespan	About 60–80 years	About 40–45 years

alternate employment, the availability of skilled human resources for climbing coconut trees has become a major concern in most coconut-growing regions. Mechanization in coconut has limited success due to inherent difficulties and the nature of plantations. Hence, the growers prefer short-statured cultivars and earliness. Therefore, breeding and adoption for short-statured palms are considered an important strategy to sustain coconut farming.

Considering the fluctuations of prices of coconut commodities, cultivation of coconut as monocrop has become less remunerative in many parts of the coconut-growing areas. Hence, coconut cultivation is promoted now as a 'cropping system approach' to integrate many crops and livestock. Due to the differential response of coconut cultivars, it is imperative to develop varieties appropriate for different management conditions or cropping systems.

4.5 Breeding Approaches

Coconut breeding efforts have so far been limited to the use of identifying desirable cultivars for improvement or combining the targeted traits. The conventional breeding methods in coconut have several limitations; the major one is the long duration for evaluating fruit yield. Although several attempts have been made to produce pure lines from heterozygous coconut through selfing, it has not yielded desired results because of the palms' perennial nature, necessitating longer time and resources. Due to these factors, coconut breeding efforts have been largely limited to the mass selection of parental palms for desirable traits (identified through mother palm traits) and their direct use as a variety or use in making inter-varietal crosses. The mother palm selection procedures and seedling selection procedures are the effective tools being practised by the farmers and researchers. However, molecular markers are used experimentally on a limited scale to identify certain specific types. All the ongoing breeding programs follow a series of selections, i.e. mother garden selection, mother palm selection, seed selection, seedling selection, and better performing varietal selection after progeny evaluation. After the seedling selection, the cultivars are either released as varieties for commercial plantations after rigorous field evaluation or used in hybridization programs to exploit the hybrid vigor.

4.5.1 Mother Garden and Palm Selection

Mother gardens are selected based on the farmer's knowledge to identify traditionally known populations for desirable traits with higher adaptability in a particular environment. From such targeted populations, individual mother palms are selected based on specific traits and certain common traits of the population (Fig. 4.7). The mother palm selection is generally considered as the first step followed in any coconut improvement program. The general criteria followed for mother palm selection vary with cultivar, location, nature of the garden, and purpose of the selection. Generally, among tall cultivars, palms consistently yielding an annual average of 80 fruits or more with a copra yield of over 20 kg palm^{-1} year^{-1} are selected as mother palms. The mother palms ought to have attained at least 25 years of age so that the average yield performance is known. Various studies have shown that the average number of female flowers per inflorescence, number of functional leaves on the crown of the palm, crown shape, arrangement of firm bunches on the stem, total leaf production up to three years after sowing, and time taken for flowering after planting are the key parameters that contribute to the yield of fruits. Hence, these traits are given priority in any selection program.

Knowledge of floral biology of individual palms of the selected cultivar is important while selecting palms for any breeding program. Observing the status of male and female flowering phases, within the inflorescences of a palm, fully is particularly important while picking mother palms for any breeding program involving crossing experiments (Thomas et al. 2015). Generally, the male phase in an inflorescence occurs for about 18 to 25 days from inflorescence opening, whereas the female flowers of the same inflorescence remain non-receptive. The female phase, which is the receptive stage of stigma, usually lasts for a shorter period of about 3–9 days depending on the cultivar and season. The temporal differences between these two flowering phases warrant cross-pollination as the only viable option in tall cultivars. On the other hand, the time gap between both the phases is absent in most dwarf accessions, making these phases overlap either within or successive inflorescences, thus ensuring self-pollination predominantly (Nair et al. 2016).

4.5.2 Seed and Seedling Selection

The seed nuts are to be selected from the harvested lot of fruits from the identified mother palms. To ensure the quality seed nuts, the fruit bunches are to be harvested at the stage where at least one nut in the bunch starts drying, which

Fig. 4.7 Palms showing good mother palm traits in the population

generally occurs after 11–12 months of fruit set. The matured fruit bunches are lowered to the ground from tall palms by tying with a rope to avoid any mechanical damage to the husk. Each bunch will have fruits of varying sizes and shapes, although some basic traits remain same. This happens because of the inherent maturity differences of fruits developed from the first or last female flower, the position of fruits in the bunch, fruit setting per cent, etc. The medium-sized fruits of a cultivar are selected as seed nuts and used for sowing. Seed nuts with severe pest and disease infestation, which are malformed and do not confine to the size and shape of fruits of the selected cultivars, are avoided. The sowing conditions of different genotypes are uniform, either in poly bags or in nursery beds. Fully dried seed nuts, exhibiting the splashing sound of water inside, are sown. Completely dried nuts without water, rotten seed nuts, and seed nuts with secondary fungus infection are rejected. The seed nuts, which germinate within six months after sowing, are retained in the nursery for studying the seedling parameters for appropriate seedling selection procedures.

At 12 months after sowing, the well-grown seedlings are compared for a set of morphological traits such as rate of growth, time required for splitting of leaf lamina into leaflets, vigor of seedling manifested by girth at collar region, seedling height measured as the length of longest leaf, leaf production, and hue of the petiole. These traits have been reported to be positively correlated with better feat in terms of earliness and fruit yield at the adult stage of the palm. Even though these traits may vary according to varieties and growing nursery environment, generally, seedlings with 6–7 leaves at 12 months of age, with at least one split spindle leaf, seedling collar girth of over 10–12 cm and free

Fig. 4.8 A coconut seedling with desirable traits

from pests and diseases are regarded as quality seedlings (Fig. 4.8). The color of petiole and height of seedlings are important traits that should exhibit true-to-parental palms selected for breeding work. Hence, in any coconut breeding program, the final seedling selection occurs about 24 months from the commencement of pollination in the identified mother palms. The seedling selection procedure is also followed among hybrid progenies, as they tend to segregate for seedling traits due to the heterozygous nature of selected parental palms. The use of molecular markers is suggested as an effective tool in differentiating the true-to-type descendants (Rajesh et al. 2013).

4.5.3 Evaluation of Cultivars and Selection

The earliest selection of coconut would have been made for large-sized, round-shaped fruits that could have provided uncontaminated and sweet nut water for travellers during long voyages. Later, the fruit yield with more copra out turn for higher edible oil output had become the priority when coconuts were increasingly exploited for oil purposes. Presently, the selection is followed for many useful traits such as precocity in bearing, dwarfness, stress tolerance, or resistance to biotic and abiotic factors and traits associated with product diversification as coconuts are now increasingly being used for several other purposes meeting the market-driven demand and competitiveness. As explained above, the mass-selected planting material is used directly for either cultivation or further exploitation in breeding programs for specific objectives.

Multi-location evaluation of introduced cultivars along with locally selected desirable varieties is effectively followed for coconut improvement. It has helped to recognize several better performing cultivars for a particular region and to develop new varieties with enhanced yield potential. In some countries, selective introductions were made to test against the specific needs such as disease tolerance and the selections from the commercially exploited populations. Several such improved selections made from local tall and dwarf collections and exotic introductions have been released in India for commercial cultivation after field evaluation. It includes selections from Laccadive Ordinary Tall (Chandrakalpa), Cochin China Tall (Kalpa Pratibha), Andaman Giant Tall (Kalpa Dhenu), Java Tall (Kalpa Mitra), Tiptur Tall (Kalpatharu), Malayan Orange Dwarf (Kalpa Surya), Malayan Yellow Dwarf (Kalpa Jyothi), Malayan Green Dwarf (Kalparaksha, Kera Madhura), Chowghat Orange Dwarf, Chowghat Green Dwarf (Kalpasree), Kulasekharam Green Tall (Kalpa Haritha), Niu Leka Green Dwarf (CIARI Annapurna), Hari Papua Orange Dwarf (CIARI Surya), Nikkore Orange Dwarf (CIARI Chandan), Niu Oma Dwarf (CIARI Omkar), Philippines Ordinary Tall (Kera Chandra), Fiji Tall (Kera Bastar), San Ramon Tall (Kalpa Shatabdi), Jamaica Tall (Kalyani Coconut 1), Gangabondam Green Dwarf (Gouthamiganga), Seychelles Tall (Kera Sagara), Federated Malay States Tall (Kalpa Ratna), Benawali Green Round (Pratap), Andaman Ordinary Tall (VPM 3), and West Coast Tall (Kera Keralam). These selections were made for different desirable traits such as higher fruit yield, tender nut traits, dwarfness, earliness, disease and pest resistance/tolerance, and abiotic stress tolerance (Niral et al. 2016; Jerard et al. 2017).

Besides, trait-specific accessions have been identified for several promising/desirable and novel traits such as soft endosperm, sweet endosperm, aromatic water, pink husked coconuts, compact crown, and smallest fruits (Jerard et al. 2017).

Similarly, in the Philippines, selections were made from promising talls and dwarfs like Tacunan Green Dwarf, Catigan Green Dwarf, Aromatic Green Dwarf, Malayan Red Dwarf, Rennell Island Tall, and Baybay Tall and used for multiplication for the establishment of seed gardens (Gerardo and Rivera 1998). In Indonesia, beginning from 1970, local ecotypes with high copra yield were identified, viz. Mapanget, Tenga, Bali, Palu, Sawarna, Riau, and Igo Daku, and mass selection from high-yielding blocks of these tall coconut populations was adopted for the production of quality planting material (Novarianto et al. 1998). In Thailand, selection from Maphrao Yai or Thai Tall, a commercially grown cultivar for copra yield and a toddy type, for its high sugar content in the sap, was performed (Chulapan and Anupap 1998). In Vietnam, the cultivars, viz. Catigan, Sri Lanka Green Dwarf, Malayan Yellow Dwarf, West African Tall, San Ramon Tall, and Hijo Tall, are reported to perform better (Vo Van Long 1998). Selections from Fiji Tall, Rotuman Tall, and Rennell Tall are recommended for cultivation in Fiji (Joeli 1998). The palms from Vanuatu Tall population are selected for cultivation and hybridization programs considering their tolerance to coconut foliar decay disease. However, it produces smaller sized fruits with mediocre productivity (Gerard 1998). Hence, the selection process varies depending upon the priority and the prevailing biotic and abiotic stresses at the local conditions. When tackling disease or pest incidence is the prime concern, the research priority involves introgressing the gene(s) conferring resistance/tolerance trait(s) and combining them with moderate yield levels, thus averting complete yield loss. In India, despite the lesser or moderate productivity potential, a selection from CGD (Kalpasree) has been recommended in the root (wilt) disease prevalent tracts considering its disease resistance trait. In Cote d'Ivoire, West African Tall is the predominantly grown cultivar and the mother palms selected from the population are used for seed production (Bourdeix et al. 1998). In Tanzania, the East African Tall and Pemba Red Dwarf are the widely grown cultivars (Mkumbo and Kullaya 1998). Hence, the selections are always meant for meeting the local demand, using coconuts in a particular growing region, according to the available resources.

4.5.4 Exploitation of Hybrid Vigor

The morphological and performance variations existing between the tall and dwarf coconut cultivars and the prospect of combining the desirable traits of precocity with high yield and sturdiness through the exploitation of hybrid vigor resulted in the development of coconut hybrids. The first coconut hybrid developed through artificial pollination was made between Malayan Red Dwarf (MRD) and Niu Leka Dwarf (NLAD), both dwarf types at Sigatoka Research Station of Fiji (Marechal 1928). The first coconut hybrid in India was between West Coast Tall (WCT) and Chowghat Green Dwarf (CGD) made at the Nileshwar Coconut Research Station during the 1930s to combine the copra quality of the tall and the higher yield cum earliness of the dwarf. This report on hybrid vigor in coconut palms (Patel 1937) is considered as the significant milestone in the history of coconut varietal improvement and subsequently led to many successful coconut breeding attempts worldwide, including India, the Philippines, Indonesia, Sri Lanka, Côte d'Ivoire, Jamaica, and other countries. In Fiji, where the first hybrid was produced, subsequent generations of hybrid derivatives of MRD × NLAD were planted for genetic studies. Over the years, the number of local cultivars selected from the germplasm evaluation program has been exploited to develop several newer crosses and to further evaluate yield and resistance/tolerance to diseases. Most of the combinations of dwarf × tall, tall × dwarf, and tall × tall have exhibited superiority of those hybrids for precocity in bearing, higher productivity in copra, and oil yield per palm than the

Fig. 4.9 A dwarf × tall hybrid palm showing early and regular production of bunches

local tall cultivars (Fig. 4.9). Bourdeix et al. (1998) reported a significant yield difference between hybrids and the West African Tall (WAT), and the best hybrids yielded twice the economic produce than the local WAT. The hybrid PB121, involving MYD and WAT, was identified as superior and sent for demonstration or multi-location testing or commercial cultivation in many growing regions besides planted in Côte d'Ivoire. In Sri Lanka, the multi-location trials have confirmed the superiority of the hybrids Green Dwarf × Tall and Yellow Dwarf × Tall (Jayasekera et al. 1993).

After the earlier successful attempts to develop better performing hybrids, several hybrid combinations have been developed using different local dwarf and tall cultivars of respective coconut-growing regions. Yield improvement of about 30–70% for fruit, copra, and oil was recorded from a dwarf × tall hybrid named Kalpa Samrudhi over the local tall and other hybrids from India (Jerard et al. 2015). The improved and better performing hybrids identified and released from ICAR-Central Plantation Crops Research Institute (ICAR-CPCRI) in India (Fig. 4.10a, b) included Chandra Sankara (COD × WCT), Kera Sankara (WCT × COD), Chandra Laksha (LCT × COD), Kalpa Sankara (CGD × WCT), Kalpa Samrudhi (MYD × WCT), and Kalpa Sreshta (MYD × TPT) (Niral et al. 2016). Besides, several hybrids have been released from trials conducted at various All India Coordinated Research Project Centres on Palms (AICRP on Palms) from different locations within India. The hybrids such as COD × WAT, MYD × EAT 32, and COD × LCT are some of the most promising combinations identified for specific traits in addition to fruit and copra yield. The locally produced hybrids in Lakshadweep Islands, such as Laccadive Yellow Dwarf × LCT and Laccadive Green Dwarf × LCT, were better for the yield of fruits and copra out turn. Presently, over 150 hybrids produced through the crossing of selected tall and dwarf accessions are under various evaluation stages in India at multi-location trials under ICAR-CPCRI.

In Indonesia, hybrids KHINA-1 (GKN × Tenga Tall or DTA), KHINA-2 (GKN × Bali Tall or DBI), and KHINA-3 (GKN × Palu Tall or DPU) were developed using Nias Yellow Dwarf (GKN) as the female parent for precocity in flowering and released (Novarianto et al. 1998). Performance evaluation for fruit yield in the Philippines has identified hybrids, CAT × LAG (PCA 15–1); Malayan Red Dwarf or MRD × TAG (PCA 15–2); and MRD × BAY (PCA 15–3) for replanting program (Gerardo and Rivera 1998). In Thailand, two hybrids, viz. Chumphon Hybrid 60 (THT × WAT) and Chumphon Hybrid No 2 (MYD × HIT), were released for cultivation (Chulapan and Anupap 1998). In Vietnam, the hybrids PB121, PB141, JVA1, JVA2, Tam Quan × Hijo Tall, MYD × Ta Tall, MYD × Palu Tall, and MYD × Rennell Tall are produced utilizing the pollen of Palu Tall and Rennell Tall reportedly obtained from Indonesia (Vo Van Long 1998). The potential of cultivating hybrid coconut combinations has been demonstrated based on the performance of MAREN hybrid (Malayan Yellow Dwarf × Rennell Tall) in Papua New Guinea for yield and

Fig. 4.10 Dwarf × tall hybrid palms of Kalpa Samrudhi (**a**) and Kalpa Sreshta (**b**)

precocity, although it was susceptible to beetle attacks (Thompson 1981; Brook 1985). In Fiji Islands, MRD × FJT was found better for earliness and higher yield and was promoted for cultivation in Fiji (Joeli 1998).

In Jamaica, hybrid combination MAYPAN (Malayan Dwarf × Panama Tall) was selected and released for commercial planting due to larger nut size, greater wind storm tolerance, and ability to sustain the production under marginal/low input conditions (Charles 1961; Harries and Romney 1974).

Selection of parental palms of cultivars for crossing program based on the testing of combining ability has been one of the strategies to produce superior hybrid progenies. In 1972, a set of 36 crosses in the 9 × 9 diallel mating system was established to study the combining ability in coconut, which resulted in the identification of Laccadive Ordinary Tall (LCT) and Gangabondam Dwarf (GBGD) as the best general combiners for fruit yield. The hybrid LCT × GBGD was identified as to exhibit better specific combining ability effects for the traits copra out turn and nut yield (Nampoothiri et al. 1999). The experiments conducted in Côte d'Ivoire with the production and evaluation of MAWA (Malayan Dwarf × West African Tall) hybrid also highlighted the potential for combining ability tests performed through mating designs such as line × tester, diallel for selection of parents. In Côte d'Ivoire, the Rennel Island Tall (Pacific) and the West African Tall (Côte d'Ivoire and Benin) were identified as possessing good combining ability between themselves as well as with other dwarf and tall ecotypes (Bourdeix et al. 1991). Malayan Yellow Dwarf (MYD), Malayan Red Dwarf (MRD), West African Tall (WAT), Polynesian Tall (PYT), and Rennell Tall (RIT) have been used in combining ability tests with the local germplasm in Indonesia to determine the best parental combination (Novarianto et al. 1998).

Tall × tall cross-combinations have been utilized to develop improved varieties to increase the yield of the quality copra and impart the biotic and abiotic stress tolerance traits. Although the tall × tall hybrids are recorded to be late flowering like their tall parents, their yield potential is good. Two tall × tall hybrids, viz. WCT × SSAT and LCT × CCNT, have been released for commercial cultivation in Kerala and Tamil Nadu states of India for higher copra yield. Besides, several tall × tall hybrids are presently being evaluated for drought tolerance. Four tall × tall hybrids, namely KB-1, KB-2, KB-3, and KB-4 (Balitka 1989), were released for cultivation in Indonesia.

Although the earliest hybrid production in coconut was reported from a combination of two distinct dwarf types, such combinations of dwarfs are yet to be exploited as variety. In Côte d'Ivoire, a hybrid between the Malayan Yellow and Red Dwarfs yielded 3.8 t of copra ha^{-1} during the adult phase, at a planting density of 170 palms ha^{-1} (Le Saint et al. 1987). In the same trial, Malayan Yellow Dwarf, the control, produced one ton of copra less than the hybrid, the yield of which is comparable to the dwarf × tall materials (IRHO-CIRAD 1989). It was suggested (Bourdeix et al. 1991) that dwarf × dwarf hybrids should be developed for accumulating desirable genes of precocity and a higher rate of bunch production. Using promising dwarf cultivars, twenty-one different dwarf × dwarf hybrids were developed during 2003 to evolve elite hybrids suitable for tender nut purposes (ICAR-CPCRI 2004). The evaluation revealed the dwarf × dwarf hybrids as early flowering, with significant and positive heterosis for growth and yield traits (ICAR-CPCRI 2014). The traits earliest flowering (15 months after planting) and regular bunch production were observed in the cross involving MYD × CGD. The hybrid MYD × NLGD (Niu Leka Green Dwarf) showed several desirable traits (Fig. 4.11) such as a compact crown with closely arranged leaves, high number of fruits per bunch, higher kernel content, tender nut water, and stout stem with closely arranged leaf scars manifesting slow vertical growth. Another hybrid COD × GBGD was identified for early flowering and the number of fruits per bunch. These potential examples of dwarf hybrids increased the scope of higher density planting and ease of crop management.

4.6 Status and Achievements of Breeding for Specific Traits

4.6.1 Climate Resilience

A perennial crop like coconut is exposed to numerous stresses during the growing season and over the years, depending on nature's vagaries. While higher temperature directly interferes with photosynthesis and accumulation of photosynthates, low-temperature conditions cause tissue damage and indirectly alter the metabolic process, thereby deteriorating growth and development (Kumar et al. 2008; Hatfield and Prueger 2015). Further, in the coconut palm, the adverse climatic factors would negatively impact growth and fruit yield for the subsequent 2–3 years, considering that the initiation of inflorescence primordia happens nearly 36 months before the emergence of the inflorescence from the leaf axils (Rajagopal and Kasturi Bai 2002).

It has been estimated that coconut palms exhibit luxuriant growth and yield better when they receive about a monthly rainfall of 150 mm. However, such good rainfall or moisture

Fig. 4.11 Bunches of MYD × NLGD hybrid

availability is not ensured in most traditional coconut-growing areas worldwide. The palms are exposed to scanty or poorly distributed rainfall, longer dry period, etc., which severely reduces their productivity. Hence, the development of drought-tolerant hybrids/varieties is an important requirement in coconut improvement. Palms with a greater number of roots and a better root density have been reported to be less affected by drought (Cintra et al. 1993). Traits such as stomatal frequency, stomatal index, chlorophyll fluorescence, amount of leaf epicuticular wax, leaf water potential, activities of lipases and proteases, glutamate oxaloacetate transaminase, and acid phosphatase were useful for characterizing and differentiating drought-tolerant coconut cultivars (Rajagopal et al. 1988; Repellin et al. 1994; Gomes and Prado 2007; Kasturi Bai et al. 2009). Drought tolerance has also been linked with the relative level of lipid peroxidation (Chempakam et al. 1993). In general, dwarf cultivars have been observed to use more water due to the high number of stomata unit leaf area^{-1}, low level of leaf epicuticular wax content, and poor stomatal regulation (Rajagopal et al. 1990).

In contrast, the tall cultivars were reported to exhibit a conservative water utilization pattern (Voleti et al. 1993). In India, the cultivars, viz. Federated Malay States Tall, Java Tall, Fiji Tall, Laccadive Ordinary Tall, and Andaman Giant Tall, were categorized as drought-tolerant (Fig. 4.12) based on morphological and physiological traits, and they have been further exploited in the crop improvement initiatives. The better performing hybrids, Laccadive Ordinary Tall × Chowghat Orange Dwarf (released by ICAR-CPCRI as Chandra Laksha) and Malayan Yellow Dwarf × West Coast Tall (released by ICAR-CPCRI as Kalpa Samrudhi), are categorized as drought-tolerant hybrids (Kasturi Bai et al. 2006; Rajagopal et al. 2005). Similarly, the hybrid between West Coast Tall × Chowghat Orange Dwarf (released by ICAR-CPCRI as Kera Sankara) was also found suitable for cultivation under rainfed conditions (Niral et al. 2016). A comparative study of three dwarf genotypes, viz. Malayan Yellow Dwarf (MYD), Malayan Red Dwarf (MRD), and Brazilian Green Dwarf (BGD), indicated better response for water stress in MYD with higher stomatal sensitivity, better leaf cooling, nutrient uptake capacity, higher water uptake efficiency, and a relatively higher rate of leaf and inflorescence emergence even during the dry period (Passos et al. 1999). The evaluation of identified drought-tolerant types based on physiological traits in actual drought-prone areas for their performance may require a longer time. Naresh et al. (2006) have indicated the way to utilize the existing plantations in the drought-prone areas for identification of putative drought-tolerant palms based on the phenotype and the physiological parameters which may shorten the breeding cycle. Study of *in situ* drought-tolerant palms in drought-affected areas of India revealed vigorous and uniform seedling traits among the progenies of the putative drought-tolerant palms of East Coast Tall population of Tamil Nadu, indicating the potential of such selection (Jerard et al. 2013).

Further at ICAR-CPCRI, the selected talls are crossed to produce tall × tall hybrids and also dwarf × tall combinations for further evaluation. Preliminary results indicated significant differences among tall × tall hybrids for the fruit yield under moisture stress conditions. The hybrid West Coast Tall × Fiji Tall appears to be better over other hybrids (ICAR-CPCRI 2015).

Temperatures above 31 °C cause a decrease in nut and copra, whereas oil content in copra increases (Kumar 2005). Exposure to increased temperature of 3 °C above 31 °C was reported to have adverse effects on palm growth and development because of reduction in photosynthesis, leaf area, and chlorophyll content, and at the same time enhances higher deposition of leaf epicuticular wax (Hebbar et al. 2013). Genotypic variations in coconut have also been observed in the cardinal temperatures (T_{min}, T_{opt}, and T_{max}) for pollen germination and pollen tube growth. The tall genotypes such as West Coast Tall (WCT), Laccadive Ordinary Tall (LCT), Federated Malay States Tall (FMST), dwarf cultivar Chowghat Orange Dwarf (COD), and hybrids (COD × WCT and MYD × WCT) showed better adaptability to high temperature. In contrast, the dwarf

Fig. 4.12 A drought-tolerant Fiji Tall palm amidst drying palms

genotype Malayan Yellow Dwarf (MYD) was less adaptable (Hebbar et al. 2018). Based on pollen germination and high-temperature tolerance levels, coconut hybrids could be grouped into different heat-tolerant clusters in a study in Sri Lanka, wherein the Sri Lankan Green Dwarf × Sri Lankan Tall and Sri Lankan Brown Dwarf × Sri Lankan Green Dwarf were identified as high-temperature stress-tolerant (Ranasinghe 2019).

Being a tropical palm, coconut is highly susceptible to low temperatures, with flowering and fruit set getting adversely affected below 13 °C. Symptoms such as spear leaf damage, leaf drying, and wrinkled kernel are associated with cold injury in coconut (Mao 1986). Flower drying, poor fruit set, and persistent inflorescence peduncle were observed in the cold regions of the Coorg region in India (ICAR-CPCRI 2009). In China, Hainan Tall is reported to possess cold tolerance (Mao and Lai 1993). In India, cold-tolerant palms have been identified among the populations of East Coast Tall and the Assam Green Tall. The released variety Kamrupa of India also showed better cold tolerance (Chowdhury et al. 2001). Fiji Tall was observed to be more climate-resilient, showing better adaptability across the coconut-growing regions.

In contrast, Andaman Ordinary Tall showed better consistency in yield levels under input deficient conditions (Khan et al. 2002). Further, *in situ* putative cold-tolerant palms have been identified by ICAR-CPCRI from the Sub-Himalayan terai region of West Bengal (ICAR-CPCRI 2009), Karnataka (ICAR-CPCRI 2009), Assam and Meghalaya (ICAR-CPCRI 2010), and Bihar (ICAR-CPCRI 2011) for conservation as well as utilization in the crop improvement work. Observations on seedling progenies of 15 putative cold-tolerant lines from West Bengal under evaluation at ICAR-Research Centre, Mohitnagar, indicate fewer cold injury symptoms than the WCT control obtained from ICAR-CPCRI, Kasaragod (ICAR-CPCRI 2017).

Another important aspect of climate change is the increasing frequency of typhoons/cyclones and associated damage to coconut plantations (Fig. 4.13), implying an emerging need to develop cyclone-resilient varieties. While no specific variety has been developed for typhoon resistance, a few observations are available from different countries. Marty (1986) observed a high percentage of felling in dwarfs (32.2–57.2% felling) compared to talls (5.6–10.4%) and hybrids (4.6–14.6% felling). Based on the observations on mortality caused by felling or uprooting on 13 dwarf varieties, 10 tall varieties, and 28 dwarf × tall hybrids, Labouisse et al. (2007) stated that in general, dwarfs are more susceptible than the talls and hybrids, and substantial differences exist between the varieties of the same group. The Tacunan Green Dwarf from the Philippines was most resistant, followed by the Vanuatu Red Dwarf and the Samoan Yellow, while the Malayan Yellow Dwarf was the most susceptible. Among the talls, Vanuatu Tall proved to be the most resistant, along with varieties originating from the South Pacific (Rennell Island Tall, Rotuman Tall, New Caledonia Tall, Tonga Tall), the Philippines (Baybay Tall, Tagnanan Tall), and Malaysia (Malayan

Fig. 4.13 Cyclone damaged coconut plantations

Tall) displaying appreciable resistance compared to the varieties from Papua New Guinea (Gazelle Peninsula Tall and Karkar Tall) and West Africa (West African Tall). Among the hybrids, the high-yielding Malayan Red Dwarf × Rennell Tall was highly susceptible to felling. Subsequently, the Philippine Coconut Authority (PCA 2016) identified the Tacunan Green Dwarf as an outstanding variety that can withstand strong winds and typhoons and attributed it to the sturdiness of the palm, with physical characteristics of a thick and short stem and well-anchored root system.

Similarly, the compact dwarfs (Fig. 4.14) from Fiji, such as the Niu Leka Dwarf, and the hybrid derivatives of MRD × Niu Leka are likely to be more resistant to typhoons, owing to the robust trunk and slow height increment. Field observation in Fiji by the author revealed the stout but the short stem and strong leaves with compact crown among the hybrid derivatives of the Niu Leka crosses compared to Niu Leka parental palms and Fiji Tall populations. Therefore, these genotypes are preferred for planting in-house compounds in cyclone-prone areas of Fiji.

Fig. 4.14 Compact crown hybrid palm

4.6.2 Disease Resistance

Coconuts in different growing regions are infected by several fungal and phytoplasmal diseases. The important ones are lethal yellowing, root (wilt) disease, *Phytophthora* bud rot, stem bleeding, basal stem rot/Ganoderma wilt, and leaf spot/leaf blight. Among these, lethal yellowing disease (LYD), a virulent phytoplasma disease, is the most serious disease which causes quick mortality of palms devastating entire plantations in the disease prevalent tracts. The disease was first reported from Jamaica, in the Caribbean region, and is now reported across coconut-growing regions, viz. East and West Africa (Kenya, Tanzania, Nigeria, Ghana, Togo, Mozambique, Côte d'Ivoire), North and South America (Florida—USA, Mexico, Cuba), and the Caribbean (Jamaica, Honduras, Bahamas, St Kitts, Nevis, Dominican Republic, Vanuatu), Indonesia (Natuna Islands), and Papua New Guinea (Madang). Thus, LYD is a devastating serious disease in the coconut-growing countries of the Atlantic region. Synonymous with LY in the Caribbean is the Ghanian form of LYD referred to as Cape St Paul Wilt disease (CSPWD), Maladie de kaincope in Togo and Dahomey, and Kribi disease in Cameroon. In Sri Lanka, the coconut phytoplasmal disease is referred to as Weligama leaf wilt disease.

In India, the phytoplasma-associated root (wilt) disease (RWD) is the most serious disease, affecting coconut palms in the southern state of Kerala and adjoining districts in the neighbouring state of Tamil Nadu. In the absence of effective control measures against phytoplasmal diseases, evolving resistant varieties are of utmost importance.

Screening of coconut germplasm to identify sources of resistance against LYD was first initiated in Jamaica and identifying resistant sources of Malayan Dwarfs—yellow, green as well as orange dwarfs (Natman and Robert 1955; Harries 1974a), and Panama Tall (Harries 1974b). Subsequently, the Maypan hybrids, produced by crossing the Malayan Dwarfs with Panama Tall, were field planted in Jamaica (Harries and Romney 1974). However, Malayan Dwarfs, as well as Maypan hybrid, were found to be susceptible to the disease in Ghana. The long-term field trials in Jamaica, Badouin et al. (2009), concluded that the Maypan, as well as the Malayan Dwarfs and Panama Tall, cannot be considered as disease-resistant lines. LYD resistance of Niu Leka/Fiji Dwarf (Flores 2008), Sri Lankan Dwarf, Indian Dwarf, and King Coconut (Been 1981) was also reported. Quiacoe et al. (2009) reported resistance in Sri Lankan Green Dwarf and Vanuatu Tall (VTT) to the Ghanaian lethal yellowing CSPWD, while Danyo and Dery (2011) reported higher disease susceptibility of West African Tall (WAT) compared to the hybrid MYD × VTT. Sangare et al. (1992) also reported the susceptibility of WAT to LYD in Africa. On the other hand, Mpunami et al. (1992) reported a relatively high resistance of East African Tall from Tonga and the East African Tall × Pemba Red Dwarf hybrid. Considering that resistance of accessions/cultivars varies across regions/countries, it is postulated that varietal resistance is influenced by the environmental factors, viz. strains of phytoplasma (Eden-Green 1997), temperature, vector population, and sub-optimal growing conditions (Harries 1998).

The RWD, which causes huge production losses in India, is a non-lethal, debilitating malady caused by a phytoplasma. Although it is confined to the state of Kerala and adjoining places in Tamil Nadu state, the production loss has been estimated to be very high. Characteristic bending of the leaflets termed 'flaccidity', yellowing of leaves, marginal necrosis of leaflets, and progressive decline in fruit yield are the key features of the infection. Phytoplasma has been suggested to be associated with the symptoms of the disease (Solomon et al. 1983), and hence, development and utilization of resistant varieties have been considered as viable option for the disease management. Evaluation of coconut germplasm against RWD in India identified RWD resistance/tolerance (Thomas et al. 2018). In the early breeding programs, screening the available coconut germplasm in disease-affected areas was adopted. A high degree of resistance to both leaf rot disease and RWD in Andaman Ordinary Tall and New Guinea Tall was established (Radha 1961). During 1972, large-scale screening trials at ICAR-CPCRI, Kayamkulam, and in farmers' gardens revealed the susceptibility of all the evaluated cultivars and hybrids. Subsequently, Kenthali Orange Dwarf (Jacob et al. 1998), Chowghat Green Dwarf (CGD) (Nair et al. 2004), and Malayan Green Dwarf (MGD) (Nair et al. 2007) were reported to show a higher level of resistance to RWD. Through sustained efforts, WCT, CGD, and COD palms have been identified in the RWD disease 'hotspots' in the state of Kerala with better performance for fruit yield and which were free from RWD when phenotypically and serologically tested. Those palms were used to produce RWD-tolerant planting material and the development of RWD-tolerant varieties. A comprehensive breeding program was implemented at ICAR-CPCRI since 1988 for evolving resistant/tolerant coconut varieties (Nair et al. 2010) against root (wilt) disease. The high-yielding, disease-free palms amidst heavily diseased palm population in disease hotspots of the root (wilt) prevalent areas were utilized for the breeding program. The evaluation trials have led to the development and release of three coconut varieties with varying levels of resistance/tolerance to root (wilt) disease. Observations in farmers' gardens and evaluation trials involving 10 varieties have revealed that Chowghat Green Dwarf (CGD) exhibited the highest resistance level (Nair et al. 2004). The selection from CGD, named Kalpasree, was released for growing in homesteads of the root (wilt) disease prevalent areas. Studies carried out at a seed production plot at the Coconut Development Board (CDB) Farm at Neriamangalam, Kerala State, have helped in the identification and development of another resistant variety, Malayan Green Dwarf (Nair et al. 2007), which was later released as Kalparaksha for commercial cultivation. In the field conditions, the predominant local cultivar WCT showed 84% disease incidence, whereas CGD showed 19.9%, followed by MGD with 22.4% of disease incidence.

Further 70% of the hybrids derived from the cross CGD × WCT were affected within 18 years of planting, and they recorded an average yield of 84 nuts palm^{-1} year^{-1}, highlighting the tolerant nature of the hybrid (Nair et al. 2006). Considering this impressive performance of CGD × WCT in the RWD prevalent areas, the hybrid combination was subsequently released as Kalpa Sankara for cultivation. Anatomical studies in dwarf accessions at ICAR-CPCRI indicated high lower cuticle thickness in MGD (10.27 μm). In comparison, younger leaves of CGD showed higher values for leaf thickness (1077.65 μm), parenchyma (between the vascular bundle and epidermis) width (47.29 μm), and larger distance between the stomata and phloem tissues (88.25 μm). Hence, these morphological/structural features are postulated to contribute to the higher level of resistance reported in CGD and MGD (ICAR-CPCRI 2009).

Coconut genetic resources have also been evaluated to identify genotypes resistant to bud rot caused by *Phytophthora palmivora*, stem bleeding disease caused by *Thielaviopsis paradoxa*, leaf spot/leaf blight caused by a host of fungi, viz. *Pestalotia palmarum* Cooke & Grevillea, *Bipolaris incurvata*, *Helminthosporium* sp., and basal stem rot caused by *Ganoderma sp*. Preliminary studies indicate differences in disease intensity among accessions, though no disease-resistant varieties have been identified so far. In India, Cochin China Tall, Gangabondam Dwarf, and Laccadive Ordinary Tall showed lesser intensity of stem bleeding infection (Jerard et al. 2017). In Indonesia, Novarianto et al. (1998) reported that the hybrid MYD × PYT exhibited resistance. The accessions PYT, RLT, DJP, and DBI are more resistant and the hybrid PB 121 and WAT as highly susceptible to bud rot of coconut. Generally, the dwarf palms with erect leaves, such as COD, are susceptible to bud rot disease as the rupturing or bending or breakage of the spindle due to moderate to heavy winds facilitates the entry of the fungal pathogen into the bud. The spindle leaves damaged by rhinoceros beetles are also prone to bud rot infection (Jerard et al. 2017).

Apart from the breeding initiative against root (wilt) disease in India, identification of disease-resistant or disease-tolerant or susceptible types against stem bleeding, bud rot, and leaf spot diseases was initiated. Cochin China Tall, GBGD, and Laccadive Ordinary Tall were identified as having a lesser incidence of stem bleeding disease in varietal evaluation trials. In contrast, Malayan Yellow Dwarf and its hybrids were observed to be susceptible to the stem bleeding disease.

The cadang-cadang disease, a destructive disease of coconut caused by the Cadang-Cadang Viroid (CCVd), causes death of many palms in the Bicol Region of the Philippines, and there have been no genetic sources of resistance reported. The coconut foliar decay (CFD) disease is prevalent in Pacific Ocean Islands such as Vanuatu. The tolerance to disease was observed among the local talls. Almost all the introduced ecotypes were susceptible except Rennell Tall, which was the least susceptible exotic cultivar. Its hybrid with Vanuatu Tall (VTT) was almost free of CFD (Gerard 1998). Hence, the strategy for planting material production for Vanuatu is focussed on the exploitation of CFD tolerance among the local ecotypes.

4.6.3 Pest Resistance

A few pests of coconut, namely eriophyid mite (*Aceria guerreronis*), rhinoceros beetles (*Oryctes rhinoceros*), and red palm weevil (*Rhynchophorus ferrugineus*), cause major economic loss across the most coconut producing nations.

Infestation by eriophyid mite reduces nut yield as well as the quality of nuts (Fig. 4.15). Considering the extreme difficulty in eradicating the pest through conventional plant protection measures, development or identification of mite-resistant varieties assumes greater importance. Genetic variability for eriophyid mite tolerance among the coconut cultivars was reported. Many researchers observed a strong correlation of tightness of perianth with lesser mite infestation. It was postulated that the rounded morphology of fruits facilitates tighter appression of the perianth (tepal) onto the fruit surface, restricting the entry of the coconut mite beneath the perianth and reducing its ability to establish a colony on the fruit (Mariau 1977; Julia and Mariau 1979; Moore and Alexander 1990; Howard and Rodriguez 1991; Aratchige et al. 2007). Arunachalam et al. (2013) reported a highly significant correlation between mite damage score with color and weight of the tepal. Studies have also indicated that the lesser incidence of the pest could be correlated with the traits such as fruit color, shape, and tepal arrangement in the female flowers. The regular arrangement of inner tepals of female flowers makes the gap very narrow or virtually absent, thereby restricting the incidence of the mite compared to the irregularly arranged inner tepals. The mites' entry underneath the tepals depends on tepals' tightness to the fruit surface at the early stages of fruit development. Whenever the fruit–tepal gap is narrow or absent, reduced mite damage was recorded. The tepals are tightly wound around the nuts of round-shaped fruit types than elongated and angled fruit types. However, this kind of tepal aestivation varies within the palm from flower to flower. A Cambodian variety with tight tepals was observed with a high degree of mite tolerance by Mariau (1986). In another study, round and dark green fruits exhibited higher tolerance to the eriophyid mite damage than the elongated fruits and fruits of other colors (Moore and Alexander 1987).

Fig. 4.15 Eriophyid mite damage in coconut

On the other hand, Varadarajan and David (2003) reported a fewer population of mites in orange-colored fruits with small-sized inner tepals arranged with cleft angle > 136°. Aratchige et al. (2007) reported a higher level of susceptibility of the Sri Lankan Green Dwarf, having elongated, and small-sized fruits. Kulasekharam Green Tall (KGT) was documented to have significantly lesser mite infestation (Niral et al. 2014). A selection from KGT was released as a variety named Kalpa Haritha (Fig. 4.16) characterized with less mite incidence compared to West Coast Tall cultivar. In India, Kenthali Orange Dwarf (Ramaraju et al. 2000) and Chowghat Orange Dwarf were also identified with lesser mite damage, while the Laccadive Micro Tall and Straits Settlement Apricot Tall exhibited higher mite infestation (Nair 2000). Subsequently, many researchers have recorded the natural mite infestation levels in different accessions and hybrids and reported moderate resistance in British Solomon Islands Tall, Philippines Ordinary Tall, Spicata, and the hybrids, viz. Philippines Ordinary Tall × San Blas Tall, Cochin China Tall × Philippines

Ordinary Tall (Muthiah and Natrajan 2004). Girisha and Nandihalli (2009) reported a significantly lesser mite population in Gangabondam Green Dwarf (GBGD). Sujatha et al. (2010) recorded a lower mite damage index in Laccadive Ordinary Tall (LCT) and ECT × GBGD. In contrast, the highest damage index was recorded in Laccadive Micro Tall (LMT) and LMT × GBGD. At ICAR-CPCRI, observations during December 2018 on the incidence of eriophyid mite on 70 random palms indicated infestation of 60.84% of nuts with a damage score of 2.17 (based on ICAR-CPCRI scale). The differential varietal response was observed among the palms studied. While Kalpa Haritha recorded the least eriophyid mite incidence, relatively low mite incidence was found in Navasi Tall, Gangapani Tall, Jamaica Tall, and East Coast Tall, among the accessions studied (ICAR-CPCRI 2019).

Variations in the susceptibility levels of the coconut cultivars have been reported for the pests; leaf-eating caterpillar (*Opisina arenosella*); and rhinoceros beetle damage; however, there were no resistant varieties (Kapadia 1981; Sumangala Nambiar 1991). The hybrids with COD as pollen parent have been reported to be more susceptible (Nambiar 1988). At ICAR-CPCRI, CGD is found to be susceptible to red palm weevil infestation. Falerio and Rangnekar (2001), based on the observations of ovipositional preference of red palm weevil in Goa, India, reported higher egg-laying capacity in the dwarf accessions, CGD and COD, and in a tall accession Benaulim Tall. In India, Sosamma et al. (1988) screened a few accessions for tolerance to burrowing nematode (*Radopholus similis*) and observed Java Tall, Andaman Giant Tall, Klapawangi Tall, and Kenthali Orange Dwarf to be more tolerant.

4.6.4 Product Diversification, Value Addition, and Novel Traits

Traditionally, coconut is considered as an oil crop, and hence the breeding focus has been on evaluation for oil yield and quality. With growing competition from oil palm and other vegetable oils, there has been a steady decline in the use of coconut oil for consumption.

Coconut is now actively promoted as a food crop with myriad of uses. Arunachalam et al. (2001) have listed mutant coconut types with desirable traits and highlighted their potential in breeding programs to improve profitability,

Fig. 4.16 Kalpa Haritha variety with very less incidence of mite damage

increase consumer preference, promote product diversification, and reduce the drudgery of coconut climbing. A spontaneous mutant *spicata* with many flowers on the unbranched inflorescence (Arunachalam et al. 2014), the *Thairu thengai* mutant with soft endosperm (Jerard et al. 2013), the Kaithathali with edible husk at tender stage (Satyabalan 1997), sweet kernel coconut (Samsudeen et al. 2006), flavored tender and matured coconut (ICAR-CPCRI 2009), pink husked coconuts (Jerard and Niral 2016), coconut palms with a compact crown (Bourdeix 1998), and the occurrence of haploid seedlings and bulbil producing palms (Jerard et al. 2014a) are a few instances of novel types having enormous potential in coconut improvement programs.

The tender coconut water served as a mineral drink with therapeutic properties that help regain the vitality of the human body has become popular as an emerging, natural, and healthy product. It has also been used as a potable water source by sailors in the Asia-Pacific region on long sea voyages. In the recent past, the awareness and popularity of tender nut water as a refreshing health drink are gaining momentum. Tender nut water quality has received greater attention with a growing emphasis on product diversification and to sustain/increase the income of coconut growers. Brazil is the world's largest market for packaged coconut water, accounting for 67% of juice volume sales in 2010, up from 47% in 2005 and 21% in 2003. It is reportedly growing at the expense of orange juice. Considering the business implications, several multinational companies have established their presence in the rapidly growing packaged coconut water market in major coconut-growing countries. The tender nut water products are rapidly gaining popularity in traditional and non-traditional areas, thus opening new vistas for enterprising coconut entrepreneurs to capture the widening tender nut market.

The water of the tender coconut fruit is technically the liquid endosperm of the coconut fruit. The quantity and quality of the tender coconut water vary with genotype, agro-ecological conditions, age of the palm, and progress of nut maturity. In general, tender nut water content reduces during nut maturation, coupled with reduced potassium and increased sodium content. Overall, the seven-month-old coconut is recommended for harvest as tender nuts considering the water volume and taste of nut water and sugar and sodium–potassium content.

The quantity of the tender coconut water in the fruit is correlated with the size of the coconut fruit and the endosperm cavity. It varies from negligible to more than 1000 mL. The sweetness of the tender nut water and mineral content also varies and organoleptically is graded as very good, good, average, or poor. In India, the conserved germplasm and experimental hybrids are evaluated for the quantity and quality of tender nut water. In addition to organoleptic evaluation, the sugar, amino acid, sodium, and potassium content in the tender nut water are also estimated.

Soft endosperm types have been reported from almost all the major coconut-growing countries (Arunachalam and Rajesh 2008) and named differently, such as *Makapuno* (Philippines), *Dikiri Pol* (Sri Lanka), *Kelapa Kopyor* (Indonesia), *Maphrao Kathy* (Thailand), *Coco Gra* (Seychelles), *Niu Garuk* (Papua New Guinea), *Pia* (Polynesia), *Dong Kathy* (Cambodia), and *Dua Sap* (Vietnam). This soft kernel mutant has been effectively exploited for product diversification after multiplication through controlled pollination among *Makapuno* bearing palms and in vitro embryo culture. In Indonesia, Kopyor Dwarf coconut, having a higher percentage of kopyor fruits per bunch (15–20) as compared to the tall kopyor types (5–10), has been reported (Maskromo et al. 2013). In India, palms bearing fruits with soft endosperm, called *Thairu thengai*, have been observed from the parts of Andaman and Nicobar Islands (Jerard et al. 2013) among the open-pollinated Andaman Ordinary Tall population. Such soft kernel-type palms have also been recorded in the WCT population in a sporadic manner (ICAR-CPCRI 2014).

The sweet kernel coconut (Mohachao Narel) is another invaluable mutant form reported from India (Samsudeen et al. 2013). The fruits yield fiberless or little fibered sweet kernel, suitable for table purpose as a salad. An open-pollinated population, raised at ICAR-CPCRI from Mohachao Narel palms, reveals the genetic inheritance of trait. A similar sweet kernel type, named *Ngot* (sweet), has been reported from Vietnam (Vo Van Long 1998).

The sweet husked type, called Kaithathali coconut, has been reported from Lakshadweep Islands. Similar types have been reported from Andaman and Nicobar islands also. The sweet husked types are very common in Rotuma Island of Fiji, as observed by the author. The husks of very tender fruits are sweeter in these palms than the tender fruits from other palms. The husk fibers turn brown upon maturity in normal coconut, whereas in the sweet husked types the husk becomes pale in color and loosely attached with the shell. This makes the nuts easily separable from the husk, unlike the normal coconuts. Incorporation of this trait in common cultivars could be explored so as to ease the process of dehusking in processing industries.

As the coconut palms grow tall over the years, they pose challenges while climbing for crop management activities such as harvesting, spraying, and cleaning. Hence, dwarf statured palms combined with other desirable traits of talls are aimed in breeding programs. Niu Leka Dwarf accession of Fiji origin, conserved at ICAR-CPCRI, possesses the traits of tall coconuts for fruit components coupled with a conspicuously short stem and compact crown. A selection made from Niu Leka Dwarf has been registered for its unique trait of dwarfness combined with high copra content among dwarf coconut accessions (Jerard et al. 2014b). Palms similar to the Niu Leka Dwarf but of varied nut color and bigger bole have been reported as compact dwarfs in Fiji as observed by the author and could have been the hybrid derivatives of MRD × Niu Leka hybrid. The occurrence of such compact dwarfs at varying degrees of compactness, crown size, nut size, and color, along with typical Niu Leka palms with green-colored nuts, was observed during a field survey in Viti Levu Island. In contrast, only Niu Leka palms were observed in Vanua Levu and Taveuni islands of Fiji. The utilization of these compact dwarfs is an important area in future coconut breeding. Breeding of compact dwarf coconut may facilitate ease of climbing the palm and also accommodates more palms per unit area, eventually increasing the productivity. Besides, these compact dwarfs in Fiji are relatively less affected by cyclonic storm (unpublished) and the report of tolerance/resistance of Niu Leka Dwarf to lethal disease (Merrow et al. 2003) warrants the exploitation of these dwarfs in breeding programs.

The aromatic or scented coconuts are becoming popular in tender coconut markets, and Thailand is a leader in the aromatic coconut industry. The mild aroma similar to that of basmati rice or *Pandanus aromaticus* leaves with a sweeter tinge could be easily realized during consumption of freshly cut tender nut. The sweet flavor of water appears to increase the palatability of tender coconut water. In Thailand, the aromatic coconut is called Nam Hom, and in Vietnam, it is called Dua (aromatic). Similar aromatic types were collected and conserved by ICAR-CPCRI from a coconut enthusiast, and farmer's garden in Andhra Pradesh (ICAR-CPCRI 2012) and the progeny palms are being evaluated at ICAR-CPCRI. Among the conserved accessions in India, Klapawangi Tall was identified to have the aroma trait. Similar types have been observed in Lakshadweep tall coconut populations from Minicoy and Nicobar Island (unpublished). With the limited number of palms exhibiting aroma trait, crosses with seven dwarf cultivars have been developed, and the hybrid offspring are under evaluation (ICAR-CPCRI 2015).

Natural haploid seedlings at the nursery stage have been reported (Whitehead and Chapman 1962), which have an immense potential for developing dihaploid breeding lines. Dihaploids could be utilized for the production of highly heterotic hybrids coupled with a high degree of homogeneity. The recent developments in molecular biology techniques in coconut offer scope for screening a large number of coconut seedlings in a short period to distinguish naturally occurring haploids as similar approaches

Fig. 4.17 Pink coconut type showing pinkish female flowers (**a**) and pink husk (**b**)

(a)

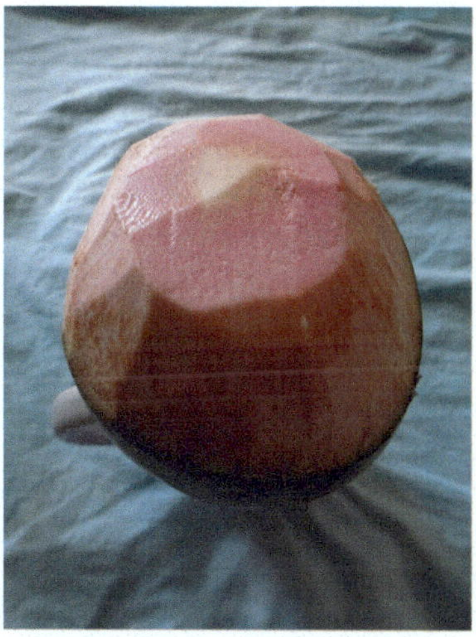

(b)

have paid fruitful results in oil palm (Dunwell et al. 2010). The outcome will generate 100% homogeneous F_1 seed from intercrosses between DH/mixoploids once they develop female inflorescences.

Generally, fruit color in coconut cultivars exhibit shades of green, yellow, red, or orange and brown. Most tall populations produce progenies, mostly with a mixture of greens and browns, whereas the dwarf cultivars produce palms with definite colors of green, orange or red, yellow, and brown. Some individual palms have been reported to produce an attractive pink color surrounding the tepals among these populations. The Pilipog Green Dwarf from the Philippines produces such palms in which the female flowers are seen with pink-tinged tepals (Rivera et al. 2010). The cross section of the fruits also revealed the pink color in the husk. This feature has been reported from fruits of individual palms (Fig. 4.17a, b) belonging to different coconut populations such as yellow dwarf and East Coast Tall, West Coast Tall, and San Ramon (Jerard and Niral 2016).

In the pink husked coconuts, when the tender fruits are trimmed, husk fibers also exhibit the color ranging from intense to light pink from the outer to the inner side, giving an attractive appearance. The pink color also appeared on the inner cavity surface of the young fruits of 3–5 months age, the cut surface of bunches, the cut surface of leaves, flowers, etc. It was observed that the palm produces two types of male flowers having anther filaments of either dark pink or yellow. This occurrence of two types of male flowers in the same palm with different colored anthers has potential application in marker-assisted selection in coconut for breeding varieties with attractive pink husks. This kind of palms could be potentially used to develop a range of pink husked lines that will have better marketability when compared to the other varieties. In Fiji, a green dwarf with a pink husk was identified in which the sprout and root tips were pink in color. Among the progenies, the seedlings with pink tipped roots could be easily distinguished, highlighting the potential of this type in breeding programs.

4.7 Conclusion

Considering the need for product diversification, beyond copra and oil purpose, to increase profitability, there exists a great scope for utilizing the novel mutant types having traits such as soft endosperm, aroma, sweet kernel, and pink husk in the improvement programs. Besides, breeding for oil quality and fatty acid composition could be another area of research warranting attention in order to make coconut oil competitive with other edible oils and suitable for several industrial applications. Breeding for climate resilience along with biotic stress tolerance needs much more focus as many emerging pests pose challenges in sustaining the production at several coconut-growing pockets. The selected populations or cultivars in the field gene banks have to be purified to minimize the experimental errors in obtaining the desirable traits while breeding varieties and hybrids. A comprehensive maintenance breeding strategy needs to be evolved, combining conventional and molecular approaches. As the evaluation of identified varieties and hybrids in multi-location trials consume time and resources, a concurrent farmer participatory breeding program mechanism could be ideal for hastening crop improvement in coconut. Considering the achievements made, the challenges ahead, and the avenues available, the strategies for future breeding programs include the development of improved high-yielding varieties with traits suitable for coconut processing, development of climate-resilient varieties to sustain or expand the coconut cultivation, development of inbreds and homogenous lines, exploitation of novel traits for developing new niche products are suggested. To fulfil the requirements of these strategies, concerted efforts are needed from time to time for collection, conservation, and exchange of diverse coconut types and the knowledge sharing in a globally coordinated approach among the coconut breeders around the world so that the breeding cycle could be hastened in efficient manner.

References

Aratchige NS, Sabelis MW, Lesna I (2007) Plant structural changes due to herbivory: do changes in *Aceria* infested coconut fruits allow predatory mites to move under the perianth? Exp Appl Acarol 43:97–107

Arunachalam V, Augustine Jerard B, Elain Apshara S, Jayabose C, Subaharan K, Ravikumar N, Palaniswami C (2013) Digital phenotyping of coconut and morphological traits associated with eriophyid mite (*Aceria guerreronis* Keifer) infestation. J Plantn Crops 41(3):417–424

Arunachalam V, Jerard BA, Elangovan M, Ratnambal MJ, Dhanapal R, Rizal SK, Damodaran V (2001) Unexploited diversity in coconut palm (*Cocos nucifera* L.). Plant Genet Resour Newslett 127:39–43

Arunachalam V, Rajesh MK (2008) Breeding of coconut palm. CAB Rev Perspect Agric Veterinary Sci Nutrition Natural Resour 3(053):1–12

Arunachalam V, Rajesh MK, Jerard BA, Jayabose C, Sairam CV (2014) Characterization of a spicata mutant of coconut palm in India. J Plantn Crops 42 (1):108–112

Balitka (1989) Coconut hybrids. DOK 018/1/1989. Research Institute for Coconut. Manado, Indonesia

Baudouin L, Philippe R, Quaicoe R, Dery S, Dollet M (2009) General overview of genetic research and experimentation on coconut varieties tolerant/resistant to lethal Yellowing. Oléagineux 16(2):127–131

Beccari O (1916) *Cocos nucifera* Linn *forma palmyrensis* Becc. In: Rock JF Palmyra Island with a description of its flora, College of Hawaii Bull, pp 1–58

Been BO (1981) Observations on field resistance to lethal yellowing in coconut varieties and hybrids in Jamaica. Oleagineux 36:9–12

Bourdeix R (1988) Etude du déterminismegénétique de la couleur du germe chez le cocotier Nain. Oléagineux 43(10):371–373

Bourdeix R, Meunier J, N'Cho YP (1991) Unestrategte de selection du cocotier *Cocos nucifera* L. II. Amelioration des hybrides Grand x Grand. Oleagineux 46:367–382

Bourdeix R, N'Choand YP, Sangare A (1998) Coconut breeding programme in Cote d'Ivoire. In: Batugal PA, Ramanatha Rao V (eds) Coconut Breeding. Papers presented at a Workshop on Standardization of Coconut Breeding Research Techniques, 20–25 June 1994, Port Bouet, Cote d'Ivoire, IPGRI-APO, Serdang, Malaysia

Brook RM (1985) Early yields from dwarf x tall coconut experiment. Harvest 11:66–70

Charles AE (1961) Selection and breeding of the coconut palm. Trop Agric 38(4):283–296

Chempakam B, Kasturi Bai KV, Rajagopal V (1993) Lipid peroxidation and associated enzyme activities in' relation to screening for drought tolerance in coconut (*Cocos nucifera,* L) PI. Physiol Biochem 20 (1):5–10

Chowdhury D, Nath JC, Mohan NK (2001) 'Kamrupa'—a newly released coconut variety by Assam Agricultural University. Indian Coconut J 31:12–13

Chulapan P, Anupap T (1998) Coconut breeding programme in Thailand. In: Batugal PA, Ramanatha Rao V (eds) Coconut Breeding. Papers presented at a Workshop on standardization of coconut breeding research techniques, 20–25 June 1994, Port Bouet, Cote d'Ivoire, IPGRI-APO, Serdang, Malaysia

Cintra FLD, Passos EEM, De Leal LS (1993) Evaluation of root system distribution in tall coconut cultivars. Oleagineux 48:453–461

Copeland EB (1931) The Coconut. Macmillan and Co., London

Danyo G, Dery SK (2011) Overview of the epidemiology of coconut Lethal Yellowing Disease (LYD) in Malayan yellow dwarf × Vanuatu tall hybrids in the Western and central regions of Ghana. J Ghana Sci Assoc 13(1):30–38

Davis TA (1969) Prospects of the clonal propagation of the coconut. Ceylon Coconut Planters Rev 6:1–5

Dunwell JM, Wilkinson MJ, Nelson S, Wening S, Sitorus AC, Mienanti D, Alfiko Y, Croxford AE, Ford CS, Forster BP, Caligari PDS (2010) Production of haploids and doubled haploids in oil palm. BMC Plant Biol 10:218

Eden-Green SJ (1997) History, world distribution and present status of lethal yellowing disease of palms. In: Eden-Green SJ, Offori F Elmina (eds) Proceedings of International Workshop on Lethal Yellowing Like Diseases of Coconut. Ghana, November 1995, pp 9–26

Faleiro JR, Rangnekar PA (2001) Ovipositional preference of red palm weevil *Rhynchophorus ferrugineus* Olive. to coconut cultivars. Indian Cocon J 32 (6):22–23

Flores A (2008) Fiji Dwarf sets new durability standards in coconuts. Agri Res (May-June): 17

Gerard D (1998) Vanuatu national coconut breeding programme. In: Batugal PA, Ramanatha Rao V (eds) Coconut Breeding. Papers presented at a Workshop on Standardization of Coconut Breeding Research Techniques, 20–25 June 1994, Port Bouet, Cote d'Ivoire. IPGRI-APO, Serdang, Malaysia

Gerardo AS, Rivera RL (1998) Coconut breeding programme of the Philippines. In: Batugal PA, Ramanatha Rao V (eds) Coconut Breeding. Papers presented at a Workshop on Standardization of Coconut Breeding Research Techniques, 20–25 June 1994, Port Bouet, Cote d'Ivoire. IPGRI-APO, Serdang, Malaysia

Girisha RC, Nandihalli BS (2009) Seasonal abundance and varietal reaction of coconut perianth mite, *Aceria guerreronis* Keifer in Dharwad area. Karnataka J Agric Sci 22(3):606–608

Gomes FP, Prado CH (2007) Ecophysiology of coconut palm under water stress. Braz J Plant Physiol 19 (4):377–391

Harries HC (1974) Selection and breeding of coconuts for resistance to diseases such as lethal yellowing. Indian J Gen 34A:66–74

Harries HC (1974) The performance of F_1 hybrid coconuts in Jamaica. J Plantn Crops 2(2):15–20

Harries HC (1998) Breeding phytoplasma disease resistant coconuts: alternative strategies for field exposure trials. In: Proceedings of International Cashew and Coconut Conference: Trees for Life—the Key to Development, Dar es Salaam, Tanzania, 17–21st February 1997, pp 375–379

Harries HC, Romney DH (1974) MayPan: an F_1 hybrid for commercial production in Jamaica. World Crops 26:110–111

Hatfield JL, Prueger JH (2015) Temperature extremes: effect on plant growth and development. Weather Clim Extreme 10:4–10

Hebbar KB, Rose HM, Nair AR, Kannan S, Niral V, Arivalagan M, Gupta A, Samsudeen K, Chandran KP, Chowdappa P, Prasad PV (2018) Differences in *in vitro* pollen germination and pollen tube growth of coconut (*Cocos nucifera* L.) cultivars in response to high temperature stress. Environ Exp Bot 153:35–44

Hebbar KB, Sheena TL, ShwethaKumari K, Padmanabhan S, Balasimha D, Berwal MK, Thomas GV (2013) Response of coconut seedlings to elevated CO_2 and high temperature in drought and high nutrient conditions. J Plantn Crops 41:118–122

Howard FW, Rodriguez EA (1991) Tightness of the perianth of coconuts in relation to infestation by coconut mites. Fla Entomol 74(2):358–361

ICAR-CPCRI (2004) Annual report 2003–04. ICAR-Central Plantation Crops Research Institute, Kasaragod, 152p

ICAR-CPCRI (2009) Annual report 2008–09. ICAR-Central Plantation Crops Research Institute, Kasaragod, 145p

ICAR-CPCRI (2010) Annual report 2009–10. ICAR-Central Plantation Crops Research Institute, Kasaragod, 150p

ICAR-CPCRI (2011) Annual report 2010–11. ICAR-Central Plantation Crops Research Institute, Kasaragod, 146p

ICAR-CPCRI (2012) Annual report 2011–12. ICAR-Central Plantation Crops Research Institute, Kasaragod, 128p

ICAR-CPCRI (2014) Annual report 2013–14. ICAR-Central Plantation Crops Research Institute, Kasaragod, 139p

ICAR-CPCRI (2015) Annual report 2014–15. ICAR-Central Plantation Crops Research Institute, Kasaragod, 126p

ICAR-CPCRI (2017) Annual report 2016–17, ICAR-Central Plantation Crops Research Institute, Kasaragod – 671124. Kerala, India, p 236p

ICAR-CPCRI (2019) Annual report 2018–19, ICAR-Central Plantation Crops Research Institute, Kasaragod – 671124. Kerala, India, p 164p

IRHO-CIRAD (1989) Rapport d'activities Recherches de la Station de Recherches "Marc Delorme", annees 1988, 361p

Jack HW, Sands WN (1922) The dwarf coconut in Malaya. Malayan Ag J 10:4–12

Jacob PM, Nair RV, Rawther TSS (1998) Varietal resistance. In: Nampoothiri KUK, Koshy PK (eds) Coconut root (wilt) disease. CPCRI, Kasaragod, pp 97–104

Jayasekera KS, Wahala WMPB, Jayasekera C, Peries RRA (1993) Water use in coconut (*Cocos nucifera* L.). In: Proceedings of the Annual Sessions of the Sri Lanka Association for the Advancement of Science (SLAAS), Peradenlya

Jerard BA, Damodaran V, Niral V, Samsudeen K, Rajesh MK, Sankaran M (2013) Conservation and utilization of *Thairu Thengai*–soft endosperm coconut accession from Andaman Islands. J Plantn Crops 41 (1):14–21

Jerard BA, Niral V, Chowdappa P (2017) Breeding. In: Chowdappa P, Niral V, Jerard BA, Samsudeen K (eds) Coconut. Daya Publishing House, A Division of Astral International Pvt Ltd, New Delhi, India. pp 131–174

Jerard BA, Niral V, Dhanapal R, Damodaran V, Saran Kumar Rizal, Arunachalam V, Sankaran M, Nair RV,

Ratnambal MJ, Bhaskara Rao EVV, Koshy PK, Thomas GV (2014b) Plant Germplasm Notice 14. IND 099—NiuLeka Green Dwarf (EC0415218; INGR13065), a coconut (*Cocos nucifera*) germplasm of short statured palm but possessing the advantageous characters of talls. Indian J Plant Genet Resour 27 (1):66–89

Jerard BA, Niral V, Samsudeen K, Nair RV, Jayabose C, George V Thomas (2015) Development of a Dwarf x Tall coconut hybrid 'Kalpa Samrudhi'. J Plantn Crops 43(1):46–52

Jerard BA, Rajesh MK, Elain SE, Sajini KK, Rahman S, Fayas TP, Karun A (2014) Scope of novel and rare bulbiferous coconut palms (*Cocos nucifera* L.). Genetic Resour Crop Evol 61:1–6

Jerard BA, Niral V (2016) Pink husked coconut. CPCRI News letter. ICAR-Central Plantation Crops Research Institute, Kasaragod

Joeli K (1998) National coconut breeding programme in Fiji. In: Batugal PA, Ramanatha Rao V (eds) Coconut Breeding. Papers presented at a Workshop on Standardization of Coconut Breeding Research Techniques, 20–25 June 1994, Port Bouet, Cote d'Ivoire. IPGRI-APO, Serdang, Malaysia

Julia JF, Mariau D (1979) New research on the coconut mite *E. guerreronis*(K), in the Ivory Coast. Oléagineux 34:181–189

Kapadia MN (1981) Susceptibility of three cultivated coconut varieties to *Nephantis serinopa* Meyrick. Indian Cocon J 12(4):4–5

Kasturi Bai KV, Naresh Kumar S, Rajagopal V (2009) Abiotic stress tolerance in coconut. CPCRI, Kasaragod, India

Kasturi Bai KV, Rajagopal V, Arunachalam V (2006) Assessment of diversity on coconut varieties for drought responsive physiological traits. J Plant Crops 34:118–120

Khan HH, Elangovan M, Arunachalam V, Nambiar PTN, Reddy RVKS, Prasanna Kumar, Nagawekar DO, Suresh S, Giridharan S, Natrajan C, Nampoothiri KUK, Parthasarathy VA (2002) Stability of coconut yield in coordinate multi location trials. In: Plantation crops research and development in the new millennium, pp 300–302

Kumar SN (2005) Variability in content and composition of fatty acids in coconut oil due to genetic and environmental factors. AP cess fund project. Final report submitted to ICAR, New Delhi

Kumar SN, Bai KK, Rajagopal V, Aggarwal PK (2008) Simulating coconut growth, development and yield with the InfoCrop-coconut model. Tree Physiol 28 (7):1049–1058

Labouisse JP, Sileye T, Hamelin C (2007) New observations on the resistance of coconut cultivars to tropical cyclones in Vanuatu. CORD 23(2):28–38

Le Saint JP, de Nuce de Lamothe M (1987) Les hybrides de cooctiersnains: performance et interet. Oleagineux 42(10):353–362

Mao Z (1986) An investigation on meteorological indices for coconut cultivation in China. Oleagineux 41:119–128

Mao Z, Lai Y (1993) The coconut germplasm of Hainan Island, China. Plant Genet Res Newsl 91(92):53–57

Marechal H (1928) Observations and preliminary results on the coconut palm with a view to developing improved seed nuts for Fiji. Agric J Fiji 1:16–45

Mariau D (1977) *A. Guerreronis* (Eriophyes): an important pest of African and American coconut groves. Oléagineux 32:101–109

Mariau D (1986) Comportement de *Eriophye guerreronis* Keifer à l'égard de diferentes variétés de cocotier. Oléagineux 41:499–505

Marty G, Le Guen V, Fournial T (1986) Effets des dépressionscycloniques sur les plantations de cocotiers au Vanuatu. Oléagineux 41(2):63–69

Maskromo I, Novarianto H, Sukma SD, Sudarsono (2013) Productivity of three dwarf kopyor coconut varieties from Pati, Central Java, Indonesia. CORD 29(2):19–28

Menon KPV, Pandalai KM (1958) The coconut palm: a Monograph. Indian Central Coconut Committee, Ernakulam, p 384

Merrow AW, Wisser RJ, Brown JS, Kuhn DN, Schnell RJ, Broschat TK (2003) Analysis of genetic diversity and population structure within Florida coconut (*Cocos nucifera* L.) using microsatellite DNA with special emphasis on the Fiji dwarf cultivar. Theor Appl Genet 106:715–726

Mkumbo KE, Kullaya A (1998) Coconut breeding research in Tanzania. In: Batugal PA, Ramanatha Rao V (eds) Coconut Breeding. Papers presented at a Workshop on Standardization of Coconut Breeding Research Techniques, 20–25 June 1994, Port Bouet, Cote d'Ivoire. IPGRI-APO, Serdang, Malaysia

Moore D, Alexander L (1987) Aspects of migration and colonization of the coconut palm by the coconut mite *Eriophyes guerreronis* (K.) (Acari: Eriophyidae). Bull Entomol Res 63:285–288

Moore D, Alexander L (1990) Resistance of coconuts in St. Lucia to attack by the coconut mite *E. guerreronis* Keifer. Trop Agric 67:33–36

Mpunami A, Schuiling M, Kaiza D (1992) The lethal disease of coconut palm. Tanzanian Res Train Newslett (dar Es Salaam) 7(1–3):12–15

Muthiah C, Natarajan C (2004) Varietal reaction and nutrient management of coconut eriophyid mite. Planter 80(936):159–169

Nair CPR (2000) Status of coconut eriophyid mite, *Aceriaguerreronis* Keifer in India. In: Proceedings of international workshop on coconut eriophyid mite, 6–8 Jan 2000, Coconut Research Institute, Sri Lanka, pp 13–31

Nair R, Thomas RJ, Jacob PM (2010) Breeding for resistance to coconut root (wilt) disease. In: Thomas GV, Chandramohanan R, Jacob PM, Krishnakumar V (eds) Coconut Root (wilt) Management. CPCRI, Kasaragod, pp 58–71

Nair RV, Jacob PM, Ajithkumar R (2004) Screening of coconut varieties against root (wilt) disease. J Plantn Crops 32:50–51

Nair RV, Jacob PM, Thomas RJ, Pillai SP, Mathew C (2006) Performance of CGD x WCT hybrid in the root (wilt) prevalent tract. J Plantn Crops 34:15–20

Nair RV, Jerard BA, Regi J Thomas (2016) Coconut breeding in India. In: Al-Khayri JM, Jain M, Dennis V Johnson (eds) Advances in Plant Breeding Strategies: Agronomic, Abiotic and Biotic Stress Traits. Springer International Publishing, Switzerland

Nair RV, Thomas, RJ, Aravazhi E, Sasikala M, Nair CPR (2007) Malayan green dwarf, a root (wilt) resistant coconut variety with high yield suitable for the disease prevalent regions. In: Proceedings of the 94th Indian Science Congress, 3–7 Jan 2007, Annamalai University, Tamil Nadu, p 140

Nambiar SS (1988) Susceptibility of hybrid coconut varieties to *Oryctes rhinoceros* L. under rainfed conditions at Pilicode. In: Silas EG, Aravindakshan M, Jose AI (eds) Coconut breeding and management. Kerala Agricultural University, Trichur, pp 158–160

Nampoothiri KUK, Kumaran PM, Jerard BA, Ratnambal MJ, Rao EVVB, Parthasarathy VA (1999) Combining ability in coconut (*Cocos nucifera* L.). CORD XV(2):68–75

Narayana GV, John CM (1949) Varieties and forms of the coconut. Indian Coconut J 2:209–226

Naresh KS, Rajagopal V, Kasturi Bai KV (2006) Criteria for identification of coconut palms *in situ* tolerant to abiotic stresses in farmers' fields. Indian Coconut J 37:5–7

Niral V, Ananda KS, Elain Apshara S, Augustine Jerard B, Chowdappa P (2016) Improved varieties of Coconut, Arecanut and Cocoa: ICAR-CPCRI Contribution. ICAR-CPCRI Technical Bulletin No. 096, Centenary Series No. 16, ICAR-CPCRI, Kasaragod, India, p 95

Niral V, Jerard BA, Samsudeen K, Regi J Thomas, Thomas GV (2014) Coconut varieties and hybrids. CPCRI Technical Bulletin No. 87. Central Plantation Crops Research Institute, Kasaragod, Kerala, India, p 40

Novarianto H, Tine Rompas, Darwis SN (1998) Coconut breeding programme in Indonesia. In: Batugal PA, Ramanatha Rao V (eds) Coconut Breeding. Papers presented at a Workshop on Standardization of Coconut Breeding Research Techniques, 20–25 June 1994, Port Bouet, Cote d'Ivoire. IPGRI-APO, Serdang, Malaysia

Nutman FJ, Robert FM (1955) Lethal Yellowing: the unknown disease of coconut palm in Jamaica. Empire J Agri 23:257–267

Padmanabhan D (1963) Leaf development in palms. Curr Sci 32:537–539

Passos EEM, Prado CHBA, Leal MLS (1999) Condutânciaestomática, potencialhídrico foliar e emissão de folhas e inflorescênciasemtrêsgenótipos de coqueiroanão. Agrotrópica 11:147–152

Patel JS (1937) Coconut breeding. Proc Assoc Econ Biol 5:1–16

Patel JS (1938) The coconut: a monograph. Government Press, Madras, p 311

PCA (2016) (Philippine Coconut Authority). Coconut variety for commercial Buko production. http://www.pca.da.gov.ph/pdf/techno/tacunan.pdf, Accessed on February 18, 2020

Periasamy K (1965) Morphological and ontogenetic studies in palms II. Growth pattern of the leaves of *Cocos nucifera* and *Borassus flabellifer* after the initiation of plications. Australian J Bot 13:225–234

Petch T (1913) The flower of the coconut. Trop Agriculturist 41:449–455

Purseglove JW (1968) The origin and distribution of the coconut. Trop Sci 10:191–199

Quaicoe RN, Dery SK, Philippe R, Baudouin L, Nipah JO, Nkansah-Poku J, Arthur R, Dare D, Yankey EN, Pilet F, Dollet M (2009) Resistance screening trials on coconut varieties to Cape Saint Paul Wilt Disease in Ghana. Oléagineux 16(2):132–136

Radha K (1961) Leaf rot disease of coconut. In: Proceedings of the 1st Session FAO Tech Wkg Pty Coco Prod Prot and processing, Trivandrum, Kerala

Rajagopal V, Kasturi Bai KV (2002) Drought tolerance mechanism in coconut. Burotrop Bull 17:21–22

Rajagopal V, Kasturi Bai KV, Kumar N (2005) Breeding for drought tolerance in Coconut. In: Batugal P, Rao VR, Oliver J (eds) Coconut genetic resources. IPGRI, Rome, pp 282–301

Rajagopal V, Kasturi Bai KV, Voleti SR (1990) Screening of coconut genotypes for drought tolerance. Oleagineux 45:215–223

Rajagopal V, Shivshankar S, Kasturibai KV, Voleti SR (1988) Leaf water potential as an index of drought tolerance in coconut. Plant Phys Biochem 15:80–86

Rajesh MK, Jerard BA, Preethi P, Thomas RJ, Fayas TP, Rachana KE, Karun A (2013) Development of a RAPD-derived SCAR marker associated with tall-type palm trait in coconut. Sci Hortic 150:312–316

Rajesh MK, Thomas RJ, Rijith J, Shareefa M, Jacob PM (2012) Genetic purity assessment of D x T hybrids in coconut using SSR markers. Indian J Genet Plant Breed 72(4):472–474

Ramaraju K, Natarajan K, Sundara Babu PC, Palanisamy S, Rabindra RJ (2000) Studies on coconut eriophyid mite *Aceria guerreronis* Keifer in Tamil Nadu, India. In: Proceedings of the International Workshop on Coconut Eriophyid Mite, 6–8 Jan 2000, Coconut Research Institute, Sri Lanka, pp 13–31

Ranasinghe CS (2019) Climate change impacts on coconut production and potential adaptation and mitigation measures: a review of current status. In: Marambe B (ed) Proceedings of the Workshop on Present Status of Research Activities on Climate Change Adaptations Sri Lanka Council for Agricultural Research Policy, Colombo, pp 71–82

Ratnambal MJ, Nair MK, Muralidharan K, Kumaran PM, Rao EVVB, Pillai RV (1995) Coconut descriptors. Part. I. Central Plantation Crops Research Institute, Kasaragod, Kerala, India, p 198

Ratnambal MJ, Niral V, Krishnan M, Ravikumar N (2000) Coconut descriptors Part II, CD-ROM, Central Plantation Crops Research Institute, Kasaragod, Kerala, India

Repellin A, Daniel C, Zuily-Fodil Y (1994) Merits of physiological tests for characterizing the performance of different coconut varieties subjected to drought. Oleagineux 49:155–168

Rivera RL, Santos GA, Emmanuel EE, Rivera SM (2010) Pilipog Green Dwarf (PILD) in the Philippines. In: Batugal P, Bourdeix R, Oliver JT, George MLC (eds). Catalogue of Conserved Coconut Germplasm. International Coconut Genetic Resources Network (COGENT), Bioversity International, Regional Office for Asia, the Pacific and Oceania, Serdang, Selangor Darul Ehsan, Malaysia

Sampson HC (1923) The Coconut Palm. John Bale, Sons and Danielsson Ltd, London

Samsudeen K, Jacob PM, Niral V, Kumaran PM, Salooja R, Moosa H (2006) Exploration and collection of coconut germplasm in Kadmat and Amini Islands of Lakshadweep, India. Genetic Resour Crop Evol 53:1721–1728

Samsudeen K, Rajesh MK, Nagwaker DD, Reshmi R, Kumar PA, Devadas K, Anitha K (2013) Diversity in Mohachao Narel, a sweet endosperm coconut (*Cocos nucifera* L.) population from Maharashtra, India. Natl Acad Sci Lett 36(3):319–330

Sangare A, de Taffin G, de Franqueville H, Arkhust ED, Pomier M (1992) Coconut lethal yellowing in Ghana. Initial results on planting material performance in the field. Oleagineux 47(12):699–704

Satyabalan K (1997) Coconut varieties and cultivars—their classification. Asian Pacific Coconut Community, Jakarta, Indonesia, p 105

Satyabalan K (1993) The Coconut palm-botany and breeding. Asian Pacific Coconut Community, Jakarta, Indonesia, p 214

Solomon JJ, Govindankutty MP, Nieuhaus F (1983) Association of mycoplasma-like organisms with the coconut root (wilt) disease in India. Z Pflkrankhpflschutz 90:295–297

Sosamma VK, Koshy PK, BhaskaraRao EVV (1988) Response of coconut cultivars to the burrowing nematode *Radopholus similis*. Indian J Nematol 18 (1):136–137

Sujatha A, Chalam MSV, Kalpana M (2010) Screening of coconut germplasm against coconut eriophyid mite, *Aceria guerreronis* Keifer in Andhra Pradesh. J Plantn Crops 38(1):53–56

Sumangala Nambiar S (1991) Susceptibility of hybrid coconut varieties to *Oryctes rhinoceros* under rain fed conditions at Pilicode. In: Silas EG, Aravindakshan M, Jose AI (eds) Coconut Breeding and Management. Kerala Agricultural University, Thrissur, pp 158–160

Swaminathan MS, Nambiar MC (1961) Cytology and origin of dwarf coconut palm. Nature 192:85–86

Thomas RJ, Rajesh MK, Jacob PM, Jose M, Nair RV (2015) Studies as genetic uniformity studies of Chowghat Green Dwarf and Malayan Green Dwarf varieties of coconut using molecular and morphometric methods. J Plantn Crops 43(2):89–96

Thomas RJ, Shareefa M, Nair RV (2018) Varietal resistance in coconut. In: Nampoothiri KUK, Krishnakumar V, Thampan PK, Achuthan Nair M (eds) The Coconut Palm (*Cocos nucifera* L.)—Research and Development Perspectives. Springer Nature Singapore Pte Ltd, Singapore, pp 113–156

Thompson P (1981) Hybrid coconuts in Papua New Guinea. Harvest 7(3) (Papua New Guinea)

Tomlinson PB (1961) Anatomy of the Monocotyledons II Palms. Cocoid Palms: Cocos, pp 195–198

Varardarajan MK, David PMM (2003) Effect of ground vegetation and nut characters on the severity of infestation by *Aceria guerreronis*: in coconut. Entomon 28:361–365

Venkatanarayana G (1957) On certain aspects of the development of the leaf of *Cocos nucifera*. Phytomorphol 7:297–305

Vo Van Long (1998) Coconut breeding programme of Vietnam. In: Batugal PA, Ramanatha Rao V (eds) Coconut Breeding. Papers presented at a Workshop on Standardization of Coconut Breeding Research Techniques, 20–25 June 1994, Port Bouet, Cote d'Ivoire. IPGRI-APO, Serdang, Malaysia

Voleti SR, Kasturi Bai KV, Rajagopal V (1993) Water potential in the leaves of coconut (*Cocos nucifera* L.) under rainfed and irrigated conditions. In: Nair MK, Khan HH, Gopalasundaran P, Bhaskara Rao EVV (eds) Advances in Coconut Research and Development. Oxford & IBH Publishing, New Delhi, India, pp 243–245

Whitehead RA, Chapman GP (1962) Twinning and haploidy in *Cocos nucifera*. Nature 195:1228–1229

Characterization of Genetic Diversity Using Molecular Markers

5

Lalith Perera and R. Manimekalai

Abstract

The coconut palm (*Cocos nucifera* L.) is grown throughout the tropical world and virtually every part of the palm possesses some human use. It is a monotypic species in the genus *Cocos* and belongs to the Arecaceae family. Genetic diversity in coconut is mainly compartmentalized between the tall and dwarf palm types. The coconut crop is spread across the globe; hence, different agro-climatic regions contribute to its rich germplasm diversity. Besides morphological descriptors, molecular markers, namely RAPD, ISSR, ISTR, SSR, AFLP, etc. have been widely used to characterize the genetic diversity of coconut originating from different regions. With the expansion of the knowledge of the coconut genomic resources including its genome sequences, DNA sequence-based markers are also being widely utilized presently to characterize the genetic diversity and population structure. Further, genomics and transcriptomics approaches are being utilized to unravel the gene diversity and gene-function relationships. This chapter elaborates the principles and applications of diversity analysis of coconut using various molecular markers and other genomic approaches of coconut that are in vogue.

5.1 Introduction

Coconut (*Cocos nucifera* L.) is monotypic genus which belongs to the family Arecaceae (Palmaceae) and the subfamily Cocoideae. The diploid genome of coconut has 32 chromosomes ($2n = 2x = 32$). The diversity of coconut genotypes is enormous; multiple varieties and morphological forms of the palm have been described (Narayana and John 1949; Gangolly et al. 1957; Menon and Pandalai 1958 and references there in; Liyanage 1958) mainly based on their morphology and breeding habit (Ohler 1984). In its varietal classification, two principal types of coconuts have been identified worldover, namely the tall (Typica) group and the dwarf (Nana) group. In addition, several intermediate type coconut populations have also been identified in certain countries. Tall coconuts, besides possessing tall stature, display larger crowns and an enlargement at the base of the stem. As the classification suggests, the dwarf coconuts are short in stature, produce a smaller crown and are having a slender stem from the base to crown. The dwarf coconuts are further distinguished into four groups based on their fruit color, i.e. green dwarfs, yellow dwarfs, red

L. Perera (✉)
Coconut Research Institute, Lunuwila, Sri Lanka
e-mail: lalithperera1234@yahoo.com

R. Manimekalai
ICAR-Sugarcane Breeding Institute, Coimbatore, Tamil Nadu, India

M. K. Rajesh et al. (eds.), *The Coconut Genome*, Compendium of Plant Genomes,
https://doi.org/10.1007/978-3-030-76649-8_5

dwarfs and brown dwarfs. The intermediate coconut forms/varieties possess mixtures of tall and dwarf characters. There are two genetically diverse groups within the tall coconuts; they are the coconuts originating from South East Asia and Pacific region and the coconuts from South Asia and Africa region. Evolutionary analysis suggests that the dwarf forms of the palm could have emerged from their tall counterparts belonging to the South East Asian/Pacific region (Perera et al. 2000). Genetically, the dwarfs are highly homozygous exhibiting more than 90 per cent homozygosity for simple sequence repeat (SSR) alleles and thus can be considered as pure lines. When a population of dwarfs is developed as a result of self-pollination, the resultant population could be called 'clones'. Nevertheless, the tall morphoforms are found to be naturally cross-pollinating, and hence, the populations are highly heterogeneous. However, different accessions of dwarfs display varying degrees of heterozygosity. Consequently, the talls are rich in alleles, while of the dwarfs have diminished allelic richness. Similarly, coconut exhibits great morphological diversity for each of its parts, for instance, fruits vary for size, shape and color and the weight proportions of various fruit component such as nut water, kernel, husk, or shell.

Recent research findings have accorded a number of health benefits of coconut and its products, and hence, coconut is considered as a wonder food commodity. Since many of the health benefits are attributed to the coconut and its products, there is a growing demand for coconut in the world market today, and consequently, an increased production and supply of coconut is warranted. Thus, the collection and evaluation of coconut germplasm and breeding coconut for high yield and specialty traits have become a priority more than ever in many of the national research programs.

The genetic diversity information about a crop forms a basis for its systematic exploitation through multiple breeding programs. Coconut hybrids, derived by crossing talls and dwarfs, showed 80 to 100% increase in the yield levels of economic parts compared to talls alone (Batugal et al. 2009). In addition, naturally occurring 'super palms' having superior yield, yield stability and drought tolerance have been identified (Wickramaratne 1987). Nevertheless, crop improvement programs of coconut are hindered by several issues such as its narrow genetic base, long juvenile phase, cross-pollinating reproductive biology, lack of any vegetative propagation methods, highly heterogeneous phenotypes, and the long intervals between the generations. Also, the monotypic nature of the coconut is one of a major limiting factors in breeding coconut varieties suitable for cultivation in diverse ecological situations. Many of these genetic constraints have to be overcome by non-conventional techniques to achieve considerable genetic improvement in coconut.

In this context, the characterization of coconut germplasm has become an important step to exploit the genetic diversity. Towards this, several biomarkers including biochemical, physiological and phenotypic markers along with molecular markers of coconut have been investigated. Among these, polymorphisms existing in protein or DNA-based molecular markers have been extensively utilized in resolving phylogeny, taxonomy and ecological questions pertaining to plant breeding.

Molecular markers are quite robust and potent genetic tools used in the crop improvement programs and in assisting conventional breeding of many crops. The incorporation and the utilization of molecular marker techniques have improved the breeding efficiency by estimation of genetic diversity in coconut accessions, categorizing the genetic groups of different accessions, characterization and organization of germplasm, developing linkage mapping and in characterization of novel QTLs for marker-assisted selection (MAS). These markers would also help in reducing the selection of crosses to be tested in a coconut hybrid breeding program. Identification of molecular markers linked to economically useful traits will strengthen and hasten the breeding programs, reduce costs, improve the efficiency and curtail the length of the selection cycles.

To decode the genetic architecture of an individual by detecting the molecular polymorphisms, many techniques, comprising biochemical markers and DNA markers, have been utilized. Markers greatly aiding molecular characterization include techniques such as restriction fragment length polymorphism (RFLP), randomly amplified polymorphic DNA (RAPD), amplified fragment length polymorphism (AFLP), inverse sequence-tagged repeat (ISTR), simple sequence repeats (SSRs) and inter-simple sequence repeats (ISSR). In the recent context, genome sequence-based molecular markers add to the ever-growing repertoire of tools useful for molecular characterization.

5.2 DNA Markers

Despite the wide usage of protein markers, a major limitation of isozyme-based analysis is its limited standard assay conditions for wide range of studies. DNA-based molecular markers are frequently used in different disciplines of botany such as in embryology, genetic modification and in resolving taxonomic questions. Furthermore, DNA-based markers are being continuously altered and their performance enhanced in order to automate the genome analysis processes.

5.2.1 Features of Ideal DNA Markers

i. Should occur frequently in the genome being investigated
ii. Well distributed all over the genome
iii. Should be neutral devoid of any masking or pleiotropic effects
iv. Abundantly polymorphic (moderate to high level of polymorphism is desired)
v. Should show co-dominant inheritance pattern so that homozygous and heterozygous individuals of an organism could be differentiated
vi. Easy access, easy to perform and rapid assay conditions
vii. Should possess relatively high reproducibility
viii. Unambiguous assignment of alleles
ix. Exchange of marker data between the laboratories should be less cumbersome
x. Should be economical for the development of marker and assay performance

Considering the difficulties in satisfying all these criteria by a single molecular marker, a marker which satisfies some or most of these characteristic features finds an application in crop genomics. Different types of marker technologies can be applied for the characterization the crop genome.

5.2.2 Restriction Fragment Length Polymorphism (RFLP)

Further extending the application of Southern hybridization technique, the RFLP-based DNA marker system was developed. This molecular marker system is specifically suitable for a single clone/restriction enzyme combination and it exploits sequence variations in homologous DNA sequences. RFLP has multiple advantages over other traditional marker systems: (i) RFLPs are co-dominant and are not influenced by the environment conditions; (ii) any DNA starting material could be used; (iii) multiple markers could be assigned or mapped in a population without any interference from the phenotypic mutations. Hence, RFLP markers are widely finding their use in forensic science, discrimination or resolution of parentage conflicts, and in identification of appropriate genotypes. Lebrun et al. (1998) applied the RFLP technique in coconut to assay the genetic diversity in 100 palms derived from diverse geographical conditions. Relatively high quality genetic maps have been constructed utilizing locus-specific RFLP markers (Tanksley et al. 1989). Results from the RFLP studies furnished the historical dispersion pattern of coconut. Considerable ecotype diversity was recorded from Far East and Pacific regions, since these regions are suggested as putative centres of origin of the coconut. The West African ecotypes were related to the ecotypes of South Asia, especially India and Sri

Lanka, suggesting recent extension of the palm along the length of the Atlantic Coasts of Africa from the coconuts that have their origin from the Indian Ocean. The tall ecotypes exhibited high polymorphism compared to the dwarfs. However, on the downside, RFLP assays are in general time demanding, involves tedious experimental procedures, and requires good quality DNA.

5.2.3 Random Amplified Polymorphic DNA (RAPD)

The RAPD is a DNA-based molecular marker which works on the principle of differential PCR amplicons resulting from the same set of template DNA and short primers (oligonucleotides) used in a typical PCR reaction. The PCR amplicons are derived from the known DNA source but unknown sequence information utilizing a single, short arbitrarily designed primers. Thus, this technique does not require prior sequence information of template DNA. RAPD is a less expensive technique, efficiently provides large number of molecular markers in a relatively short time, using a simple PCR station; however, great concerns still exist with its reproducibility. RAPD have a wide range of applications including the genetic identification of varieties, cultivars and clones, used in the field of genetic mapping, marker-assisted selection of agronomically important traits and in studies such as population genetics and molecular systematics.

Though application of RAPD as a marker technology was initiated in the early 1990s, it was adopted late in coconut when Upadhyay et al. (2004) reported use of eight polymorphic RAPD primers to analyse genetic diversity among the twenty coconut (15 Indian and 5 exotic) accessions. This study suggested that the tall accessions exhibited relatively high polymorphism and hence are genetically diverse and more heterozygous. Based on the RAPD marker analysis, moderate occurrence of (60%) genetic variability within the populations was reported. The relatively low inter-population diversity was attributed to the phenomenon of genetic drift and a possible bottleneck in evolutionary past of the species. However, RAPD molecular markers linked to specific set of populations were also identified.

Several others studies also have reiterated the utility of RAPD techniques to delineate the extent of genetic diversity among the coconut accessions. RAPD assay has revealed that the coconut populations derived from the seeds of open pollinated palms are the major source of greater genetic variability of accessions found outside East Java (Dewi Hayati et al. 2000). Cardena et al. (2003) identified RAPD markers that are putatively linked to lethal yellowing disease of the coconut palm utilizing three diverse coconut populations, viz. West African Tall (susceptible), Malayan Yellow Dwarf (resistant) and Atlantic Tall palms (resistant).

The genetic diversity and relatedness among three dwarf coconut accessions, namely Malayan Yellow Dwarf (MYD), Kulashekaram Yellow Dwarf (KYD) and Andaman Yellow Dwarf (AYD) were decoded using the RAPD markers. A close association of the two Indian yellow dwarfs (AYD and KYD) with the exotic yellow dwarf (MYD) was observed, indicating the likelihood of AYD and KYD evolving from a common progenitor, the MYD. Manimekalai et al. (2006a) characterized a set of coconut accessions and reported the congruence between geographic origin and clustering by RAPD markers. Rajesh et al. (2015) used RAPD-derived SCoT markers to characterize the coconut accessions.

A major demerit of RAPD markers is they are dominant hence have limited utility in the field of population genetics of cross-pollinated crops like coconut. Hence, the gene frequency estimates are less reliable compared to that of any co-dominant markers. RAPD technique is very sensitive to slight changes in the PCR conditions, which seriously affects the experimental outcomes.

5.2.4 Inverse Sequence-Tagged Repeat (ISTR)

The ISTR has been a powerful tool for genotype identification and differentiation, but do not hold much significance for the analysis of genetic relatedness (Anzizar et al. 1998). The ISTR analysis, as proposed by Rohde et al. (1995), was used for the detection of DNA polymorphism in coconut. ISTR primers were designed and utilized to amplify the repetitive elements such as *copia*-like retroposons in the coconut genome (Rohde et al. 1992) offering DNA polymorphism survey among a set of select coconut genotypes originating from various regions of the world. Further, this technique was extended to analyse East African Tall coconut populations (Duran et al. 1997). Thus, ISTR represents a valuable tool for identification of genotype, genetic variability analysis in germplasm and its use in coconut breeding. It corroborates the view that retro-elements in the genome of coconut could have played a considerable role in providing genetic diversity of coconut.

5.2.5 Amplified Fragment Length Polymorphism (AFLP)

Amplified fragment length polymorphism (AFLP) technology (Zabeau and Vos 1993) represents a resourceful combination of RFLP and PCR techniques. This high-multiplex method of DNA profiling (Vos et al. 1995) offers a unique potential to generate a huge polymorphic genetic loci. AFLP technology could be utilized to study all organisms where prior genome sequence information is not available as it generates very informative fingerprints. Thus, AFLP has become a methodology of choice for DNA polymorphism studies. Features unique for AFLP technology are: (i) PCRs can be performed with random primers, as no prior sequence information is needed; (ii) robustness and high reproducibility; (iii) a combination of AFLP primers could be used to generate unique set of DNA fingerprints.

Perera et al. (1998) performed AFLP assay (comprising eight primer combinations) to study 42 genotypes of coconut in Sri Lanka. Overall, a high genetic variation was observed in talls than in intermediate or dwarf accessions. Also, the intermediate accessions were more genetically related to the dwarfs than to the talls and the duplicate accessions of intermediate forms were culled out. Moreover, the cross-pollinated tall forms showed relatively high genetic variations between various forms and among themselves.

5.2.6 Simple Sequence Repeats (SSR)

SSRs or short tandem repeats (STRs), also called microsatellites, consist of tandemly reiterated (2–8 bp size), polymorphic short DNA sequences. They are co-dominant markers and therefore can be used to study gene duplication, deletion, genetic diversity, and genetic and comparative mapping analysis. Genotypes differ in number of repeats present and in turn in the size of the repeated region. Hence, flanking primers are designed targeting the repeat regions and site-specific amplification of repeats is performed to generate sequence-tagged microsatellite markers.

In plants, RFLP-based fingerprinting using the short probes identified the microsatellites and it was suggested that microsatellites in plants are less abundant compared to the animals. Later, large data sets have explored the abundance of microsatellites in plants than previously anticipated. Many studies have reported that tri-, tetra- and penta-nucleotide motifs are found less compared to mono- and di-nucleotide repeats. Generally, tri-nucleotide repeat types are found predominantly occurring in exonic regions, whereas mono-, di-, tetra-, penta- and hexa-nucleotide repeats are rare in genic regions. Microsatellite-based markers were used to investigate the coconut germplasm genetic diversity. Eight SSR primers, derived from coconut genomic library, were reported by Rivera et al. (1999), and these were utilized in the genetic diversity analysis of 40 genotypes from coconut gene bank of the Philippine Coconut Authority. It revealed a total of 64 alleles and on an average of eight alleles per SSR.

The genetic diversity among 130 coconut individuals (comprising 75 talls and 55 dwarfs) was investigated utilizing eight sets of SSR primers by Perera et al. (2000). The template DNA was adequately representing various coconut ecotypes (94) of the world and thereby allowed the genetic discrimination of 116 of 130 individuals. A phylogenetic tree was constructed based on the genetic distance yielding five main clusters each having either talls or dwarfs. Similarly, 33 Sri Lankan tall populations were subjected to genetic analysis using SSR primers with an objective of devising suitable germplasm collection strategies and to aid the selection of appropriate parents for crop improvement programs (Perera et al. 2001). A considerable level of genetic diversity was detected in all the populations.

A microsatellite kit has been developed at French Agricultural Research Centre for International Development (CIRAD), France, to identify the coconut cultivars (Baudouin and Lebrun 2002). The kit was developed by gathering more than 600 DNA samples belonging to 113 reference populations adequately representing the coconut genetic diversity worldwide. The study has produced a set of 80 polymorphic microsatellite markers, followed by screening them on a reduced set of coconut genotypes, testing the remaining 14 markers on the 600 coconut genotypes. Suitable protocols and statistical methods were also devised to identify the reference population.

Meerow et al. (2003) used 15 SSR/microsatellite DNA loci to study the genetic architecture of eight coconut accessions held at two locations in South Florida. Further analysis of putative parents of the cultivar 'Fiji Dwarf' was performed using these markers. It has revealed that the Yellow and Green Malayan Dwarfs are genetically different from the Red Malayan Dwarfs. The variant 'Red Spicata' was found to be closely related to 'Fiji Dwarf'.

Eight SSR primers were used in the delineation of genetic diversity of 45 select Indian and exotic coconut accessions (Devakumar et al. 2006). Application of these eight SSR primers revealed that the primer sets could distinguish 48 alleles and in accordance with their breeding behavior genetic diversity values of tall accessions were high compared to the dwarfs. An exception to this rule was an indigenous dwarf, Kulasekharam Orange Dwarf, which exhibited more genetic diversity than talls (Devakumar et al. 2006).

Application of microsatellite markers for gene mapping investigations showed that lethal yellowing (LY) tolerant genotypes (Vanuatu Tall and Sri Lanka Green Dwarf) are genetically distinct from the susceptible West African Tall (Konan et al. 2007) as this study reported 58 different microsatellite alleles suggesting relatively high genetic polymorphism.

Manimekalai and Nagarajan (2007) used SSR markers to study the genetic diversity among the worldwide collections of germplasm accessions. Genetic diversity of 10 Brazilian Tall coconut accessions was evaluated by Ribeiro et al. (2010) using 13 SSR loci. The findings suggested the existence of a geographically confined genetic variability as the geographical vicinity of the populations defined effective sharing of greater proportion of genetic features.

Genetic diversity analysis of 10 land races of coconut (102 palms) using 14 microsatellite markers developed at CIRAD revealed that 90 alleles could be detected in the study population with an average PIC of 0.61 (Rajesh et al. 2008a). High heterozygosity was recorded in tall landraces in comparison with the dwarfs (Rajesh et al. 2008a). Using the same set of 14 primer pairs, genetic diversity of 26 coconut accessions from the Andaman and Nicobar Islands, India, was undertaken (Rajesh et al. 2008b). A total of 103 alleles were documented at an average of 7.35 alleles per locus. Further, a maximum number of rare alleles were observed in tall genotypes of Nicobar Islands.

Perera et al. (2015) evaluated Sri Lankan Yellow Dwarf (SLYD) for varietal confirmation and to study the tolerance to *Aceria* mite, using specific microsatellite marker loci. Twenty-five sample palms drawn randomly from four morphological groups based on stature and tolerance to *Aceria* mite were genotyped at seven SSR marker loci along with Sri Lanka Tall (SLT) and

Sri Lanka Green Dwarf (SLGD) as the reference varieties. Three SSR loci were found highly informative for the SLYD population and the results revealed the presence of a population structure within the SLYD coconut population in the coconut seed garden in Sri Lanka. A clear clustering was observed for tall-like and the dwarf-like groups with more variation within the tall-like group. Definite allelic identifications were established at several SSR loci for the clustering based on stature, and the markers were identified for distinguishing different groups. The study provided evidence for the need for reclassification and purification of SLYD coconut population. Further, marker associations were present for *Aceria* mite tolerance trait, suggesting the potential for marker-assisted selection.

Distribution of molecular markers within the coconut genome and the genomic or functional (ESTs) nature of the markers are important considerations for their utility in crop improvement programs. Investigations on genome mapping in coconut are still very limited and preliminary. The first genetic map of coconut (Rohde et al. 1999) was developed based on ISTR markers. It was followed by utilization of multiple molecular markers such as AFLP, ISTR, RAPD and ISSR in the mapping populations to construct an updated linkage map. These efforts led to the fixing of 382 molecular markers in the map spread over 16 linkage groups and providing information regarding six QTLs linked to early germination trait (Herran et al. 2000). Also, QTLs underlying the plant growth parameters such as leaf, stem girth and height were identified (Ritter et al. 2000). In addition, another mapping population in Côte d'Ivoire was effectively utilized to map 280 molecular markers on to the linkage groups of coconut to decipher the QTLs linked to fruit component quantitative traits such as number of nuts, bunch number and other related traits (Lebrun et al. 2001; Baudouin et al. 2006). It was also deduced that a mapping population of around 200–400 individuals of coconut is appropriate to obtain a considerable mapping resolution (Bandaranayake and Kearsey 2005).

Considering the long-term physiological adaptation of the coconut palms (Peries 1993), a collection and conservation of coconut representing different eco-geographical regions in Sri Lanka were initiated in 1991 (Perera et al. 1996). A molecular marker analysis of 33 such tall coconut accessions in Sri Lanka using SSR markers (Perera et al. 2001) revealed that there was no genetic differentiation among these accessions as the population variation accounted only less than 1%. On the other hand, high heterozygosity was reported in a population of Bali Tall (BAT) and lowest in the population of Bali Yellow Dwarf (BYD) (Mahayu and Taryono 2019) Nevertheless, these populations exhibited low heterogeneous nature. Among the populations, Sweet Green Dwarf (SGD) was found to exhibit close similarity with the population of BAT, whereas the populations of BYD and Nias Yellow Dwarf (NYD) were genetically distantly placed.

5.2.7 Inter-Simple Sequence Repeats (ISSR)

The ISSR are defined as DNA fragments located between two microsatellite regions which are in adjacently placed but in an opposite orientation. Thus, it is a cheap genotyping tool which scans the genomic variations present in the regions situated between the microsatellite repeats. Thus, ISSR fragments are PCR amplified using primers from the SSR regions. The primers are designed with or without anchored sequences (16–20 bp) from the non-repeat regions. Unlike other traditional markers, neither prior sequence information nor prior genetic studies are required for ISSR analysis; hence, it is desired in various crop improvement approaches (Prevost and Wilkinson 1999). This technique is extremely useful for the characterization of genetic diversity among the populations, DNA fingerprinting, tagging of gene(s) of agronomic importance, clonal variation detection, identification of cultivar and such other applications involving phylogeny construction, hybrid authenticity, and in the detection of genomic instability. ISSR markers produce profuse, highly repeatable, polymorphic and informative multi-locus patterns which are quite rapid to work with. Thus, ISSR markers make use of the abundant, polymorphic and ubiquitous nature of SSRs present throughout the genome. Hence, ISSR analysis is a right combination of positivities derived from AFLP, SSRs and the universal applicability of RAPD. As a result of these features and relative ease of development of ISSR markers are highly valued.

Villarreal et al. (2006) estimated the intra and inter-ecotype diversity and the genetic relationships among four Mexican Tall and five imported Tall ecotypes from a wide geographical range using 33 polymorphic ISSR loci. The genetic diversity between Mexican ecotypes is similar to that of imported ecotypes while the Malayan Yellow Dwarf (MYD) showed very low levels of diversity. Manimekalai et al. (2006b) used 19 ISSR primers and analysed the genetic diversity of 33 coconut accessions derived from the International Coconut Gene Bank (ICG) in India. A total of 154 polymorphic ISSR markers were identified among the 199 unambiguous fragments. Least similarity was observed among the Indian accessions, Nicobar Tall and Chowghat Orange Dwarf. Nevertheless, the genetic distinctness of coconut accessions obtained from the Southeast Asia, South Asia and South Pacific corroborated their origin and dispersal history.

Rasam et al. (2016) characterized five coconut varieties namely Banawali, Gangabondam Green Dwarf, East Coast Tall, Konkan Bhatye Coconut Hybrid-I and Pratap using a combination of 18 ISSR and 14 SSR markers. This investigation divulged that accessions exhibited genetic diversity both between and within the coconut varieties eventhough the range of genetic diversity within a variety was least expected. Also, the superiority of SSR markers over ISSR markers was documented. The authors inferred that it is imperative to preserve the seed purity via artificial pollination. Loiola et al. (2016) analysed the genetic relationships two Brazilian tall accessions and eight exotic accessions. The Brailzian accessions clustered with those from Africa, while those from South East Asia grouped with accessions from South Pacific, revealing their common ancestry. Geethanjali et al. (2018) analysed 79 genotypes using 48 SSR loci for deciphering the genetic diversity and population structure of the accessions. It was documented that the genotypes studied showed a relatively high quantum of genetic diversity and existence of structured population based on their geographical origins.

5.3 DNA Sequence-Based Molecular Markers

5.3.1 Single-Nucleotide Polymorphisms (SNPs)

SNPs are widely popular as a molecular marker system since the late 1990s. SNPs are the polymorphic events resulting from transition/transversion events in which the variations in the sequence features are the result of a single base substitution at a particular genomic location, which has alternative forms in the population. Within a population, SNPs can be assigned a minor allele frequency and the least frequent allele should have an abundance of at least 1%. SNPs have relative advantages over other marker systems due to the low mutation rates, and constant distribution across the genome and the availability of easy detection tools. SNPs are found to be highly abundant and frequent, but their frequency varies significantly between the different genomic regions, genomes in any species and finally among the species. Nevertheless, the development of a novel SNP marker necessitates a comprehensive genome sequencing efforts and its validation.

SNPs that are found in exons or coding regions are called exonic SNPs, or cDNA SNPs, respectively. However, a considerable portion of the SNPs are also found to be located in non-coding regions of the genome and such SNPs are referred to as non-coding SNPs (ncSNPs) of which a subset resides in the intronic regions. When SNPs are found in the exons, it may cause a significant change in the amino acid composition of the encoded protein [non-synonymous SNP (nsSNP)] or does not cause any major effects (synonymous SNP (synSNP). Promoter SNPs (pSNPs) are the ones that are found in promoter regions, thereby suggesting the high influence of SNPs in the activity of associated genes (Yates and Sternberg 2013).

SNP genotyping refers to the measurement of variations of SNPs between a species. SNP discovery in a simple plant genome is a relatively straight forward process; however, SNP analysis in a complex genome is cumbersome due to latter's highly repetitive sequences. The simplest approach to discover SNP in the targeted genome is by direct DNA sequencing using PCR. This approach is very expensive for genome-scale study since it requires large locus-specific primers. Herrera et al. (2006) evaluated 15 accessions of coconut, representing six cultivars, using SNPs. The analysis was performed through single-strand conformation polymorphism (SSCP) to detect the SNP-containing alleles. Manju et al (2011) computationally predicted SNPs/InDels from coconut WRKY gene sequences. SNPs were seen at a frequency of 2.84/100 bp and two novel RKY-SNP markers were reported in the study. Further, SNP-based molecular markers have a great potential to serve as anchor markers so that the currently available genotype-specific maps could be converted into a general reference maps for the crops such as coconut and oil palm (Bai et al. 2018).

5.3.1.1 SNP Genotyping

Plant breeders generally use SNPs for quantitative trait loci (QTL) analysis and in association studies. In crop genetics, the researchers aim to discover the SNPs in the cultivars to find the variants with desirable characteristics. Due to their abundance, the genetic maps that are generated based on SNP markers are very dense and saturated. These maps greatly assist in the conduct of MAS and also in the studies involving quantitative genetics. The interest in the discovery of these SNPs has led to development of efficient genotyping methods which is revolutionizing the crop improvement programs of twenty-first century (Thompson 2014).

Recent developments in the next-generation sequencing (NGS) and single-nucleotide polymorphism (SNP) sequencing have accelerated the crop improvement targets. Automated high-throughput SNP genotyping has more advantages than other marker systems. The whole genome sequencing utilizing NGS technologies have led to the discovery of SNP pools to select informative markers in different sets of germplasm having wide ramifications for marker-assisted selection techniques that have significant impact on modern plant breeding.

5.3.1.2 SNP Genotyping Methods

Presently, SNP markers are available for most of the crops. SNP-based genotyping methods have also grown. For instance, the utilization of fluorescent tagged allele-specific TaqMan probes from Applied Biosystems are available for SNP genotyping (Schleinitz et al. 2011). Besides, the cleaved amplified polymorphic sequence (CAPS) marker approach was also employed for SNP genotyping (Thiel et al. 2004). Another method utilizes endpoint fluorescence detection in order to differentiate the alleles in Kompetitive allele-specific PCR (KASP) (Semagn et al. 2013; He et al. 2014). Among the various genotyping methodologies available, the TaqMan and KASP-based SNP assays are frequently employed owing to their relatively low cost, and medium to high-throughput and improved sensitivity features. Also, these techniques greatly rely on the high-resolution melting (HRM) curve analysis feature present in the real-time PCR machines. The rhAmp method is a variant that employs RNase H2 to trigger the successfully bound primers and it generates sequence-specific reaction (Dobosy et al. 2011). RAD-sequence tags were used for studying the Brazilian dwarf coconut populations and the report concluded that SNP markers have immense power of detection of genetic variation (Santos et al. 2020). Using SNPs markers, Muñoz-Peréz et al. (2019) conducted genetic diversity and population structure analysis of cultivated varieties of northern South America and identified five loci as candidates for domestication targets, including genes related to vital functions such reproduction, respiration, and defence. Recently, SNP markers have been reported as useful marker in analysing coconut leaf beetle *Brontispa longissima* (Chen et al. 2020).

5.4 Genomics Approaches

Genomic studies entail analysis of the structure and function of the entire genome of a living organisms, and applications of genome-scale technologies for the crop improvement programs. It is aimed at understanding the role of genes, expression patterns and characteristics of gene products and thus is involved in mapping, sequencing and functional analysis of genomes. Comparative genomic analysis offers great opportunity to investigate the evolutionary features of genome structure and provides molecular evolutionary genomic information with relative ease. In addition, genomics studies offer approaches to identify and characterize novel genes and other functional or genomic elements and provide crucial resources for the annotation of completed genomes and enlists tool box for the isolation and characterization of genes in non-model species.

5.4.1 Advantages of Genome Sequencing

Genomics research has already made a considerable footprint in life sciences which further spurred rapid progresses in the crop science. The various advantages of genome sequencing efforts are the relative ease with which molecular relationships between the genes could be unravelled and with great confidence. It also offers evolutionary insights into the genome organization, utilizing various functional genomics approaches, thereby contributing to developmental biology and generating information like SNPs/InDels for use in crop breeding.

The genome sequence information will enable in-depth appreciation of genes encoding for yield and yield attributing components so as to develop agronomically improved plant phenotypes. Further identification and characterization of novel biomarkers offers great potential for the agriculture and related industries. For instance development of designer oils, crops possessing durable resistance to diseases and appreciable tolerance to abiotic stresses among others. Also, these markers have a greater role in identification of plant genotypes that are good at assimilating the plant nutrients.

Xiao et al. (2013) scanned 57,304 coconut unigenes for di-, tri-, tetra-and hexa-nucleotide repeat sequences and an additional 30 microsatellites markers have been developed. These markers have greatly helped in assessing the genetic relatedness of coconut accessions obtained from China and SE Asia. It was concluded that coconut populations of China have its origins from South East Asia based on population structure analysis and historical records. It was also inferred that human dispersal and sea currents could have played a major role in bringing coconuts to Yunnan and Hainan provinces, respectively.

The recent advances in the genomic research have shifted the exploitation of gene-targeted, functional markers in lieu of random DNA markers. One such novel marker system is the start codon targeted polymorphism (SCoT), which was utilized for assessment of genetic relationship and diversity among 23 coconut accessions (10 talls and 13 dwarfs), representing different geographical regions (Rajesh et al. 2015).

Alsaihati et al. (2014) reported the size of coconut genome to be around 2.6 GB and containing 50–70% repeat sequences. The assembly of the raw sequence data generated an estimated 0.97–2.47 GB (37–94.5%) of coconut genome sequence coverage. Indonesian Palm Genome Project sequenced 'Tenga Tall' and 'Pati Kopyor Green Dwarf' a representative tall and dwarf palm genomes. The coconut genome sequencing project resulted in 148,109,277,288 basepairs (bp) for the dwarf and 99,306,500,882 bp for the tall coconut palm, respectively. The total GC and AT content of Dwarf coconut were 38% and 62% while for tall it was reported as 37% and 63%

(Lestari et al. 2016). The generation of a reference genome should be of vital importance for the molecular breeding of coconut in the future. Xiao et al. (2017) performed de novo whole genome sequence of coconut. A total of 419.67 gigabases (GB) of raw reads were generated by the Illumina HiSeq 2000 platform representing 90.91% of the genome (2.42 GB, variety Hainan Tall). It was followed by whole genome sequencing of indigenous dwarf cultivars of the Philippines and India, by Lantican et al. (2019) and Rajesh et al. (2020), respectively. These genome sequence information could be used to accelerate the breeding process. A complete chloroplast genome of coconut (Huang et al. 2013) has also been made available.

5.5 Transcriptomics to Unravel Genes and Their Functions

Transcriptome and Expressed Sequence Tags (EST) sequences generated from the different tissues, development stages or cultivar with resistance/susceptibility to a disease of coconut, may be vital to identify the genes involved in various metabolic pathways. Transcriptome and RNA-sequencing (RNA-seq) data generation, coupled with identification of differentially expressed (DE) genes, would immensely help in such breakthrough studies. Various RNA-seq studies of coconut have illustrated the possibilities of such applications (Fan et al. 2013; Huang et al. 2014; Liang et al. 2014; Rajesh et al. 2016). For instance, Fan et al. (2013) identified 57,304 unique genes of coconut, Liang et al. (2014) identified and annotated differentially expressed genes in pulp development, Nejat et al. (2015) identified genes in response to yellow decline phytoplasma and Rajesh et al. (2018) identified genes in response to root (wilt) disease. Punzalan et al. (2019) developed transcriptome of Laguna tall variety which is useful for the annotation of coconut genes and Reynolds et al. (2019) identified genes involved in lipid biosynthesis and Ramesh et al. (2020) characterized the transcriptome response of coconut seedlings to water deficit stress.

5.6 Conclusions

All these molecular marker, transcriptomic and genomic studies have generated an enormous quantum of molecular, DNA, and RNA sequence data and most of these data are available in publicly repositories such as TROPgeneDB (http://tropgenedb.cirad.fr/tropgene/JSP/index.jsp) and NCBI GenBank Database (http://www.ncbi.nlm.nih.gov/nucleotide). These research endeavours should be further strengthened in the future to harness the needs of the coconut breeders. State-of-the-art research facilities and capacity building should be provided to almost all coconut growing countries to carry out their own molecular breeding research in the future. Multi-nation research collaborations are needed to commercialize these new technologies for application in the field.

Acknowledgements This chapter has updated information of previous book chapters and original research articles of the authors along with the information from other publications. The authors would like to thank all those who have directly or indirectly contributed to this book chapter.

References

Alsaihati B, Liu W, Lin Q, Al-Mssallem IS (2014) Coconut genome *de novo* sequencing. In: XXII conference of plant and animal genome. San Diego, CA. Available at http://www.intpag.org. Accessed May 2015

Anzizar I, Herrera M, Rohde W, Santos A, Dowe JL, Goikoetxea P, Ritter E (1998) Studies on the suitability of RAPD and ISTR for identification of palm species (Arecaceae). Taxon 47:635–645

Bai B, Wang L, Zhang YJ, Lee M, Rahmadsyah R, Alfiko Y, Ye BQ, Purwantomo S, Suwanto A, Chua NH, Yue GH (2018) Developing genome-wide SNPs and constructing an ultra high-density linkage map in oil palm. Sci Rep 8(1):691

Bandaranayake CK, Kearsey MJ (2005) Genome mapping, QTL analysis and MAS: importance, principle, constraints and application in coconut. Int Plant Genet Resour Newslett 142:47–54

Batugal P, Bourdeix R, Baudouin L (2009) Coconut breeding. In: Jain SM, Priyadarshan PM (eds) Breeding plantation tree crops: tropical species. Springer, New York, NY

Baudouin L, Lebrun P (2002) The development of a microsatellite kit for use with coconuts. IPGRI, Rome, p 52

Baudouin L, Lebrun P, Konan JL (2006) QTL analysis of fruit components in the progeny of a Rennell Island Tall coconut (*Cocos nucifera* L.) individual. Theor Appl Genet 112(2):258–268

Cardena R, Ashburner GR, Oropeza C (2003) Identification of RAPDs associated with resistance to lethal yellowing of the coconut (*Cocos nucifera* L.) palm. Sci Hortic 98:257–263

Chen Z, Wang G, Li M, Peng Z, Ali H, Xu L, Gurr GM, Hou Y (2020) Development of single nucleotide polymorphism (SNP) markers for analysis of population structure and invasion pathway in the coconut leaf beetle *Brontispa longissima* (gestro) using restriction site-associated DNA (RAD) genotyping in Southern China. Insects 11(4):230

Devakumar K, Jayadev K, Rajesh MK, Chandrasekar A, Manimekalai R, Kumaran PM, Parthasarathy VA (2006) Assessment of genetic diversity of Indian coconut accessions and their relationship to other cultivars using microsatellite markers. Plant Genet Resour Newslett 145:38–45

Dewi Hayati PK, Hartana A, Suharsono (2000) International Conference on Science and Technology For Managing Plant Genetic Diversity in the 21" Century, June 12–16, Kuala Lumpur, Malaysia

Dobosy JR, Rose SD, Beltz KR, Rupp SM, Powers KM, Behlke MA, Walder JA (2011) RNase H-dependent PCR (rhPCR): improved specificity and single nucleotide polymorphism detection using blocked cleavable primers. BMC Biotechnol 11:80

Duran Y, Rohde W, Kullaya A, Rohde W, Kullaya A (1997) Molecular analysis of East African Tall coconut genotypes by DNA marker technology. J Genet Breed 51:279–288

Fan H, Xiao Y, Xia YY, W, Mason AS, Xia Z, Qiao F, Zhao S, Tang H, (2013) RNA-Seq analysis of *Cocos nucifera*: transcriptome sequencing and *de novo* assembly for subsequent functional genomics approaches. PLoS ONE 8(3):e59997

Gangolly SR, Satyabalan K, Pandalai KM (1957) Varieties of coconut. Ind Coconut J X: 3–28

Geethanjali S, Rukmani AJ, Rajakumar D, Kadirvel P, Viswanathan PL (2018) Genetic diversity, population structure and association analysis in coconut (*Cocos nucifera* L.) germplasm using SSR markers. Plant Genet Resour 16(2):156–168

He C, Holme J, Anthony J (2014) SNP genotyping: the KASP assay. Methods Mol Biol 1145:75–86

Herran A, Estioko L, Becker D, Rodriguez MJ, Rohde W, Ritter E (2000) Linkage mapping and QTL analysis in coconut (*Cocos nucifera* L.). Theor Appl Genetic 101:292–300

Herrera M, Meerow AW, Borrone JW, Kuhn DN, Schnell RJ (2006) Ten informative markers developed from WRKY sequences in coconut (*Cocos nucifera* L.). Mol Ecol Notes 6(3):904–906

Huang YY, Matzke AJM, Matzke M (2013) Complete sequence and comparative analysis of the chloroplast genome of coconut palm *Cocos nucifera*. PLoS ONE 8(8):e74736

Huang YY, Lee CP, Fu JL, Chang BC, Matzke AJ, Matzke M (2014) *De novo* transcriptome sequence assembly from coconut leaves and seeds with a focus on factors involved in RNA-directed DNA methylation. G3 4:2147–2157

Konan JNK, Koffi KE, Konan JL, Lebrun P, Dery SK, Sangare A (2007) Microsatellite gene diversity in coconut (*Cocos nucifera* L.) accessions resistant to lethal yellowing disease. Afr J Biotechnol 6:341–347

Lantican DV, Strickler SR, Canama AO, Gardoce RR, Mueller LA, Galvez HF (2019) *De novo* genome sequence assembly of dwarf coconut (*Cocos nucifera* L. 'Catigan Green Dwarf') provides insights into genomic variation between coconut types and related palm species. G3: Genes Genomes Genet 9(8):2377–2393

Lebrun P, Baudouin L, Bourdeix R, Konan JL, Barker JHA, Aldam C, Herran A, Ritter E (2001) Construction of a linkage map of the Rennell Island Tall coconut type (*Cocos nucifera L.*) and QTL analysis for yield characters. Genome 44:962–970

Lebrun P, N'cho YP, Seguin M, Grivet L, Baudouin L (1998) Genetic diversity in coconut (*Cocos nucifera* L.) revealed by restriction fragment length polymorphism (RFLP) markers. Euphytica 101:103–108

Lestari et al (2016) Pemanfaatan Teknologi Genomika dan Bioteknologi untuk Mendukung Percepatan Pemuliaan Palma Penting di Indonesia. KKP3S Project Annual Report. IAARD. Ministry of Agriculture, Republic of Indonesia

Liang Y, Yuan Y, Liu T, Mao W, Zheng Y, Li D (2014) Identification and computational annotation of genes differentially expressed in pulp development of *Cocos nucifera* L. by suppression subtractive hybridization. BMC Plant Biol 14:1–17

Liyanage DV (1958) Varieties and forms of coconut palms grown in Ceylon. Ceylon Coconut Quarterly 9:1–10

Loiola CM, Azevedo AON, Diniz LEC, Aragao WM, Azevedo CD, Santos PH, Ramos HC, Pereira MG, Ramos SR (2016) Genetic relationships among tall coconut palm (*Cocos nucifera* L.) accessions of the International Coconut Genebank for Latin America and the Caribbean (ICG-LAC), evaluated using microsatellite markers (SSRs). PLoS ONE 11(3): e0151309

Mahayu WM, Taryono (2019) Coconut (*Cocos nucifera* L.) diversity in Indonesia based on SSR molecular marker. In: AIP conference proceedings 2099, 020013. https://doi.org/10.1063/1.5098418

Manimekalai R, Nagarajan P (2006) Inter-relationship among coconut (*Cocos nucifera* L.) germplasm accessions using RAPD technique. Genet Resour Crop Evol 53:1137–1144

Manimekalai R, Nagarajan P (2006) Assessing genetic relationships among coconut (*Cocos nucifera* L.) accessions using inter simple sequence repeat markers. Sci Hortic 108:49–54

Manimekalai R, Nagarajan P (2007) Use of simple sequence repeat markers for estimation of genetic diversity in coconut (*Cocos nucifera* L.) germplasm accessions. J Plant Biochem Biotechnol 16:29–33

Manju KP, Manimekalai R, Arunachalam V (2011) Microsatellites in palm (Arecaceae) sequences. Bioinformation 7(7):347–351

Meerow AW, Wisser RJ, Brown JS, Kuhn DN, Schnell RJ, Broschat TK (2003) Analysis of genetic diversity and population structure within Florida coconut (*Cocos nucifera* L.) germplasm using microsatelite DNA, with special emphasis on the Fiji Dwarf cultivar. Theor Appl Genetic 106:715–726

Menon KPV, Pandalai KM (1958) The coconut. A Monograph. Indian Central Coconut Committee, pp 86–102

Muñoz-Pérez M, Cañas GP, López L, Arias T (2019) Genome-wide diversity of northern South America cultivated Coconut (*Cocus nucifera*) uncovers diversification times and targets of domestication of coconut globally. bioRxiv. https://doi.org/10.1101/825398

Narayana GV, John CM (1949) Varieties and forms of coconut. Madras Agric J 36:349–366

Nejat N, Cahill DM, Vadamalai G, Ziemann M, Rookes J, Naderali N (2015) Transcriptomics-based analysis using RNA-Seq of the coconut (*Cocos nucifera*) leaf in response to yellow decline phytoplasma infection. Mol Genet Genomics 290:1899–1910

Ohler JG (1984) Coconut, Tree of Life, Plant production and protection Paper 57. Food and Agricultural Organisation of the United Nations, Rome

Perera L (2001) Origin, domestication and genetic diversity of coconut: DNA Information CORD XVII (1): 35–51

Perera L, Kamaral CJ, Fernando WBS (2015) Molecular assessment of *Cocos nucifera* L. Var. Sri Lanka yellow dwarf for genetic purity and aceria mite tolerance. Intl J Mol Evol Biodivers 5(1):1–5

Perera L, Peries RRA, Fernando WMU (1996) Collection and conservation of coconut biodiversity in Sri Lanka. Int Plant Genet Resour Newslett 106:1–4

Perera L, Russel JR, Provan J, McNicol JW, Powell W (1998) Evaluating genetic relationships between indigenous coconut (*Cocos nucifera* L.) accessions from Sri Lanka by means of AFLP profiling. Thoer Appl Genet 96:545–550

Perera L, Russell JR, Provan J, Powell W (2000) Use of microsatellite DNA markers to investigate the level of genetic diversity and population genetic structure of coconut (*Cocos nucifera* L.). Genome 43:15–21

Perera L, Russell JR, Provan J, Powell W (2001) Levels and distribution of genetic diversity of coconut (*Cocos nucifera* L., var. Typica form typica) from Sri Lanka assessed by microsatellite markers. Euphytica 122:381–389

Peries RRA (1993) Improving the efficiency of the low input coconut agroecology through conservation and utilisation of biodiversity. In: Proceedings of the 10th IFOAM Conference, Lincoln University, New Zealand, p 122

Prevost A, Wilkinson MJ (1999) A new system of comparing PCR primers applied to ISSR fingerprinting of potato cultivars. Theor Appl Genet 98(1):107–112

Punzalan R, Bautista A, Ernesto Emmanuel E, Rivera R, Rivera S, Saloma C (2019) Transcriptome of the traditional coconut variety Laguna Tall. Philippine J Sci 148(S1):153–164

Rajesh MK, Nagarajan P, Jerard BA, Arunachalam V, Dhanapal R (2008) Microsatellite variability of coconut accessions (*Cocos nucifera* L.) from Andaman and Nicobar Islands. Curr Sci 94:1627–1631

Rajesh MK, Arunachalam V, Nagarajan P, Lebrun P, Samsudeen K, Thamban C (2008) Genetic survey of ten Indian coconut landraces by simple sequence repeats (SSRs). Sci Hortic 118:282–297

Rajesh MK, Fayas TP, Naganeeswaran S, Rachana KE, Bhavyashree U, Sajini KK, Karun A (2016) *De novo* assembly and characterization of global transcriptome of coconut palm (*Cocos nucifera* L.) embryogenic calli using Illumina paired-end sequencing. Protoplasma 253(3):913–928

Rajesh MK, Rachana KE, Kulkarni K, Sahu BB, Thomas RJ, Karun A (2018) Comparative transcriptome profiling of healthy and diseased Chowghat Green Dwarf coconut palms from root (wilt) disease hot spots. Eur J Plant Pathol 151(1):173–193

Rajesh MK, Sabana AA, Rachana KE, Rahman S, Jerard BA, Karun A (2015) Genetic relationship and diversity among coconut (*Cocos nucifera* L.) accessions revealed through SCoT analysis. 3 Biotech 5 (6):999–1006

Rajesh MK, Chowdappa P, Behera SK, Kasaragod S, Gangaraj KP, Kotimoole CN, Nekrakalaya B, Mohanty V, Sampgod RB, Banerjee G, Das AJ, Niral V, Karun A, Mahato AK, Gaikwad K, Singh NK, Keshava Prasad TS (2020) Assembly and annotation of the nuclear and organellar genomes of a dwarf coconut (Chowghat Green Dwarf) possessing enhanced disease resistance. OMICS: J Integr Biol 24 (12):726–742

Ramesh SV, Hebbar KB, Rajesh MK, Abhin SP, Gangaraj KP, Athul Bobby (2020) Transcriptome analysis of *Cocos nucifera* L., seedlings having contrasting water use efficiency (WUE) under water-deficit stress: molecular insights and genetic markers for drought tolerance. In: Proceedings of The 1st International Electronic Conference on Plant Science. https://doi.org/10.3390/IECPS2020-08853

Rasam DV, Gokhale NB, Sawardekar SV, Patil DM (2016) Molecular characterisation of coconut (*Cocos nucifera* L.) varieties using ISSR and SSR markers. J Hortic Sci Biotechnol. https://doi.org/10.1080/14620316.2016.1160544

Reynolds KB, Cullerne DP, Tahchy A, Rolland V, Blanchard CL, Wood CC, Singh SP, Petrie JR (2019) Identification of genes involved in lipid biosynthesis through *de novo* transcriptome assembly from *Cocos nucifera* developing endosperm. Plant Cell Physiol 60(5):945–960

Ribeiro FE, Baudouin L, Lebrun P, Chaves LJ, Brondani C, Zucchi MI, Vencovsky R (2010) Population structures of Brazilian tall coconut (*Cocos nucifera* L.) by microsatellite markers. Genetic Mol Biol 33 (4):696–702

Ritter E, Rodriguez MJB, Herran A, Estioko L, Becker D, Rohde W (2000) Analysis of quantitative trait loci (QTL) based on linkage maps in coconut (*Cocos nucifera* L.). In: Arencibia A (ed) Plant genetic engineering towards the third millennium. Elsevier Science B.V. Amsterdam, The Netherlands, pp 42–48

Rivera R, Edwards KJ, Arnold GM, Arnold GM, Ayad G, Hodgkin T, Karp A (1999) Isolation and characterisation of polymorphic microsatellites in *Cocos nucifera* L. Genome 42(4):668–675

Rohde W, Salamani F, Ashburner R, Randles JW (1992) An *Eco*RI repetitive sequence family of the coconut palm (*Cocos nucifera* L.) shows sequence homology to copia-like elements. J Genet Breed 46:391–394

Rohde W, Becker D, Kullaya A, Rodriguez J, Herran A, Ritter E (1999) Analysis of coconut germplasm biodiversity by DNA marker technologies and construction of a first genetic linkage map. In: Oropeza C, Verdeil JL, Ashburner GR et al (eds) Current advances in coconut biotechnology. Kluwer Academic Publishers, Dordrecht, The Netherlands, pp 99–120

Rohde W, Kullaya A, Rodriguez J, Ritter E (1995) Genome analysis of *Cocos nucifera* L. by PCR amplification of spacer sequences separating a subset of copia-like 16RI repetitive elements. J Genetic Breed 49:179–186

Santos PHAD, Venaˆncio TM, dos Santos PHD, Ramos HC, Arêdes FA, Azevedo AO, Boechat MS, de Souza Filho GA, Ramos SR, Mirisola LA, Aragão WM (2020) Genotyping-by-sequencing technology reveals directions for coconut (*Cocos nucifera* L.) breeding strategies for water production. Euphytica 216:45. https://doi.org/10.1007/s10681-020-02582-1

Schleinitz D, Distefano JK, Kovacs P (2011) Targeted SNP genotyping using the TaqMan(R) assay. Methods Mol Biol 700:77–87

Semagn K, Babu R, Hearne S, Olsen M (2013) Single nucleotide polymorphism genotyping using Kompetitive Allele specific PCR (KASP): overview of the technology and its application in crop improvement. Mol Breed 33:1–14

Tanksley SD, Young ND, Paterson AH, Bonierbale MW (1989) RFLP mapping in plant breeding: new tools for an old science. Biotechnol 7:257–264

Thiel T, Kota R, Grosse I, Stein N, Graner A (2004) SNP2CAPS: a SNP and INDEL analysis tool for CAPS marker development. Nucleic Acids Res 32(1):e5

Thomson MJ (2014) High-throughput SNP genotyping to accelerate crop improvement. Plant Breed Biotechnol 2(3):195–212

Upadhyay A, Jayadev K, Manimekalai R, Parthasarathy VA (2004) Genetic relationship and diversity in Indian coconut accessions based on RAPD markers. Sci Hortic 99:353–362

Villarreal DZ, Rodriguez MR, Hugh H, Colunga-GarcíaMarín P (2006) Population genetics, lethal yellowing disease, and relationships among Mexican and imported Coconut ecotypes. Crop Sci 46(6):2509–2516. https://doi.org/10.2135/cropsci2005.12-0462

Vos M, Hogers R, Bleeker M, Reijans M, Lee TV, Hornes M, Friters A, Pot J, Paleman J, Kuiper M, Zabeau M (1995) AFLP: a new technique for DNA fingerprinting. Nucleic Acids Res 23:4407–4414

Wickramaratne MRT (1987) Breeding coconut for adaptation to drought. COCOS 6:16–23

Xiao Y, Xu P, Fan H, Baudouin L, Xia W, Bocs S, Xu J, Li Q, Guo A, Zhou L, Li J (2017) The genome draft of coconut (*Cocos nucifera*). Gigascience 6(11):1–11. https://doi.org/10.1093/gigascience/gix095.PMID:29048487;PMCID:PMC5714197

XiaoY LY, Yang Y, Fan H, Xia W, Mason AS, Zhao S, Sager R, Qiao F (2013) Development of microsatellite markers in *Cocos nucifera* and their application in evaluating the level of genetic diversity of *Cocos nucifera*. Plant Omics J 6:193–200

Yates CM, Sternberg MJ (2013) Proteins and domains vary in their tolerance of non-synonymous single nucleotide polymorphisms (nsSNPs). J Mol Biol 425:1274–1286

Zabeau M, Vos P (1993) Selective restriction fragment amplification: a general method for DNA fingerprinting. European Patent Office. Publication 0 534 858 A1 bulletin 93/13

Quantitative Trait Loci (QTL) and Association Mapping for Major Agronomic Traits

6

M. K. Rajesh, S. V. Ramesh, Lalith Perera, and A. Manickavelu

Abstract

The enormous potential of DNA-based molecular markers has greatly assisted crop improvement programs of various crops, including non-model species like plantation crops. Despite the perennial nature, analysis of genomic loci governing quantitative traits of economic importance in coconut has gained impetus due to rapid strides in utilization of DNA markers and development of linkage maps. Although various molecular markers have been used to develop linkage maps and to delineate trait-marker association studies, genome-wide association analysis or linkage disequilibrium analysis-based mapping is less explored in coconut. The advent of next-generation sequencing technologies and the availability of whole-genome sequences of coconut genotypes would spur the genome-wide trait-marker association analysis and fine-mapping studies. This chapter discusses the perspective of QTL mapping investigations in coconut and proposes a way forward in the context of developments in genomics-assisted breeding.

M. K. Rajesh (✉) · S. V. Ramesh
ICAR-Central Plantation Crops Research Institute, Kasaragod, Kerala, India
e-mail: rajesh.mk@icar.gov.in

S. V. Ramesh
e-mail: ramesh.sv@icar.gov.in

L. Perera
Coconut Research Institute, Lunuwila, Sri Lanka
e-mail: kanthaperera@yahoo.com

A. Manickavelu
Central University of Kerala, Kasaragod, Kerala, India
e-mail: amanicks@cukerala.ac.in

6.1 Introduction

Cocos nucifera L., (called 'Tree of life'), is a crop of tropics and sub-tropics importance found in more than 90 countries in the coastal and island ecosystems. It belongs to the monotypic genus *Cocos*, of tribe Cocoeae, subfamily Arecoideae. It is a diploid crop ($2n = 2x = 32$) with an estimated genome size of ~2.6 Gb. Coconut mainly consists of two morphoforms, 'talls' and 'dwarfs,' showing varying polymorphism levels within and between them. The tall types are characterized by outcrossing behavior and display more polymorphism than dwarfs, which predominantly show inbreeding behavior (Menon and Pandalai 1958). *De novo* genome sequencing estimated the coconut genome size as 2.6 Gb (Alsaihati et al. 2014). Genome sizes of selected coconut accessions estimated using flow cytometry revealed that the mean coconut genome size is 2C = 5.57 pg (average tall genome size 5.59 pg and average dwarf genome size 5.55 pg) which corresponds to 2.72 Gb/haploid set (Freitas Neto et al. 2016). Interestingly, the 2C genome sizes of related *Elaeis* sp. are small, ranging from 3.64 to 4.10 pg.

M. K. Rajesh et al. (eds.), *The Coconut Genome*, Compendium of Plant Genomes,
https://doi.org/10.1007/978-3-030-76649-8_6

The little differences in the genome sizes of talls and dwarf forms of coconut disproved the hypothesis that dwarfs are tall variants that originated due to mutations (Freitas Neto et al. 2016).

The major traits of economic importance are mainly governed by genetic factors and the influence of environmental factors. Hence, genetic dissection of traits will pave way for their greater understanding and manipulation. Traits are generally governed by genes either quantitatively or qualitatively. In general, characters such as the number of fruits, the weight of fruits, size of fruits, traits conferring resistance/tolerance to stressors, etc., are quantitative. In other words, these traits are governed by multiple genes and are called 'quantitative' or 'polygenic' or 'multifactorial' or 'complex' traits. The genomic regions that harbor the genes and genetic elements governing these quantitative traits are called quantitative trait loci (QTL). The developments in the field of molecular markers, in general, and rapid strides in the advancements of DNA-based molecular markers, in particular, have dramatically revolutionized classical plant breeding by way of identification of QTLs.

DNA markers that exhibit close linkage with the inheritance of agronomically or economically important gene(s) could be effectively utilized as a tool for marker-assisted selection (MAS) in crop improvement programs. Significant applications of MAS in monocots, especially cereals [rice (McCouch and Doerge 1995), wheat (Koebner and Summers 2003), maize (Tuberosa et al. 2003), barley (Thomas 2003) etc.] have led to the development of potentially improved genotypes. In coconut, various DNA molecular markers such as inverse sequence-tagged repeat (ISTR) (Rohde et al. 1995; Duran et al. 1997), randomly amplified polymorphic DNA (RAPD) (Ashburner et al. 1997; Duran et al. 1997), restriction fragment length polymorphism (RFLP) (Lebrun et al. 1998), amplified fragment length polymorphism (AFLP) (Perera et al. 1998), simple sequence repeats (SSR) (Perera et al. 1999; Rivera et al. 1999; Teulat et al. 2000; Meerow et al. 2003; Rajesh et al. 2008; Ribeiro et al. 2010) have been effectively utilized mainly to assess the genetic diversity of germplasm.

Bi-parental mapping is a straight forward strategy to identify a particular QTL by developing the mapping population with two parents possessing contrasting traits. Though it is considered a gold standard approach to identify a particular chromosome locus, developing bi-parental population in a crop like coconut is a cumbersome approach because of its perennial nature. The use of traditional linkage mapping principles in perennials is beset with several problems such as developing a requisite mapping population, such as recombination inbred lines (RILs), with adequate pedigree information. Development of RILs, in perennials such as coconut, is extremely difficult. The concept of linkage disequilibrium mapping thus came to rescue of perennial or tree species. Linkage disequilibrium or genome-wide association studies (GWAS) employ the genetic principle of linkage disequilibrium in a set of crop genotypes to identify QTLs; thus, it exploits the phenomenon of historical recombination within populations of unrelated individuals to trace the sequence variations contributing to the phenotypic variation. Also, to perform trait-marker association analysis or association mapping studies, it is imperative to have a germplasm collection with excellent trait variability and devoid of linkage disequilibrium among the unlinked loci (Geethanjali et al. 2018). Furthermore, genome-wide identification and characterization of QTLs in economically important perennial trees were impeded due to the non-availability of high-density linkage maps. Nonetheless, the advent of next-generation sequencing (NGS) technologies and use of tools such as genotyping-by-sequencing (GBS) (Elshire et al. 2011) and restriction-site-associated DNA sequencing (RAD-seq) (Peterson et al. 2012) have greatly assisted in the rapid discovery of molecular markers which in turn has helped in QTL mapping of traits of agronomic and economic importance (Torkamaneh et al. 2016). In the related palm species, oil palm, QTL mapping of yield components (Seng et al. 2016) and genome-wide association studies have been performed (Ithnin et al. 2017).

In coconut, though considerable improvements have been made utilizing molecular markers, conventional breeding programs have encountered severe challenges due to long breeding cycle, difficulties in selecting traits as attributed to yield, and traits involved in the manifestation of disease/pest resistance, etc. To overcome some of the shortcomings of traditional breeding approaches, it is imperative to develop marker-assisted selection (MAS) procedures utilizing DNA markers tightly associated with QTL to help in screening phenotypes. Analysis of coconut QTLs would help in breeding for valuable traits that show low to moderate heritability and are difficult to improve through traditional breeding techniques.

6.2 Linkage Maps and QTL Identification-Coconut Perspective

Genetic linkage map of an organism forms an essential component for efficient MAS. DNA-based molecular markers have greatly helped construct linkage maps-a map of genes on chromosomes-based not on their physical distances; rather, their relative positions are determined by recombination frequency. Molecular markers and statistical tools that analyze such markers and their inheritance pattern have made it possible to precisely locate the chromosome loci of an organism to identify a QTL. Genetic linkage maps are saturated with adequate polymorphic markers, and it allows for QTL mapping, estimating the QTL number, their phenotypic effects and mode of gene action. It provides chromosome-wise order of molecular markers and their genetic distance from each other. This process of developing linkage maps and identifying and characterizing genomic regions and their relationship with the expression of quantitative traits is known as QTL mapping. Hence, mapping and identifying QTLs would improve our knowledge of genetic mechanism governing the traits and design effective selection procedures to hasten the breeding programs. Identifying agronomically important trait-linked markers and its co-segregation with QTLs could help genomics-assisted breeding programs in coconut.

The first molecular linkage maps, involving the coconut genotypes, Malayan Yellow Dwarf (MYD) and Laguna Tall (LAGT), were developed by Rohde et al. (1999) and Herrán et al. (2000). The ISTR-based linkage map was constructed using 52 F_1 plants from the cross involving MYD20 × LAGT07. This mapping population was used for genetical dissection of coconut, which led to the identification of 51 polymorphic markers, of which 43 were placed into 12 linkage groups (Rohde et al. 1999). It was followed by a linkage map developed by Herrán et al. (2000) who used the F_1 mapping populations from the controlled crosses involving LAGT and MYD, genotype local to the Philippines and widely used in breeding programs worldwide, respectively. In this study, a combination of four molecular markers (including AFLP, ISSR, ISTR, and RAPD) was used to develop the linkage map, consisting of 16 linkage groups. The linkage groups comprised a minimum of six to a maximum of 21 markers varying in their lengths from 99.3 and 173.2 cM having a total map length of 2226 cM. On the other hand, the map of MYD linkage groups was constructed with 4–32 markers of total map length of 1266 cM, and the linkage groups had varying sizes of 35.4–148.3 cM (Herrán et al. 2000). Analysis of common molecular markers of linkage groups of both the parents had led to the identification of nine homologous chromosomes. Also, six QTLs linked to precocious germination (between 20.5 and 26 days of reduced germination time) were identified. Notwithstanding the small mapping population (52 F_1 progenies), six QTLs linked with early germination and yield, which are important traits for coconut breeding, were identified (Herrán et al. 2000).

This pioneering work was followed by the investigation by Lebrun et al. (2001), who constructed a coconut linkage map comprising 16 linkage groups with a map length of 1971 cM utilizing 227 AFLP and SSR markers. Half-sib progenies obtained from the crosses involving a set of 12 Cameroon Red Dwarf (CRD) genotypes and a common male parent Rennell Island

Tall (RIT) were used as the mapping population. A linkage map comprising 16 linkage groups, having 227 molecular markers, was created. The total map length for RIT corresponds to 1971 cM with 5–23 markers per linkage group, and the length of the individual linkage groups varied from 52.1 and 191.7 cM. Three and two QTLs were detected in two different evaluation periods, respectively, linked to yield attribute, number of bunches, explaining between 7.1 and 21.1% of the total phenotypic variance (Lebrun et al. 2001). Similarly, three more QTLs governing number of nuts were detected explaining total phenotypic variance between 11.8 and 21.3% accounting for about 4.7 and 14 nuts per genotype. Thus, this investigation identified nine QTLs for yield and yield attributing characters such as bunch number and nut number of coconut along with other traits such as early germination (Herrán et al. 2000), girth height, and leaf production (Ritter et al. 2000).

Further, improving the linkage map published by Lebrun et al. (2001), Baudouin et al. (2006) utilized the same mapping population to map QTLs linked to fruit component characters in a first of its kind attempt in coconut. To construct this improved linkage map, one AFLP primer combination, eight SSR markers derived from the relative species oil palm, and 48 novel coconut SSRs were used. A linkage map of RIT parent of length 1849.8 cM, comprising of 16 linkage groups sharing lengths from 51.9 to 181.8 cM, was established. The linkage group harbors around 274 molecular markers, and the updated map had 47 additional markers, including 44 SSRs, which are co-dominant and highly polymorphic. Around 48 QTLs have been identified to be involved in the expression of fruit component traits, and 34 among them were found to be located in six clusters, demonstrating the pleiotropic effect. These QTLs, located in the cluster, very well explained the magnitude of their effects compared to other non-cluster or 'scattered' QTLs. The percentage of phenotypic variance exemplified by these QTLs within the cluster varied from 9 to 28%. Thus, these six QTL clusters act upon the whole fruit development, whereas 14 other QTLs appear to be involved in expressing other fruit characteristic features. Similarly, four other QTLs known to govern the fruit endosperm humidity explained 11–22% of genetic variance. Also, authors suggested that QTLs identified for coconut meat, endosperm humidity, and fruit production could be linked to copra yield since those QTLs showed very high regression coefficients of $R^2 = 99\%$ to the copra yield and associated traits (Baudouin et al. 2006). An interesting finding of this study is that various QTLs for fruit component traits, such as weight and endosperm humidity, were located in different genomic regions of coconut. Hence, marker-assisted selection (MAS) for yield traits in coconut requires careful selection of multiple QTLs, each governing individual yield and yield attributing traits.

Considering the significance of leaf cuticular wax in conferring resistance against abiotic stresses in coconuts, Riedel et al. (2009) have attempted to study the qualitative and quantitative variations in wax components and to identify QTL linked to this trait utilizing the mapping population developed from the cross East African Tall (EAT) and Rennell Island Tall (RIT). A series of 241 AFLP primers, 64 SSR primer combinations, and 22 heterologous SSR primers (from oil palm) were studied in the mapping population. An integrated linkage map of EAT/RIT was developed (of 2739 cM map length) containing 1026 molecular markers. Among the mapped markers, a greater proportion belongs to SSRs (53 coconut and eight oil palm SSR loci) and AFLP markers. This study yielded a total of 46 QTLs associated with wax content and components traits (a QTL for iso-skimmiwallinol acetate and eight QTLs for skimmiwallinol acetate—major components of epicuticular wax of coconut).

In many cases, these QTLs are found to be co-located, suggesting their pleiotropic effect. This study also cautioned for further validation of the QTLs identified utilizing more progeny genotypes to avoid 'false positive' QTLs. The summary of the mapping, QTL identification, and marker-trait association studies in coconut is presented in Table 6.1.

Table 6.1 Summary of the mapping, QTL identification and marker-trait association studies in coconut

Sl. No.	Mapping population	Molecular markers	Total map length	No. of QTLs	References
1.	Malayan Yellow Dwarf (MYD) × Laguna Tall (LAGT)	Inverse sequence-tagged repeat (ISTR)	–	–	Rohde et al. (1999)
2.	MYD × LAG	Combination of AFLP, ISSR, ISTR and RAPD	LAGT: 2226 cM; MYD 1266 cM	Six QTLs linked to precocious germination and yield	Herrán et al. (2000)
3.	Cameroon Red Dwarf (CRD) × Rennell Island Tall (RIT)	AFLP and SSRs	1971 cM	Nine QTLs for yield and yield attributing characters such as bunch number and nut number	Lebrun et al. (2001)
4.	CRD × RIT	AFLP and SSRs derived from oil palm and novel coconut SSRs	1849.8 cM	48 QTLs for fruit component traits	Baudouin et al. (2006)
5.	East African Tall (EAT) × RIT	241 AFLP, 64 SSRs and 22 heterologous SSR primers (from oil palm)	2739 cM	46 QTLs associated with wax content and components traits	Riedel et al. (2009)
6.	West African Tall (WAT)- MYD- Atlantic Tall (AT)	RAPD	–	12 RAPD markers putatively associated with lethal yellowing disease resistance	Cardeña et al. (2003)
7.	Multiple accessions	SSRs and RAPD	–	$OPE18_{600}$, $CnCirE2_{151}$ and $OPP16_{1200}$ markers linked to mite resistance	Shalini et al. (2007)
8.	Bulked DNA of Talls and Dwarfs	RAPD converted into a SCAR marker	–	OPA09 linked to tall morphoform	Rajesh et al. (2013)
9.	Bulked DNA of Talls and Dwarfs	RAPD	–	Validate parents and hybrids	Rajesh et al. (2014)
10.	Aromatic and non-aromatic coconut accessions	Gene specific primers	–	CnBadh2-m1, and CnBadh2-m2, selectively amplify aromatic and non-aromatic coconut	Vongvanrungruang et al. (2016)
11.	Aromatic and non-aromatic coconut accessions	Gene specific primers using *de novo* transcriptome approach	–	Primers to detect biallelic SNP variation (G/C)	Saensuk et al. (2016)

(continued)

Table 6.1 (continued)

Sl. No.	Mapping population	Molecular markers	Total map length	No. of QTLs	References
12.	Seventy-nine genotypes across the world	SSRs	–	Linkage disequilibrium-based mapping identified SSR locus CnCir73 linked to fruit component traits	Geethanjali et al. (2018)
13	Back-cross MYD × (MYD × WAT)	SNPs	2365 cM	Marker density per chromosome varied from 244 (Cn15) to 904 (Cn01). Average marker density of 8.6 SNPs per Mbp	Yang et al. (2021)

6.3 Marker-Trait Association

Association studies dissecting resistance to lethal yellowing disease were performed utilizing RAPD molecular markers and three coconut populations (susceptible-West African Tall-WAT; resistant-Malayan Yellow Dwarf-MYD; and resistant-Atlantic Tall-AT) (Cardeña et al. 2003). This study assumed that RAPD linked to resistance could have a high association with MYD and AT. Less frequency of association was expected with WAT based on their phenotypic response. It yielded 82 RAPD markers that could differentiate bulked DNA samples of WAT and MYD and five of those markers having high-frequency association (>0.8) with AT plants (Cardeña et al. 2003). Bulked segregant analysis of population for contrasting traits identified 12 RAPD markers putatively associated with lethal yellowing disease resistance. These markers showed a high frequency of occurrence in MYD and AT concomitant with low frequency in WAT. Thus, this seminal study identified potent markers for utilization in marker-assisted selection (MAS) for disease resistance in coconut.

Identification of markers linked to coconut mite (*Aceria guerreronis* Keifer) resistance was carried out using separately bulked DNA of both the resistant and susceptible coconut accessions of South India (Shalini et al. 2007). Utilizing SSR and RAPD markers, a combination of five markers was identified, delineating the association of resistance trait in coconut (Shalini et al. 2007). A comprehensive survey of coconut orchards in south India identified coconut trees (40 in number) demonstrating differential susceptibility or resistance to coconut mite infestation. An SSR marker $CnCirA9_{99}$ mapped in chromosome 14 was associated with mite resistance/susceptibility trait. However, stepwise multiple regression analysis (SMRA), combining RAPD and SSR information, revealed that multiple QTLs are involved in conferring resistance to mite infestation in coconut. Accordingly, the markers $OPE18_{600}$, $CnCirE2_{151}$, $OPP16_{1200}$ were suggested to select mite resistance trait in coconut (Shalini et al. 2007).

Palm habit is an important character from the perspective of agronomy, plant protection, and harvest operations. Hence, it is extremely critical to identify genetic markers associated with plant habit that could differentiate the 'dwarfs' from 'talls' at an early stage. Differentiation of a tall from dwarf cultivars (or vice versa) of coconut relied on morphological features resulting in false characterization due to out-crossed seedlings. Rajesh et al. (2013) have utilized bulked DNA of talls and dwarf genotypes of coconut and identified an RAPD marker OPA09 yielding a unique amplicon of 260 bp exclusively in talls. This amplicon was further sequenced and characterized to develop a sequence characterized amplified region (SCAR) marker which significantly assists in selecting genuine talls from dwarf coconut genotypes independent of growth stage, genotype, and environmental factors (Rajesh et al. 2013). Similarly, utilizing the bulked DNA of tall and dwarf morph forms of coconut, Rajesh et al. (2014) identified another RAPD marker OPBA3 that was useful in screening parents and validating the hybrids emerging out of tall and dwarf cross.

Coconut genes encoding enzymes GA oxidases (*CnGA20ox1* and *CnGA20ox2*), involved in the later steps of gibberellic acid (GA) biosynthesis pathway, were characterized for sequence features and expression profile to devise suitable candidate molecular markers for differentiating the tall and dwarf forms (Boonkaew et al. 2018). The nucleotide sequence features of *CnGA20ox1* characterized from both the tall and dwarf forms are identical though three-fold differences in the gene expression were observed in talls. Base substitution in *CnGA20ox2* gene of talls and dwarfs was identified; however, gene expression was undetectable in both the kinds of accessions suggesting the *cis*-acting elements of *CnGA20ox1* could be a determinant of plant habit (Boonkaew et al. 2018).

The fragrant ('pandan-like') or aromatic fresh young coconut is an economically important and popular trait similar to aromatic rice, pandan leaf (*Pandanus amaryllifolius* Roxb.), and *Vallaris glabra* (L.). Aromatic and non-aromatic coconut accessions were investigated for the differences in the structural and nucleotide features of the gene (*Badh2*) encoding betaine aldehyde dehydrogenase homolog 2 (BADH2). The aromatic and non-aromatic coconuts showed little differences in the expression profile of *Badh2* gene. Nevertheless, nucleotide sequence characterization revealed a single-base substitution in wild-type coconut gene, converting it into a mutant allele, causing the production of characteristic aroma (Vongvanrungruang et al. 2016). Type-specific DNA markers were developed to distinguish aromatic coconuts from non-aromatic coconuts (Vongvanrungruang et al. 2016). Concurrently, Saensuk et al. (2016) utilized *de novo* transcriptome sequencing approach and characterized rice aromatic gene ortholog *CnAMADH2* (amino aldehyde dehydrogenase 2) in coconut. A functional molecular that could distinguish the SNP variation (G/C) between the contrasting coconut accessions (non-aromatic and aromatic) was developed (Saensuk et al. 2016).

6.4 Linkage Disequilibrium-Based Mapping

Association analysis in coconut is less explored. Geethanjali et al. (2018) analyzed the population architecture by studying the representative accessions (79 genotypes) worldwide using SSR markers. Genetic diversity of the accessions revealed that the genotypes harbor two to seven alleles. Hierarchical clustering of the genotypes classified the accessions based on their geographic origin. Two major clades were observed distinguishing tall genotypes originating from Indo-Atlantic and South Asia region, from the dwarfs and some talls from Indo-Pacific and South-East Asia. Linkage disequilibrium (LD) analysis identified a very little fraction of SSR locus-pair. However, an SSR locus CnCir73 has been linked to fruit component traits in concurrence with earlier findings Baudouin et al. (2006). Recently, Zhou et al. (2020) utilized SSRs in an association mapping analysis to identity elite alleles linked to fatty acid content of coconut endosperm. The study categorized the coconut accessions into high fatty acid content and low fatty acid content sub-groups and identified a total of 10 SSR loci linked to fatty acid content. The alleles identified exhibit strong phenotypic effects as the alleles viz., CnFatB3-359 and CnFatB2-830 had positive and negative effects, respectively (Zhou et al. 2020).

6.5 Whole-Genome Sequence (WGS) and Impetus to QTL Mapping

Genomics-assisted breeding and QTL mapping will receive impetus following the availability of the whole-genome sequence of the particular plant species. In coconut, *de novo* sequence assembly of the complete draft genome (Hainan Tall variety) (HAT) was made available covering the genome length (2.42 Gb) for an estimated depth of 173×. This draft genome sequencing effort yielded a scaffold length of 2.2 Gb with N50 of 418 Kb (Xiao et al. 2017). Annotation of coconut genome revealed that over 72% of genome comprises transposable elements (TEs), and long terminal repeats (LTRs) class predominate the TEs. The number of genes predicted in coconut genome (28,039) was less compared to related palms *Phoenix dactylifera* (DPV01) and *Elaeis guineensis* (African oil palm), having 41,660 and 34,802 genes, respectively. Evolutionary analysis of 119 anti-porter genes found in coconut genome grouped the genes into 12 functional categories of *Arabidopsis* anti-porter genes. However, the expansion of characteristic gene families such as Na^+/H^+antiporters, genes coding for carnitine/acylcarnitine translocases, and potassium-dependent sodium-calcium exchangers, could be attributed to the features of coconut such as high salt tolerance, fatty acid accumulation, and nut water with high potassium ions, respectively. Whole-genome sequence of a dwarf variety 'Catigan Green Dwarf' (CATD) was obtained using a combination of multiple sequencing platforms such as Illumina Miseq, PacBio SMRT, and Dovetail Chicago technologies (Lantican et al. 2019). A total of 7139 genome-wide SSR markers, 13 gene-linked SSRs, and 58,503 SNP variants could be identified by comparing the genome sequences of Tall and Dwarf genotypes (Lantican et al. 2019). Also, 39,002 SSR loci were identified in the 'CATD' genome, along with 13 gene-linked SSR markers. Among the genes tagged with SSRs are those associated with drought response, biotic stress resistance, and genes involved in the oil biosynthesis. Further, 39, 157, and 25 SSR markers with well-defined physical distance to the drought-responsive, insect or pest resistance, and oil biosynthesis traits, respectively were identified. Recently, nuclear and organellar genomes of another dwarf cultivar Chowghat Green Dwarf (CGD), possessing enhanced resistance to root (wilt) disease, were made publicly available (Rajesh et al. 2020). Further analysis of the CGD genome divulged a total of 112 NBS–LRR gene encoding loci (Rajesh et al. 2020) though less compared to other staple monocot crops such as rice or maize. Availability of whole-genome sequence information of CGD coconut would greatly assist in developing disease-resistant genotypes in future. Further, the whole-genome sequences of coconut in the public domain form an invaluable resource to develop new molecular markers, relatively high density or marker-saturated linkage maps, identification of novel QTLs linked to traits of economic importance, and the possibility of genome-wide association mapping studies to further the efforts of molecular breeding. It lays the foundation for the exploration of potent technologies such as the development of genomic selection models in coconut to hasten genetic gains.

Availability of HAT genome led to the exploration of genotyping-by-sequencing (GBS) approach to develop a dense linkage map (8402 SNPs) (Yang et al. 2021). A backcross mapping population [MYD × (MYD × WAT)] was utilized in this endeavor, and the genome sequence-derived scaffolds were arranged on the 16 pseudochromosomes or pseudomolecules leading to the ordered arrangement of 77% of coconut genes onto 16 linkage groups (Yang et al. 2021). It was deduced that chromosome Cn15 exhibited segregation distortion as it presented a strong heterozygote deficit (27.5% average percentage of heterozygosity against the expected 50%) which was attributed to the strong selection among pollen grains favoring maternal allele. Evolutionary analysis point toward the large scale expansion of transposable elements in the coconut genome induced due to the alterations of sea level during the glaciations at Pleistocene. Also, RNA-seq analysis further sheds light on the osmotic adjustment signaling pathways of coconut that eliminates reactive oxygen species, but brings Na^+/K^+ homeostasis and stomatal closure in response to salt stress (Yang et al. 2021). The chromosome-scale assembling of coconut genome exploiting the close synteny between the coconut and oil palm is step forward in establishing a valuable resource for furthering the genomics and breeding efforts.

6.6 Conclusion

Marker-assisted selection has been a valuable tool for hastening the genetic improvement of perennials in general and economically important palm species in particular. Hence, it is imperative to have molecular markers, linkage maps, and mapping of QTLs on the whole genome for furthering the genomics-assisted breeding in coconut. It is also important to construct genetic linkage maps with co-dominant markers such as SSRs/SNPs. Previous linkage maps of coconut were constructed with a combination of dominant and co-dominant molecular markers. Also, fine mapping of QTLs to assist breeding in coconut is crucial by saturating the linkage map with more DNA markers so that fine QTL mapping in the seedling stage could be performed for early selection. Identifying marker-trait association with LD mapping is having great scope in coconut, which will help find the loci naturally fixed. Literature mining in coconut revealed that only one study was carried out taking 79 genotypes or accessions, which is relatively low to comprehend the LD in coconut.

Hence, it is necessary to have more molecular markers in coconut that are tightly linked to the gene(s) of agronomic and economic importance and utilize map-based cloning of targeted traits. Utilization of restriction-site-associated DNA sequencing (RAD-seq) approach and genotyping-by-sequencing to construct high density, high-quality genetic linkage maps based on SNP markers and investigate the genetic basis of traits such as oil biosynthesis, fruit yield, etc., could help in avoiding destructive phenotyping. Also, the publications of whole-genome sequence of tall (Xiao et al. 2017) and dwarf coconuts (Lantican et al. 2019; Rajesh et al. 2020) are crucial resources for further developing useful genomic information to assist the crop improvement programs. Identification of QTL linked to important traits such as plant height, drought tolerance, and disease/pest resistance, followed by sequence characterization of the respective genomic scaffold region, may go a long way in improving coconut breeding programs and understanding gene-function relationship.

References

Alsaihati B, Liu W, Lin Q, Al-Mssallem IS (2014) Coconut genome *de novo* sequencing. XXII conference of plant and animal genome. San Diego, CA. Available at [http://www.intpag.org]

Ashburner GR, Thompson WK, Halloran GM (1997) RAPD analysis of south Pacific coconut palm populations. Crop Sci 37:992–997

Baudouin L, Lebrun P, Konan JL, Ritter E, Berger A, Billotte N (2006) QTL analysis of fruit components in the progeny of a Rennell Island Tall coconut (*Cocos nucifera* L.) individual. Theor Appl Genet 112:258–268

Boonkaew T, Mongkolsiriwatana C, Vongvanrungruang A, Srikulnath K, Peyachoknagul S (2018) Characterization of *GA20ox* genes in tall and dwarf types coconut (*Cocos nucifera* L.). Genes Genom 40:735–745. https://doi.org/10.1007/s13258-018-0682-4

Cardeña R, Ashburner GR, Oropeza C (2003) Identification of RAPDs associated with resistance to lethal yellowing of the coconut (*Cocos nucifera* L.) palm. Sci Hortic 98:257–263

Duran Y, Rohde W, Kullaya A, Goikoetxea P, Ritter E (1997) Molecular analysis of East African Tall coconut genotypes by DNA marker technology. J Genet Breed 51:279–288

Elshire RJ, Glaubitz JC, Sun Q, Poland JA, Kawamoto K, Buckler ES, Mitchell SE (2011) A robust, simple genotyping-by-sequencing (GBS) approach for high diversity species. PLoS ONE 6: e19379

Freitas Neto M, Pereira TN, Geronimo IGC, Azevedo AON, Ramos SRR, Pereira MG (2016) Coconut genome size determined by flow cytometry: tall versus dwarf types. Genet Mol Res 15 (1):15017470. https://doi.org/10.4238/gmr.15017470

Geethanjali S, Rukmani JA, Rajakumar D, Kadirvel P, Viswanathan PL (2018) Genetic diversity, population structure and association analysis in coconut (*Cocos nucifera* L.) germplasm using SSR markers. Plant Genet Resour 16:156–168. https://doi.org/10.1017/S1479262117000119

Herrán A, Estioko L, BeckerD Rodriguez MJB, Rohde-Wand Ritter E (2000) Linkage mapping and QTL analysis in coconut (*Cocos nucifera* L.). Theor Appl Genet 101:292–300

Ithnin M, Xu Y, Marjuni M, Serdari NM, Amiruddin MD, Low ETL, Tan YC, Yap SJ, Ooi LCL, Nookiah R, Singh R (2017) Multiple locus genome-wide

association studies for important economic traits of oil palm. Tree Genet Genomes 13(5):103
Koebner RMD, Summers RW (2003) 21st century wheat breeding: plot selection or plate detection? Trends Biotechnol 21:59–63
Lantican DV, Strickler SR, Canama AO, Gardoce RR, Mueller LA, Galvez HF (2019) *De novo* genome sequence assembly of dwarf coconut (*Cocos nucifera* L.'Catigan Green Dwarf') provides insights into genomic variation between coconut types and related palm species. G3: Genes Genomes Genet g3-400215
Lebrun P, N'cho YP, Seguin M, Grivet L and Baudouin L (1998) Genetic diversity in coconut (*Cocos nucifera* L.) revealed by restriction fragment length polymorphism (RFLP) markers. Euphytica 101:103–108
Lebrun P, Baudouin L, Bourdeix R, Konan JL, Barker JH, Aldam C, Herrán A, Ritter E (2001) Construction of a linkage map of the Rennell Island Tall coconut type (*Cocos nucifera* L.) and QTL analysis for yield characters. Genome 44:962–970
McCouch SR, Doerge RW (1995) QTL mapping in rice. Trends Genet 11:482–487
Meerow AW, Wisser RJ, Brown JS, Kuhn DN, Schnell RJ, Broschat TK (2003) Analysis of genetic diversity and population structure within Florida coconut (*Cocos nucifera* L.) germplasm using microsatellite DNA, with special emphasis on the Fiji Dwarf cultivar. Theor Appl Genet 106:715–726
Menon KPV, Pandalai KM (1958) The coconut palm: a monograph. Indian Central Coconut Committee, Government Press, Madras, India, pp 86–102
Perera L, Russell JR, Provan J, McNicol JW, Powell W (1998) Evaluating genetic relationship between indigenous coconut (*Cocos nucifera* L.) genotypes from Sri Lanka by means of AFLP profiling. Theor Appl Genets 96:545–550
Perera L, Russell CT, Provan J, Powell W (1999) Identification and characterization of microsatellite loci in coconut (*Cocos nucifera* L.) and the analysis of coconut populations in Sri Lanka. Mol Ecol 8(2):344–346
Peterson BK, Weber JN, Kay EH, Fisher HS, Hoekstra HE (2012) Double digest RADseq: an inexpensive method for de novo SNP discovery and genotyping in model and non-model species. PLoS ONE 7:
Rajesh MK, Arunachalam V, Nagarajan P, Lebrun P, Samsudeen K, Thamban C (2008) Genetic survey of 10 Indian coconut landraces by simple sequence repeats (SSRs). Sci Hortic 118:282–287
Rajesh MK, Chowdappa P, Behera SK, Kasaragod S, Gangaraj KP, Kotimoole CN, Nekrakalaya B, Mohanty V, Sampgod RB, Banerjee G, Das AJ, Niral V, Karun A, Mahato AK, Gaikwad K, Singh NK, Keshava Prasad TS (2020) Assembly and annotation of the nuclear and organellar genomes of a dwarf coconut (Chowghat Green Dwarf) possessing enhanced disease resistance. OMICS: J Integrative Biol 24(12):726–742
Rajesh MK, Jerard BA, Preethi P, Thomas RJ, Fayas TP, Rachana KE, Karun A (2013) Development of a RAPD-derived SCAR marker associated with tall-type palm trait in coconut. Sci Hortic 150:312–316
Rajesh MK, Jerard BA, Preethi P, Thomas Regi Jacob, Karun Anitha (2014) Application of RAPD markers in hybrid verification in coconut. Crop Breed Appl Biotechnol 14(1):36–41
Ribeiro FE, Baudouin L, Lebrun P, Chaves LJ, Brondani C, Zucchi MI, Vencovsky R (2010) Population structures of Brazilian Tall coconut (*Cocos nucifera* L.) by microsatellite markers. Genet Mol Biol 33 (4):696–702
Riedel M, Riederer M, Becker D, Herrán A, Kullaya A, Arana-López G, Peña-Rodríguez L, Billotte N, Sniady V, Rohde W, Ritter E (2009) Cuticular wax composition in *Cocos nucifera* L.: physicochemical analysis of wax components and mapping of their QTLs onto the coconut molecular linkage map. Tree Genet Genomes 5:53–69
Ritter E, Rodriguez MJB, Herran A, Estioko L, Becker D, Rohde W (2000) Analysis of quantitative trait locis (QTL) based on linkage maps in coconut (*Cocos nucifera* L.). In: Proceedings of the international symposium on plant genetic engineering. plant genetic engineering: towards the third millennium. Havana, Cuba, Elsevier Science Publishers, pp 42–48
Rivera R, Edwards KJ, Barker JH, Arnold GM, Ayad G, Hodgkin T, Karp A (1999) Isolation and characterization of polymorphic microsatellites in *Cocos nucifera* L. Genome 42:668–675
Rohde W, Kullaya A, Rodriguez J, Ritter E (1995) Genome analysis of *Cocos nucifera* L. by PCR amplification of spacer sequences seperating a subset of copia-like *EcoRI* repetitive elements. J Genet Breed 49:179–186
Rohde W, Becker D, Kullaya A, Rodriguez MJB, Herrán A, Ritter E (1999) Analysis of coconut germplasm biodiversity by DNA marker technologies and construction of a first genetic linkage map. In: Oropeza C, Verdeil JL, Ashburner GR, Cardega R, Santamaria JM (eds) Current advances in coconut biotechnology. Kluwer Academic Publishers, Dordrecht, The Netherlands, pp 99–120
Saensuk C, Wanchana S, Choowongkomon K, Wongpornchai S, Kraithong T, Imsabai W, Chaichoompu E, Ruanjaichon V, Toojinda T, Vanavichit A, Arikit S (2016) *De novo* transcriptome assembly and identification of the gene conferring a "pandan-like" aroma in coconut (*Cocos nucifera* L.). Plant Sci 252:324–334
Seng TY, Ritter E, Saad SHM, Leao LJ, Singh RSH, Zaman FQ, Tan SG, Alwee SSRS, Rao V (2016) QTLs for oil yield components in an elite oil palm (*Elaeis guineensis*) cross. Euphytica 212(3):399–425
Shalini KV, Manjunatha S, Lebrun P, Berger A, Baudouin L, Pirany N, Ranganath RM. Prasad DT (2007) Identification of molecular markers associated with mite resistance in coconut (*Cocos nucifera* L.). Genome 50:35–42
Teulat B, Aldam C, Trehin R, Lebrun P, Barker JHA, Arnold GM, Karp A, Baudouin L, Rognon F (2000) An analysis of genetic diversity in coconut (*Cocos*

nucifera) populations from across the geographic range using sequence-tagged microsatellites (SSRs) and AFLPs. Theor Appl Genet 100:764–771

Thomas W (2003) Prospects for molecular breeding of barley. Ann Appl Biol 142:1–12

Torkamaneh D, Laroche J, Belzile F (2016) Genome-wide SNP calling from genotyping by sequencing (GBS) data: A comparison of seven pipelines and two sequencing technologies. PLoS ONE 11: e0161333

Tuberosa R, Salvi S, Sanguineti MC, Maccaferri M, Giuliani S, Landi P (2003) Searching for quantitative trait loci controlling root traits in maize: a critical appraisal. Plant Soil 255:35–54

Vongvanrungruang A, Mongkolsiriwatana C, Boonkaew T, Sawatdichaikul O, Srikulnath K, Peyachoknagul S (2016) Single base substitution causing the fragrant phenotype and development of a type-specific marker in aromatic coconut (*Cocos nucifera*). Genet Mol Res 15(3):15038748

Xiao Y, Xu P, Fan H, Baudouin L, Xia W, Bocs S, Xu J, Li Q, Guo A, Zhou L, Li J (2017) The genome draft of coconut (*Cocos nucifera*). Giga Sci 6(11):1–11

Yang Y, Bocs S, Fan H, Armero A, Baudouin L, Xu P, Xu J, This D, Hamelin C, Iqbal A, Qadri R (2021) Coconut genome assembly enables evolutionary analysis of palms and highlights signaling pathways involved in salt tolerance. Commun Biol 4(105). https://doi.org/10.1038/s42003-020-01593-x

Zhou L, Yarra R, Cao H (2020) SSR based association mapping analysis for fatty acid content in coconut flesh and exploration of the elite alleles in *Cocos nucifera* L. Curr Plant Biol 21:100141. https://doi.org/10.1016/j.cpb.2020.100141

7 Palms in an 'Omics' Era

V. Arunachalam

Abstract

Palms, which belong to the family Arecaceae, are economically important plants next only to grasses and legumes. The science of omics in palms has witnessed considerable advances in the recent years. This chapter is a compilation of the developments in various fields of omics such as genomics, transcriptomics, proteomics, metabolomics, bioinformatics and phenomics of the palms with special reference to coconut, date palm and oil palm. Whole genome sequence resources are available only in a few of the economically important palm species. The chapter is presented under different sections in the order of abiotic stress, biotic stress, chloroplast and mitochondria, databases and web resources, evolution and phylogenomics, fruits and its parts, gender/sex determination, height/dwarfism, image analysis and other phenomics. A roadmap for future work and information of omics resources as available in the three major economic palms and related web resources are presented.

V. Arunachalam (✉)
ICAR-Central Coastal Agricultural Research Institute, Ela Old Goa, Goa, India
e-mail: v.arunachalam@icar.gov.in

7.1 Why Study Omics in Palms?

The science of omics is pivotal to improving our understanding of biological systems. 'Omics' refers to an umbrella term as it has various subdisciplines that deal with each and every biomolecule such as DNA, RNA, protein and metabolite, which are involved in biological processes. The study of structure and function of an organisms' DNA is dealt as 'genomics', that of RNA in 'transcriptomics', and the end product proteins in 'proteomics'. Metabolites synthesised by the gene action are elaborated in a field of study called 'metabolomics', and at last, the phenotypes manifested due to the concerted action of gene and environment are analysed by the science of 'phenomics'.

Palms are important economic plants next only to grasses and legumes. They play many vital roles in human welfare by supplying useful commodities such as oil (coconut and oil palm), refreshing health drink (coconut), nutrient-rich fruits (dates), furniture (rattans) and as ornamental plants. The omics of palms could uncover the genetic basis of important biological process such as climbing stem of monocots (Zhao et al. 2017), seed mass variation in the double coconut or 'Seychelles nut' (*Lodoicea maldivica*) the largest known seed weighing 17.5 kg, oil quantity and quality, metabolites of medicinal and additive chemicals such as arecoline (*Areca catechu* L.), vivipary in monocotyledons and salt

M. K. Rajesh et al. (eds.), *The Coconut Genome*, Compendium of Plant Genomes,
https://doi.org/10.1007/978-3-030-76649-8_7

tolerance of *Nypa* palm of mangrove habitat (Sahu et al. 2016). Readers are suggested to refer narrative reviews on the subject of genomics of most cultivated palms (Arunachalam 2011) such as coconut, dates and oil palm (Meerow et al. 2012; Neoh et al. 2017) in order to gain the background information. The current status of research and application of omics in palms is summarised in Table 7.1.

7.2 Abiotic Stress Omics

7.2.1 Drought and Salt Stress

Paul et al. (2010a, b) used bioinformatics and transcriptomic approaches to search for water stress-responsive genes of coconut and found major transcription factors, viz., MAP kinase, AP2, WRKY and NAC. Specific candidate genes for cold, drought and salinity stress tolerance were identified by comparative transcriptome analysis in oil palm (Xiao et al. 2014b). Proteomics analysis of the high degree of drought stress indicated up-regulation of three genes encoding for proteins such as glycolate oxidase, amino methyl transferase and the glycine cleavage system H protein 3 in *Euterpe oleracea* (de Oliveira et al. 2019). *In vitro* assays for screening coconut germplasm for drought tolerance using sodium chloride (NaCl) at 170 and 330 mM concentrations revealed the higher survival rate of drought-tolerant cultivars in comparison with drought-susceptible cultivars (Karunaratne et al. 1991). In date palm callus cultures, chlorides of sodium, potassium, calcium and polyethylene glycol (PEG) were tested for screening salinity and drought tolerance, respectively (Al-Khayri and Ibraheem 2014). Transcriptomic analysis of *Nypa fruticans* L. revealed positive selection for genes governing vivipary and salt tolerance for adaptation in the mangrove ecosystems (He et al. 2015).

Proteomic analysis indicated the significant modulations in the expression levels of the proteins namely ATP synthase and RubisCO activase during drought and salt and stress in date palm (El Rabey et al. 2015). Also, the convincing role of transketolase gene in the drought tolerance of date palm has been uncovered by El Rabey et al. (2016). Application of omics tools to decipher the salt tolerance in date palm has been reviewed by Yaish and Kumar (2015) where the authors have studied the commercial variety Khalas possessing salt tolerance trait. Review by Murugesan et al. (2017) enlists the root traits and other mechanisms of abiotic stress tolerance in genetic resources of oil palm. Composition of cell wall of date palm fruits, especially arabinosyl residues, is affected by the water deficit stress during fruit development (Gribaa et al. 2013).

7.2.2 Cuticular Wax and Proline Accumulation

Wax content in leaves of coconut palm play a major role in conferring stress tolerance to drought (Thomas et al. 2006; Riedel et al. 2009) and also to lethal yellowing disease (Arroyo-Serralta et al. 2012). Biochemically, three major compounds of epicuticular wax of coconut known include lupeol methyl ester, isoskimmiwallin and skimmiwallin (Escalante-Erosa et al. 2002). Cuticular wax forms a protective barrier of plants to overcome the harsh environment (Sharma et al. 2018). QTL analysis in coconut for marker-trait association for eight cuticular wax compounds was undertaken in coconut by Riedel et al. (2009).

Proline accumulation is observed during the imposition of abiotic stresses in palms. Oil palm seedlings subjected to water deficit stress accumulated proline when the stomatal resistance was greater than 200 s cm^{-1} (Harun 1997). Proline and other amino acids have been reported to play crucial roles in drought tolerance of coconut (Thomas et al. 2006). Proline accumulation is observed in oil palm seedlings under salt stress (Cha-Um et al. 2010), or in date palm subjected to heat and salinity stresses (Yaish 2015) and in *Nypa* palms grown at very high salt stress (EC = 57.2 dS m^{-1}) (Theerawitaya et al. 2014). The unique proline/betaine transporter gene along with promoter region characterized in oil palm (Yamada et al. 2011) further emphasizes the

Table 7.1 Selected omics resources for palm researchers

Type of omics	Coconut	Oil palm	Date palm
Whole genome sequence assembly	Hainan Tall (Xiao et al. 2017) Catigan Green Dwarf (Lantican et al. 2019) Chowghat Green Dwarf (Rajesh et al. 2020b)	AVROS pisifera (Singh et al. 2013a) Dura (Jin et al. 2016)	Khalas female (Al-Dous et al. 2011) Khalas (Al-Mssallem et al. 2013)
Chloroplast genome	Thailand Dwarf (Huang et al. 2013) Chowghat Green Dwarf (Rajesh et al. 2020b)	African oil palm (Uthaipaisanwong et al. 2012)	Khalas (Yang et al. 2010) Aseel (Khan et al. 2012) 9 Saudi Arabian cultivars (Sabir et al. 2014)
Mitochondrial genome	Tall from Oman (Aljohi et al. 2016) Chowghat Green Dwarf (Rajesh et al. 2020b)	African oil palm (57% coverage) (Uthaipaisanwong et al. 2017)	Khalas, Sukry (Fang et al. 2012) Saudi Arabian cultivars (Sabir et al. 2014)
Transcriptome	Yellow Decline infected (Nejat et al. 2015) Root (wilt) diseased (Rajesh et al. 2015, 2018) DNA metylation (Huang et al. 2014) Aromatic fruit (Saensuk et al. 2016) Gibberellic acid pathway (Rahman et al. 2016) Somatic embryogenesis (Rajesh et al. 2016a) Water-deficit stress (Ramesh et al. 2020)	Mesocarp carbon partition (Bourgis et al. 2011) Oil biosynthesis (Jin et al. 2016) Ganoderma infected roots (Tee et al. 2013) Host fungal interaction (Ho et al. 2016) Cold, drought, salt tolerance (Xiao et al. 2014b) Cold tolerance (Xiao et al. 2014a)	Red palm weevil infestation (Giovino et al. 2015) Mesocarp carbon partition (Bourgis et al. 2011)
Proteomics	Milk from coconut endosperm (D'Amato et al. 2012)	Mesocarp development (Teh et al. 2016)	Drought, salt tolerance (El Rabey et al. 2015, 2016)
Metabolomics	Coconut water (Costa et al. 2015; Kumar et al. 2021)	Mesocarp carbon partition (Bourgis et al. 2011) Mesocarp development (Teh et al. 2013, 2014) Leaf (Vargas et al. 2016) Potassium, water logging (Cui et al. 2019)	Fruit metabolites (Farag et al. 2014) Fruit nutritional quality (Hamad et al. 2015)
Phenomics	Fourier Transformed Infrared Spectroscopy of virgin coconut oil (Rohman and Che Man 2009)	*Ganoderma* disease (Izzuddin et al. 2015)	Fluorescence microscopy imaging of leaflet (Arinkin et al. 2014) Thermal imaging of red palm weevil infestation (Golomb et al. 2015)

role of stress-responsive proline accumulation. Also, the promoter region of the gene was found to be enriched with several stress-responsive *cis* elements. Proteomics and metabolomics of the proline transport and accumulation in stressed tissues of palm could further shed light on the mechanism of the amino acid proline in protecting the palms from abiotic stresses.

7.2.3 Cold, Nutrient Deficiency and Water Logging Stress

High-throughput transcriptomic analysis of cold stressed oil palm tissues revealed the role of three transcription factors namely (TFs) CBF, ICE1 and NAC (Xiao et al. 2014a). The simple sequence repeats (SSRs) mined from the transcriptome data of cold stressed tissues are promising and would serve as an excellent tool for marker-aided selection or in the introgression of cold tolerance trait in palms (Xiao et al. 2014a). Gene expression studies further corroborated the role of a unique C-repeat binding factor gene in cold tolerance of oil palm (Lei et al. 2014). The transcripts of the ribonucleoprotein coding gene, *EgRBP42*, was found to be induced under abiotic stress due to salt, cold, heat, drought and water logging (Yeap et al. 2012). Oil palm is a heavy feeder of nutrients, especially potassium. Low potassium and water logging are other major abiotic stressors in oil palm gardens located in nutrient poor soils and near river streams. The gene *EgKUP8*, responsible for potassium (K+) uptake, was characterised in oil palm (Husri and Ong-Abdullah 2018). Authors found the transcripts of *EgKUP8* were upregulated in oil palm during the potassium deficiency. Leaf metabolome and proteome analysis of the palms in response to potassium deficit and excess water conditions revealed the major role of carbon balance in these processes (Cui et al. 2019).

7.2.4 Biotic Stress

7.2.4.1 Pests

Red palm weevil (*Rhynchophorus ferrugineus* Olivier) (RPW) is a serious pest of both coconut and oil palm (Faleiro 2006) and causes severe mortality. Transcriptomics of RPW-infested palm tissues show induction of TFs like WRKY, but repression of pathogenesis-related (PR) proteins (Giovino et al. 2015). Proteomics of RPW-infested and artificially wounded date palm tissues were studied by Rasool et al. (2015) revealing the involvement of 21 differentially expressed proteins of which six were ion transport related. Proteins related to carbohydrate use, photosynthesis, protein degradation, plant stress, defence and were seem to be affected in the infested palms on further analysis (Rasool et al. 2018). Leaf metabolome of RPW-infested date palm also revealed the changes in carbohydrate and organic metabolism at the first stage of infestation and peptide and lipid metabolism pathways during the later stages of the damage (Giovino et al. 2016). Metagenomics of gut microbes revealed the predominance of organisms producing hydrolases that degrade the polysaccharides and sucrose in RPW (Jia et al. 2013) and microflora which can degrade cellulose in rhinoceros beetle (Shelomi et al. 2019). All these above investigations indicate a role for specific proteins which could be developed as biomarkers for early detection of the insect. Phenotyping using infrared thermography was employed to confirm the RPW-induced physical defects in the trunks of palms (Al-Sulaiman and Hawwa 2012) and also in spotting red palm weevil infestation (Golomb et al. 2015) in date palms.

Eriophyid mite is a serious pest of coconut palms causing serious yield losses around the World (Navia et al. 2013). DNA (RAPD and SSR) markers linked to the coconut's susceptibility to the pest were identified by Shalini et al. (2007). Round shaped and green colored fruits were documented to attract minimal damage due to the pest infestation (Moore and Alexandaer 1990). Hence, an image-based phenotyping method was employed to assess the color along with texture analysis of the fruit to discriminate the coconut varieties differing in their susceptibility to the eriophyid mite pest (Arunachalam et al. 2013).

7.2.4.2 Diseases

Basal stem rot (BSR) caused by the fungus *Ganoderma lucidum* and *Ganoderma boninense* is a major problem in coconut and oil palm cultivation, respectively. The genes associated with susceptibility of the oil palm to the basal stem rot pathogen (Ho and Tan 2015) reveal that the induced cell death and reactive oxygen species (ROS) kinetics are the major factors along with the interaction of salicylate, jasmonate and ethylene during infection. Transcriptomic analysis of oil palm infected by *G. boninense* and treated with a biocontrol agent revealed the enhanced ability of palm in the nutrient uptake and availability (Ho et al. 2016). Dehydrin gene was upregulated during *G. boninense* infection in oil palm (Tan et al. 2013). The cell wall strength was reduced due to the differences in lignin accumulation during the infection-defence process (Ho and Tan 2015). Caffeic acid O-methyl transferase protein was found to be low in roots affected by severe disease indicating the hampering of the lignin synthesis (Al-Obaidi et al. 2014). Phenolics were induced as defence mechanism during the use of biocontrol agents and chitin while controlling *Ganoderma* infection in coconut (Karthikeyan et al. 2006). Computational analysis of secretory proteins in response to *Ganoderma* infection in coconut revealed the role of hydrolases secreted during lignin degradation process in the host (Rahul et al. 2015).

Phytoplasmal diseases cause serious yield losses to coconut palms. Root (wilt) is a major disease in coconuts grown in South India. Transcriptomic analysis coupled with bioinformatics was employed to identify the Resistance Gene Analogues (RGAs) in coconut (Rajesh et al. 2015). Comparative differential gene expression analysis of healthy and root (wilt) affected coconut palms revealed that the genes related to biogenesis of cell wall, cellular transport, plant-pathogen interaction and hormone biosynthesis differ in their expression levels between healthy and diseased plants (Rajesh et al. 2018). Transcriptomics of yellow decline phytoplasma affected coconut also displayed similar set of genes related to cell rescue, defence and hormone stimulus (Nejat et al. 2015). Molecular markers (Manimekalai et al. 2012) were identified using random primers for tagging yellow leaf disease resistance in areca palm whose omics knowledge is still in infancy.

7.2.5 Chloroplast and Mitochondrial Omics

Organelle genomes of organisms play a crucial role in understanding of cytoplasmic inheritance, phylogenetic studies and evolution and adaptation of plants. Complete chloroplast genomes are available for cultivated palms such as dates (Yang et al. 2010; Khan et al. 2012), coconut (Huang et al. 2013; Rajesh et al. 2020a, b), African oil palm (Uthaipaisanwong et al. 2012), Asian palmyra (Sakulsathaporn et al. 2017), macaw palm (de Santana Lopes et al. 2018) and arecanut (Rajesh et al. 2020a). About 31 plastid genomes, mostly of members of Arecoideae tribe, have also been sequenced (Comer et al. 2015). Another set of 39 complete genomes of plastids of plants of Arecaceae and Dasypogonaceae families have also been made available (Barrett et al. 2016). Based on sequencing data of RNA of chloroplasts of different cultivars of date palm, Yang et al. (2010) identified 78 SNP sites to distinguish the cultivars. About 5 and 50% of chloroplast genome of coconut (Aljohi et al. 2016) and date palm (Sabir et al. 2014), respectively, are found to be integrated within their respective mitochondrial genomes. The genes governing photosynthesis and gene expression, plastid function are under positive selection in palms (de Santana Lopes et al. 2018).

RNA editing, the process of conversion of cytidine to uridine (C–U) or vice versa (U–C), is an important biological process in the organelles. Pentatricopeptide repeat (PPR) proteins of nuclear genome are known to specifically bind to the upstream side of sites of RNA editing (Ichinose and Sugita 2016). As high as 83 RNA editing sites in 27 genes were identified in plastid genome of coconut palms and validated by qRT-PCR experiments (Huang et al. 2013). About 32 sites of RNA editing were found in 19 plastid genes and *ndh* gene recorded the highest RNA editing sites in oil palm (Uthaipaisanwong et al. 2012). The gene content as well as the structural arrangement of plastid genomes of oil palm and date palm was found to be similar by Uthaipaisanwong et al. (2012).

Complete mitochondrial genomes of date palm (Fang et al. 2012), coconut (Aljohi et al. 2016; Rajesh et al. 2020a, b) and near complete mitogenome of oil palm (Uthaipaisanwong et al. 2017) are available for understanding the genome biology of organelles. About 734 and 600 RNA editing sites were found in mitochondrial genomes of coconut (Aljohi et al. 2016) and date palm (Fang et al. 2012), respectively. Transcriptomics analysis of mitochondrial genome indicates high level of female and male flowers, root and bud tissue specific gene expressions in date palm (Fang et al. 2012).

7.2.6 Databases, Bioinformatics and Web Resources

The science of omics generates big data sets; hence, the storage of data in online databases is neccessary so as to allow the retrieval of required data set(s) based on user queries. TropGeneDB (http://tropgenedb.cirad.fr) holds genes markers and genome mapping data for few tropical plants including the coconut and oil palm which is constantly updated (Hamelin et al. 2012; Ruiz et al. 2017). Bioinformatics tools are employed to analyse the genome sequences to mine new simple sequence repeats (SSRs) and single nucleotide polymorphism (SNP) regions and to design novel molecular markers. Transcriptome data of coconut root (wilt) diseased tissues was used to mine for genes, SSRs and other genomics features resulting in a database (Verma et al. 2017). The oil palm transcriptome sequences (Riju et al. 2007; Riju and Arunachalam 2009) were mined for SNP/InDel sites and repeat regions. A database of SSRs mined from the complete genome of oil palm is available (Babu et al. 2019). A SNP array was developed and was used along with 252 SSRs to saturate the linkage map of oil palm with genotyping data from dura, pisisfera and American oil palms and their hybrids (Ting et al. 2014). SSR markers were also discovered in mitochondrial genome of oil palm (Uthaipaisanwong et al. 2017) and NGS sequence data of sago palm (Purwoko et al. 2019) and ESTs (Manju et al. 2011) and whole genome (Manju et al. 2016) of date palm. The microsatellite markers developed for a palm species is easily cross transferrable across other palms (Anitha et al. 2008; Manju et al. 2016). Random DNA markers are employed routinely in organisms where the omics information is scanty. Converting the random markers to precise sequence characterised and amplified region (SCAR) markers using the available information by in silico tools is imperative. A bioinformatics software was developed to perform this workflow (Premkrishnan and Arunachalam 2012) and was validated with the oil palm transcriptomics data. The draft genome sequences of the oil palm are made available (http://genomsawit.mpob.gov.my) to provide genes and information on markers across the oil palm chromosomes. All the currently available web resources/databases for omics of palms are summarised in Table 7.2. Data warehouse approach is employed to access the integrated set of omics data resources in maize (Colmsee et al. 2012). Similar

Table 7.2 Internet resources for omics research on palms

	URL[a]	Reference(s)
Gene database of oil palm	http://palmxplore.mpob.gov.my/	Sanusi et al. (2018)
Genome databases of oil palm	http://genomsawit.mpob.gov.my/	
Gene and marker database- coconut, oil palm	http://tropgenedb.cirad.fr/	Ruiz et al. (2017)
Oil palm simple sequence repeats database	https://ssr.icar.gov.in/	Babu et al. (2019)
Coconut root (wilt) transcriptome database	http://webtom.cabgrid.res.in/cntdb/	Verma et al. (2017)
XCMS Online tool for metabolomic data analysis	https://xcmsonline.scripps.edu/	Tautenhahn et al. (2012)

[a]URLs links accessed on 31.07.2019

comprehensive warehouse approaches are indispensable to store the omics knowledge of palms for public use.

7.2.7 Evolution and Phylogenomics

Plastid genes have displayed slower rate of evolution in palms than grasses (Wilson et al. 1990) possibly due to their perennial nature (Smith and Donoghue 2008). Plastid gene region *trn*L—*trn*F (Baker et al. 1999) and nuclear malate synthase gene (Lewis and Doyle 2001) are some of the genomic regions used widely in phylogenetic studies of palms. Complete plastid genomes of several palms were sequenced and used to draw the phylogenetic relationship among the members palm family (Barrett et al. 2016). Plastid genomes are employed to explore the evolution in a single genus such as *Brahea* (Barrett et al. 2019b).

Phylogenomics of palms (Comer et al. 2016; Barrett et al. 2013, 2016) enhance the knowledge of gene or genome function with the help of phylogenetic studies using plastid genes or genomes or nuclear genes/genomes. Genome size/nuclear DNA content varies 33 (Barrett et al. 2019a) to 40.2-fold (Röser et al. 1997) among the members of palm family. Whole genome duplication events are reported in oil palm (Singh et al. 2013a) driving genome evolution. Chromosome number (2n) showed variation from 26 to 36 among most palms. Only *Voanioala* genus of palms recorded extremely high chromosome number of 2n = 596 (Röser et al. 1997). Hence, the genome size varies within a chromosome even amongst the related species of monocot plants (Leitch et al. 2010). Ref-Seq sequences along with whole genomes of plastids have been used to understand the phylogenomics and evolution in palms (Barrett et al. 2019a). Metabolites such as sugars, fatty acids, stilbene and rutin compounds are used as chemometric measures to understand the evolution and geographical differentiation of species of *Butia* palms (Hoffmann et al. 2017).

7.2.8 Fruits and Its Parts

Fruit or its part(s) form the major economic product of palms especially in *Acrocomia*, *Areca*, *Borassus*, *Cocos*, *Elaeis*, *Phoenix*, etc., either as source of major edible oil/fuel/industrial oil, dessert or processed fruit or chewed as masticatory. The fruit exocarp (outer surface) of coconut (Arunachalam and Rajesh 2017) and oil palm (Singh et al. 2014) display wide variations in color. In oil palm, the fruit exocarp color is under the control of the *VIRESCENS* (*VIR*) gene, which encodes a R2R3-MYB TF (Singh et al. 2014). Nearly, one-third of carotenoids found in pericarp of arecanut fruits were found to be β-carotene which forms the agro-waste of areca cultivation (Kumar et al. 2012). Color of the fruit was digitally phenotyped using image analysis techniques (Arunachalam et al. 2013).

QTL analysis of fruit component traits was carried out to identify the trait linked molecular markers in oil palm (Rance et al. 2001) and coconut (Baudouin et al. 2006). Three key genetic loci for oil per dry-matter content were identified by Genome-Wide Association Studies (GWAS) and Single Nucelotide Polymorphism (SNP) studies (Teh et al. 2016). Shell thickness trait is governed by a most important single gene (*Sh*+) which is thin in pisifera type and thick in dura type (Beinaert and Vanderweyen 1941). Tenera, having medium thick shell, showed heterosis for oil yield and is commercially exploited to enhance oil palm productivity. The shell thickness of oil palm is governed by a transcription factor SEEDSTICK and whose SNP variation was correlated with the dura, pisifera and tenera forms (Singh et al. 2013b).

Aromatic or jelly-like makapuno are two high value mutants of coconuts with commercial significance (Nguyen et al. 2016). Makapuno/curd coconut is due to the deficiency of alpha D galactosidase enzyme (Mujer et al. 1984). The makapuno trait is controlled by a recessive gene (*mm*) (Zuniga 1953) and the embryo in the nuts is unable to germinate naturally. The sequence of *CnAGal* gene, governing the mutation, was examined along with the gene expression profile, and the protein structure was predicted (Phoeurk et al. 2018). Three mutations were observed by the authors during the study which involved a long deletion of 19 amino acids and two single amino acid substitutions. Aromatic coconuts in Thailand are characteristic with a strong Pandan-like aroma due to 2-acetyl-1-pyrroline (2-AP) similar to aromatic rice. The mutation is found to be due to two different SNPs in the exon region of amino-aldehyde dehydrogenase (*CnAMADH2*) gene in aromatic green dwarf (Sanseuk et al. 2016) and yellow aromatic dwarf (Dumhai et al. 2019).

Comparison of transcriptomics and metabolomics data of oil-rich mesocarp of oil palm and non-oily mesocarp of date palm revealed the major role of the WRINKLED1 TF (WRI1) in oil biosynthesis (Bourgis et al. 2011). *WRI1* had recorded 57 times higher expression in oil palm than date palm mesocarp tissue. The role of WRI1 transcription factor in oil accumulation in mesocarp of oil palm is further endorsed by various studies (Tranberger et al. 2011; Ma et al. 2013; Dussert et al. 2013). Proteomics of oil synthesis in oil palm identified 68 proteins (Hassan et al. 2019). The role of metabolites in fruit composition in oil-rich mesocarp at different stages of development was identified in macaúba palm (*Acrocomia aculeata* (Jacq.) (Lieb et al. 2019). Metabolomics identified the changes in the composition of polyamines, phytohormones and cell wall metabolites during the development of mesocarp of oil palm (Teh et al. 2013, 2014). Pathway analysis of genes in oil palm, coconut and date palm mesocarp/endosperm tissues identified the major role of the process of conversion of pyruvate to fatty acid (Xiao et al. 2019) in fatty acid synthesis. Proteomics analysis indicates that the product of anthocyanin synthase gene accumulates during the ripening of date palm fruits (Marondedze et al. 2014). Transcriptomics and metabolomics studies of date palm fruits clearly indicate the sugar synthesis genes are expressed during the ripening stage (Yin et al. 2012). Metabolomics of fruits of date palm revealed the role of flavonoids and sugars in differentiating the varieties (Farag et al. 2014). Metabolomic fate of arecanut investigated using arecanut-fed rats revealed production of novel molecules, viz. N-methyl pectic acid and monoacylglyceride of arecaidine. The role of reduction process of carbon-carbon bond in the formation of methyl pectic acid was uncovered (Giri et al. 2006). Transcriptomics of arecanut fruits revealed the high expression of genes governing alkaloids, flavanoids and terpene molecules (Manimekalai et al. 2018).

7.2.9 Gender/Sex Determination

Gender or sex expression is an important trait governing the productivity of the cultivated palms. Dioecism, the phenomenon where the male and female flowers are borne in separate individual plants, is a challenge for farmers. *Borassus*, *Calamus*, *Phoenix* and *Lodoicea* are commonly known dioecious palms. Oil palm although monoecious in nature the degree of male, female and hermaphrodite flowers/inflorescences in a palm determines the fruit set ability. Mutant palms with extremely high proportion of either male flowers (androgena) (Arunachalam et al. 2001) or female flowers (spicata) (Arunachalam et al. 2014) are reported in coconut palm. Genomics was employed in the palms to identify sex linked molecular markers with varying degrees of success. DNA markers specific to male plants (S1443–500 bp) (Yang et al. 2005) and female-specific markers (CsMale1–509 bp) (Li et al. 2010) are reported in *Calamus*. An aldo-ketoreductose like gene (*EgAKR1*) (Somyong et al. 2015) and ACC synthase gene (*EgACCO1*) (Somyong et al. 2016) were found to control the number of female inflorescences in oil palm. Genomics of sex determination in date palm identified a XY chromosomal inheritance (Al-Dous et al. 2011). The genomic regions were physically mapped (Al-Mahmoud et al. 2012) and found to be in the telomeric region of linkage group 12 (Mathew et al. 2014). Three microsatellite markers (mPdIRDP50, mPdIRDP52, mPdIRDP80) displayed heterozygosity only in male individuals of date palm (Cherif et al. 2013).

7.2.10 Height/Dwarfism

Height is an important trait for cultivated palms as the plants grow tall as high as 15–30 m, posing severe constraints in the harvest and cultural operations. Mechanical interventions of harvesting/climbing devices and ladders are designed to overcome this. However, evolving suitable dwarf statured cultivar is one of the important objectives of breeding in coconut, arecanut and oil palm.

SNPs associated with stem height and the QTL regions on the genomes (Lee et al. 2015; Pootakham et al. 2015) and genes involved in height were identified based on the transcriptomics studies (Azni et al. 2014; Somyong et al. 2019). Major genes encoding asparagine synthase on chromosome 5 (Lee et al. 2015), GA-2 oxidase gene on chromosome 15 (Pootakham et al. 2015) and a DELLA gene (EgDELLA1) on chromosome 14 were highly correlated to stem height of oil palm. Transcriptomics experiments identified dwarfism related genes as zinc-finger protein, auxin responsive protein, circadian clock-associated protein1 and basic leucine zipper protein in oil palm (Azni et al. 2014). A genomic region OPA09-260 bp (NCBI Accession no. JX406850) was exclusively present in tall accessions of coconut (Rajesh et al. 2013). *GA20ox1*, a GA20-oxidase gene of coconut in tall and dwarf forms, although similar in sequence, differed in expression pattern in tall and dwarf palms; the expression of the gene was three-fold high in the leaves of tall palms compared to dwarfs owing to differences in *cis* elements in promoter regions (Boonkaew et al. 2018). Whole genome of Hainan Tall (Xiao et al. 2017) and Catigan Green Dwarf (Lantican et al. 2019) allowed to identify the genomic SNP variations among tall and dwarf forms of coconut (Lantican et al. 2019). Dwarf mutant of arecanut palms differs from the tall (wild type) variety by an absence of genomic region of 1.3 Kb size (KP221658) coding for the domains of a hexose carrier protein (Rajesh et al. 2016b) of palms. Phytochemical knowledge of dwarf wild relative of date palm, viz. *Phoenix roebelenii* O'Brien (Singab et al. 2015) needs comparative analysis for identification of dwarf-related molecules in *Phoenix*. Genomic differences among the species also helped to infer the geographical dispersal and species evolution of dwarf *Brahea moorei* from tall forms *B. armata* (Barrett et al. 2019b).

7.2.11 Image Analysis and Other Phenomics

Image analysis and other phenomics techniques are emerging tools for rapid omics analysis of plants. Image analysis of leaflet cross-sections was successfully utilised for the discrimination of five date palm varieties using fluorescence microscopy (Arinkin et al. 2014). Combination of 3-D imaging by magnetic resonance imaging (MRI) and metabolomics revealed varietal differences in Saw palmetto palm (*Serenoa repens*) (Jaiswal et al. 2019). Image analysis is used to understand the taxonomy and identification of orchid species (Pereira et al. 2016). Hyperspectral imaging was carried out by aerial vehicles to identify oil palm plants affected by *Ganoderma* disease (Izzuddin et al. 2015). Smartphone with appropriate image analysis software has been employed to analyse the intensity of blue color in order to indirectly estimate the tyrosinase inhibition activity of extracts of salak palm after treating with the dye dopaquinone (Moonrungsee et al. 2018). Color changes as oil palm fruits ripen were documented by image analysis techniques (Tan et al. 2010).

Growth rate estimation was carried out using radiocarbon dating in the *Oenocarpus bataua* palm (del Valle et al. 2014). Chlorophyll fluorescence is used as screening tool to identify water deficit stress tolerance in coconut (Bai et al. 2008) and oil palm (Suresh et al. 2010). A detailed account of both phenotypic variability for various traits (Arunachalam and Rajesh 2017) and composition of phytochemicals (Lima et al. 2015) in coconut palm is available for omics researchers. Fourier transformed infrared (FTIR) spectroscopy was employed to detect adulteration in virgin coconut oil (Rohman and Che Man 2009). Composition of phytochemicals (Hoffmann et al. 2017) has been worked out as a part of metabolomics in *Butia* palm. Ionomics (Baxter et al. 2007) is emerging as a tool to profile the ionic contents in plant tissues to understand the nutrient deficiency and elemental toxicity stress.

7.3 Road Map for Future Research

As the palms are entering into the new era of omics science with the advent of cost and time-efficient tools used for sequencing genomes and transcriptomes, the knowledge of biology of palms is expanding and waiting to be harnessed for various human uses. The current chapter has provided an overview of latest developments in the field and also flag the research gaps. Complete genome drafts are still unknown in economic palms such as Areca, Borassus, which need urgent attention. Comparative analysis of rapidly expanding omics knowledge in date and oil palms need to be harnessed in other palms such as coconut. Plastid genomes are extensively studied in palms, but the mitochondrial genomes are still under studied in many palms. Proteomics and metabolomics of major biotic and abiotic stress-induced changes need extensive investigations on palms.

References

Al Mahmoud ME, Al Dous EK, Al Azwani EK, Malek JA (2012) DNA based assays to distinguish date palm (Arecaceae) gender. Am J Bot 99(1):7–10

Al-Dous EK, George B, Al-Mahmoud ME, Al-Jaber MY, Wang H, Salameh YM, Al-Azwani EK, Chaluvadi S, Pontaroli AC, DeBarry J, Arondel V (2011) *De novo* genome sequencing and comparative genomics of date palm (*Phoenix dactylifera*). Nature Biotechnol 29 (6):521

Aljohi HA, Liu W, Lin Q, Zhao Y, Zeng J, Alamer A, Alanazi IO, Alawad AO, Al-Sadi AM, Hu S, Yu J (2016) Complete sequence and analysis of coconut palm (*Cocos nucifera*) mitochondrial genome. PLoS ONE 11(10):0163990

Al-Khayri JM, Ibraheem Y (2014) In vitro selection of abiotic stress tolerant date palm (*Phoenix dactylifera* L.): a review. Emir J Food Agr 26(11):921–933

Al-Mssallem IS, Hu S, Zhang X, Lin Q, Liu W, Tan J, Yu X, Liu J, Pan L, Zhang T, Yu J (2013) Genome sequence of the date palm *Phoenix dactylifera* L. Nature Comm 4(1):1–9

Anitha N, Jayaraj KL, Kumar EP, George J, Rajesh MK (2008) Assessment of cross-taxa utility of coconut microsatellite markers. Ind J Hortic 65(3): 317–321

Al-Obaidi JR, Mohd-Yusuf Y, Razali N, Jayapalan JJ, Tey CC, Md-Noh N, Junit SM, Othman RY, Hashim OH (2014) Identification of proteins of altered abundance in oil palm infected with *Ganoderma boninense*. Int J Mol Sci 15(3):5175–5192

Al-Sulaiman F, Hawwa M (2012) IC tomography and infrared tomography techniques to monitor defect sites in palm tree trunks. Adv Mat Res 445:1053–1057

ArinkinV Digel I, Porst D, Artmann AT, Artmann GM (2014) Phenotyping date palm varieties via leaflet cross-sectional imaging and artificial neural network application. BMC Bioinformatics 15(1):55

Arroyo-Serralta GA, Zizumbo-Villareal D, Escalante Erosa F, Peña-Rodríguez LM (2012) Cuticular wax composition of coconut palms and their susceptibility to lethal yellowing disease. J Mex Chem Soc 56(1):67–71

Arunachalam V (2011) Genomics of cultivated palms, p 114. Elsevier Inc

Arunachalam V, Rajesh MK (2017) In: Ahuja MR, Jain SM (eds). Biodiversity and conservation of woody plants, pp 3–36. Springer International Publishing AG

Arunachalam V, Jerard BA, Elangovan M, Ratnambal MJ, Dhanapal R, Rizal SK, Damodaran V (2001) Unexploited diversity in coconut palm (*Cocos nucifera* L.). Plant Genetic Resources Newslett 127:39–43

Arunachalam V, Jerard BA, Apshara SE, Jayabose C, Subaharan K, Ravikumar N, Palaniswami C (2013) Digital phenotyping of coconut and morphological traits associated with eriophyid mite (*Aceria guerreronis* Keifer) infestation. J Plantatn Crops 41 (3):417–424

Arunachalam V, Rajesh MK, Jerard BA, Jayabose C, Sairam CV (2014) Characterization of a spicata mutant of coconut palm in India. J Plantatn Crops 41(3):417–424

Azni INAM, Namasivayam P, Ling HC, Alwee SSRS, Abd Manaf MA (2014) Differentially expressed transcripts related to height in oil palm. J Oil Palm Res 26(4):308–316

Babu K, Rani MKL, Sahu S, Mathur RK, Kumar N, Ravichandran G, Anitha P, Bhagya HP (2019) Development and validation of whole genome-wide and genic microsatellite markers in oil palm (*Elaeis guineensis* Jacq.): first microsatellite database (OpSat-db). Sci Rep 9(1):1899

Bai KV, Kumar SN, Rajagopal V, Vijayakumar K (2008) Principal component analysis of chlorophyll fluorescence transients for tolerance to drought stress in coconut seedlings. Indian J Hortic 65(4):471–476

Baker WJ, Asmussen CB, Barrow SC, Dransfield J, Hedderson TA (1999) A phylogenetic study of the palm family (Palmae) based on chloroplast DNA sequences from thetrnL—trnF region. Plant Syst Evol 219(1–2):111–126

Barrett CF, Davis JI, Leebens-Mack J, Conran JG, Stevenson DW (2013) Plastid genomes and deep relationships among the commelinid monocot angiosperms. Cladistics 29:65–87

Barrett CF, Bacon CD, Antonelli A, Cano Á, Hofmann T (2016) An introduction to plant phylogenomics with a focus on palms. Bot J Linnean Soc 182(2):234–255

Barrett CF, McKain MR, Sinn BT, Ge XJ, Zhang Y, Antonelli A, Bacon CD (2019a) Ancient polyploidy and genome evolution in palms. Genome Biol Evol 11 (5):1501–1511

Barrett CF, Sinn BT, King LT, Medina JC, Bacon CD, Lahmeyer SC, Hodel DR (2019b) Phylogenomics, biogeography and evolution in the American genus *Brahea* (Arecaceae). Bot J Linnean Soc 190(3):242–259

Xiao Y, Xu P, Fan H, Baudouin, L., Xia W, Bocs S, Xu J, Li Q, Guo A, Zhou L, Li J (2017) The genome draft of coconut (*Cocos nucifera* L.). Giga Sci 6(11):095

BaudouinL, Lebrun P, Konan JL, Ritter E, Berger A, Billotte N (2006) QTL analysis of fruit components in the progeny of a Rennell Island Tall coconut (*Cocos nucifera* L.) individual. Theor Appl Genet 112 (2):258–268

Baxter I, Ouzzani M, Orcun S, Kennedy B, Jandhyala SS, Salt DE (2007) Purdue ionomics information management system. An integrated functional genomics platform. Plant Physiol 143(2):600–611

Beinaert A, Vanderweyen R (1941) Contribution à l'étudegénétiqueetbiométrique des variétésd' *Elaeis guineensis* Jacq. Publications de l'Institut National pour l'étudeAgronomique du Congo Belge, sériescientifique, p 27

Boonkaew T, Mongkolsiriwatana C, Vongvanrungruang A (2018) Characterization of GA20ox genes in tall and dwarf types coconut (*Cocos nucifera* L.). Genes Genomics 40(7):735–745

Bourgis F, Kilaru A, Cao X, Ngando-Ebongue GF, Drira N, Ohlrogge JB, Arondel V (2011) Comparative transcriptome and metabolite analysis of oil palm and date palm mesocarp that differ dramatically in carbon partitioning. P Natl A Sci 108(30):12527–12532

Cha-Um S, Takabe T, Kirdmanee C (2010) Ion contents, relative electrolyte leakage, proline accumulation, photosynthetic abilities and growth characters of oil palm seedlings in response to salt stress. Pak J Bot 42 (3):2091–2020

Cherif E, Zehdi S, Castillo K, Chabrillange N, Abdoulkader S, Pintaud JC, Santoni S, Salhi-Hannachi A, Glémin S, Aberlenc-Bertossi F (2013) Male specific DNA markers provide genetic evidence of an XY chromosome system, a recombination arrest and allow the tracing of paternal lineages in date palm. New Phytol 197(2):409–415

Colmsee C, Mascher M, Czauderna T, Hartmann A, Schlüter U, Zellerhoff N, Schmitz J, Bräutigam A, Pick TR, Alter P, Gahrtz M (2012) OPTIMAS-DW: a comprehensive transcriptomics, metabolomics, ionomics, proteomics and phenomics data resource for maize. BMC Plant Biol 12(1):245

Comer JR, Zomlefer WB, Barrett CF, Davis JI, Stevenson DW, Heyduk K, Leebens-Mack JH (2015) Resolving relationships within the palm subfamily Arecoideae (Arecaceae) using plastid sequences derived from next generation sequencing. Am J Bot 102(6):888–899

Comer JR, Zomlefer WB, Barrett CF, Stevenson DW, Heyduk K, Leebens-Mack JH (2016) Nuclear phylogenomics of the palm subfamily Arecoideae (Arecaceae). Mol Phylogenetics Evol 97:32–42

Costa HB, Souza LM, Soprani LC, Oliveira BG, Ogawa EM, Korres AM, Ventura JA, Romão W (2015) Monitoring the physicochemical degradation of coconut water using ESI-FT-ICR MS. Food Chem 174:139–146

Cui J, Davanture M, Zivy M, Lamade E, Tcherkez G (2019) Metabolic responses to potassium availability and waterlogging reshape respiration and carbon use efficiency in oil palm. New Phytol. https://doi.org/10.1111/nph.15751

D'Amato, A., Fasoli, E., & Righetti, P. G. (2012). Harry Belafonte and the secret proteome of coconut milk. J. Proteomics, 75(3):914–920

de Oliveira HO, de Castro GLS, Correa LO, Silvestre WVD, do Nascimento SV, da Silva Valadares RB, de Oliveira GC, Santos RIN, Festucci-Buselli RA, Pinheiro HA (2019) Coupling physiological analysis with proteomic profile to understand the photosynthetic responses of young *Euterpe oleracea* palms to drought. Photosynth Res 140(2):189–205

de Santana Lopes A, Pacheco TG, Nimz T, do Nascimento Vieira L, Guerra MP, Nodari RO, de Souza EM, de Oliveira Pedrosa F, Rogalski M (2018) The complete plastome of macaw palm [*Acrocomia aculeata* (Jacq.) Lodd.ex Mart.] and extensive molecular analyses of the evolution of plastid genes in Arecaceae. Planta 247(4):1011–1030

del Valle JI, Guarín JR, Sierra CA (2014) Unambiguous and low-cost determination of growth rates and ages of tropical trees and palms. Radiocarbon 56(1):39–52

Dumhai R, Wanchana S, Saensuk C, Choowongkomon K, Mahatheeranont S, Kraithong T, Toojinda T, Vanavichit A, Arikit S (2019) Discovery of a novel *CnAMADH2* allele associated with higher levels of 2-acetyl-1-pyrroline (2AP) in yellow dwarf coconut (*Cocos nucifera* L.). Sci Hortic 243:490–497

Dussert S, Guerin C, Andersson M, Joët T, Tranbarger TJ, Pizot M, Sarah G, Omore A, Durand-Gasselin T, Morcillo F (2013) Comparative transcriptome analysis of three oil palm fruit and seed tissues that differ in oil content and fatty acid composition. Plant Physiol 162 (3):1337–1358

El Rabey HA, Al-Malki AL, Abulnaja KO, Rohde W (2015) Proteome analysis for understanding abiotic stress (salinity and drought) tolerance in date palm (*Phoenix dactylifera* L.). Int J Genomics https://doi.org/10.1155/2015/407165

El Rabey HA, Al-Malki AL, Abulnaja KO (2016) Proteome analysis of Date Palm (*Phoenix dactylifera*

L.) under severe drought and salt stress. Int J Genomics https://doi.org/10.1155/2016/7840759

Escalante Erosa F, Gamboa-León MR, Lecher JG, Arroyo-Serralta GA, Zizumbo-Villareal D, Oropeza-Salín C, Peña-Rodríguez LM (2002) Major components from the epicuticular wax of *Cocos nucifera*. Revista de la Sociedad Química de México 46(3):247–250

Faleiro JR (2006) A review of the issues and management of the red palm weevil *Rhynchophorus ferrugineus* (Coleoptera: Rhynchophoridae) in coconut and date palm during the last one hundred years. Int J Trop Insect Sci 26(3):135–154

Fang Y, Wu H, Zhang T, Yang M, Yin Y, Pan L, Yu X, Zhang X, Hu S, Al-Mssallem IS, Yu J (2012) A complete sequence and transcriptomic analyses of date palm (*Phoenix dactylifera* L.) mitochondrial genome. PloS one 7(5):37164

Farag MA, Mohsen M, Heinke R, Wessjohann LA (2014) Metabolomic fingerprints of 21 date palm fruit varieties from Egypt using UPLC/PDA/ESI–qTOF-MS and GC–MS analyzed by chemometrics. Food Res Int 64:218–226

Giovino A, Bertolini E, Fileccia V, Al Hassan M, Labra M, Martinelli F (2015) Transcriptome analysis of *Phoenix canariensis* Chabaud in response to *Rhynchophorus ferrugineus* Olivier attacks. Front Plant Sci 6:817

Giovino A, Martinelli F, Saia S (2016) *Rhynchophorus ferrugineus* attack affects a group of compounds rather than rearranging *Phoenix canariensis* metabolic pathways. J Integr Plant Biol 58(4):388–396

Giri S, Idle JR, Chen C, Zabriskie TM, Krausz KW, Gonzalez FJ (2006) A metabolomic approach to the metabolism of the areca nut alkaloids arecoline and arecaidine in the mouse. Chem Res Toxicol 19 (6):818–827

Golomb O, Alchanatis V, Cohen Y, Levin N, Soroker V (2015) Detection of red palm weevil infected trees using thermal imaging. In: Stafford JV (Ed). Precision agriculture. Wageningen Academic Publishers, pp 643–650

Gribaa A, Dardelle F, Lehner A, Rihouey C, Burel C, Ferchichi A, Driouich A, Mollet JC (2013) Effect of water deficit on the cell wall of the date palm (*Phoenix dactylifera* 'D Negletnour', Arecales) fruit during development. Plant, Cell Environ 36(5):1056–1070

Hamad I, AbdElgawad H, Al Jaouni S, Zinta G, Asard H, Hassan S, Hegab M, Hagagy N, Selim, S (2015) Metabolic analysis of various date palm fruit (*Phoenix dactylifera* L.) cultivars from Saudi Arabia to assess their nutritional quality. Molecules, 20(8):13620–13641

Hamelin C, Sempere G, Jouffe V, Ruiz M (2012) TropGeneDB, the multi-tropical crop information system updated and extended. Nucleic Acids Res 41 (1):1172–1175

Harun MH (1997) Proline accumulation in the leaves of water stressed oil palm (*Elaeis guineensis* Jacq.) seedlings. Elaeis 9(2):93–99

Hassan H, Amiruddin MD, Weckwerth W, Ramli US (2019) Deciphering key proteins of oil palm (*Elaeis guineensis* Jacq.) fruit mesocarp development by proteomics and chemometrics. Electrophoresis 40 (2):254–265

He Z, Zhang Z, Guo W, Zhang Y, Zhou R, Shi S (2015) De novo assembly of coding sequences of the mangrove palm (*Nypa fruticans*) using RNA-Seq and discovery of whole-genome duplications in the ancestor of palms. PLoS ONE 10(12):0145385

Ho CL, Tan YC (2015) Molecular defense response of oil palm to Ganoderma infection

Ho CL, Tan YC, Yeoh KA, Ghazali AK, Yee, WY, Hoh CC (2016) *De novo* transcriptome analyses of host-fungal interactions in oil palm (Elaeis guineensis Jacq.). BMC Genom 17(1):1–19

Hoffmann JF, Carvalho IR, Barbieri RL, Rombaldi CV, Chaves FC (2017) *Butia* spp. (Arecaceae) LC-MS-based metabolomics for species and geographical origin discrimination. J Agr Food Chem 65(2):523–532

Huang YY, Matzke AJ, Matzke M (2013) Complete sequence and comparative analysis of the chloroplast genome of coconut palm (*Cocos nucifera*). PLoS ONE 8(8):74736

Huang YY, Lee CP, Fu JL, Chang BCH, Matzke AJ, Matzke M (2014) *De novo* transcriptome sequence assembly from coconut leaves and seeds with a focus on factors involved in RNA-directed DNA methylation. G3: Genes, Genomes, Genetics 4(11):2147–2157

Husri MN, Ong-Abdullah M (2018) Importance of KUP8 for K+ uptake in rooted plantlets of *Elaeis guineensis* under K+ sufficient conditions. South African J Bot 118:65–75

Ichinose M, Sugita M (2016) RNA editing and its molecular mechanism in plant organelles. Genes 8 (1):5

Izzuddin MA, Idris AS, Nisfariza NM, Ezzati B (2015) Spectral based analysis of airborne hyperspectral remote sensing image for detection of ganoderma disease in oil palm. In: Proceedings of 2015 international conference on biological and environmental science (BIOES 2015) Phuket, 1–3 Oct 2015, pp 13–20

Jaiswal Y, Weber D, Yerke A, Xue Y, Lehman D, Williams T, Xiao T, Haddad D, Williams L (2019) A substitute variety for agronomically and medicinally important *Sereno arepens* (saw palmetto). Sci Rep 9 (1):4709

Jia S, Zhang X, Zhang G, Yin A, Zhang S, Li F, Wang L, Zhao D, Yun Q, Wang J, Sun G (2013) Seasonally variable intestinal metagenomes of the red palm weevil (*Rhynchophorus ferrugineus*). Environ Microbiol 15(11):3020–3029

Jin J, Lee M, Bai B, Sun Y, Qu J, Alfiko Y, Lim CH, Suwanto A, Sugiharti M, Wong L, Yue GH (2016) Draft genome sequence of an elite Dura palm and whole-genome patterns of DNA variation in oil palm. DNA Res 23(6):527–533

Karthikeyan M, Radhika K, Mathiyazhagan S, Bhaskaran R, Samiyappan R, Velazhahan R (2006)

Induction of phenolics and defense-related enzymes in coconut (*Cocos nucifera* L.) roots treated with biocontrol agents. Brazilian J Plant Physiol 18 (3):367–377
Karunaratne S, Santha S, Kovoor A (1991) An *in vitro* assay for drought-tolerant coconut germplasm. Euphytica 53(1):25–30
Khan A, Khan IA, Heinze B, Azim MK (2012) The chloroplast genome sequence of date palm (*Phoenix dactylifera* L. cv. 'Aseel'). Plant Mol Biol Rep 30 (3):666–678
Kumar M, Moon UR, Mitra A (2012) Rapid separation of carotenes and evaluation of their *in vitro* antioxidant properties from ripened fruit waste of *Areca catechu*—a plantation crop of agro-industrial importance. Ind Crops Prod 40:204–209
Kumar M, Saini SS, Agrawal PK, Roy P, Sircar D (2021) Nutritional and metabolomics characterization of the coconut water at different nut developmental stages. J Food Compos Anal 96:103738
Lantican DV, Strickler SR, Canama AO, Gardoce RR, Mueller LA, Galvez HF (2019) *De novo* genome sequence assembly of Dwarf Coconut (*Cocos nucifera* L.'Catigan Green Dwarf') provides insights into genomic variation between coconut types and related palm species. G3: Genes Genomes Genet G3-400215. https://doi.org/10.1534/g3.119.400215
Lee M, Xia JH, Zou Z, Ye J, Alfiko Y, Jin J, Lieando JV, Purnamasari MI, Lim CH, Suwanto A, Wong L (2015) A consensus linkage map of oil palm and a major QTL for stem height. Sci Rep 5:8232
Lei X, Xiao Y, Xia W, Mason AS, Yang Y, Ma Z, Peng M (2014) RNA-seq analysis of oil palm under cold stress reveals a different C-repeat binding factor (CBF) mediated gene expression pattern in *Elaeis guineensis* compared to other species. PLoS ONE 9 (12):
Leitch IJ, Beaulieu JM, Chase MW, Leitch AR, Fay MF (2010) Genome size dynamics and evolution in monocots. J Bot. https://doi.org/10.1155/2010/862516
Lewis CE, Doyle JJ (2001) Phylogenetic utility of the nuclear gene malate synthase in the palm family (Arecaceae). Mol Phylogenetics Evol 19(3):409–420
Li M, Li F, Yang F, Yin G, Gan S (2010) A male-specific SCAR marker in *Calamus simplicifolius*, a dioecious rattan species endemic to China. Mol Breed 25:549–551
Lieb VM, Schex R, Esquivel P, Jiménez VM, Schmarr HG, Carle R, Steingass CB (2019) Fatty acids and triacylglycerols in the mesocarp and kernel oils of maturing Costa Rican *Acrocomia aculeata* fruits. NFS J 14:6–13
Lima EBC, Sousa CNS, Meneses LN, Ximenes NC, Júnior S, Vasconcelos GS, Lima NBC, Patrocínio MCA, Macedo D, Vasconcelos SMM (2015) *Cocos nucifera* (L.) (Arecaceae): a phytochemical and pharmacological review. Brazilian J Med Biol Res 48 (11):953–964
Ma W, Kong Q, Arondel V, Kilaru A, Bates PD, Thrower NA, Benning C, Ohlrogge JB (2013) *Wrinkled1*, a ubiquitous regulator in oil accumulating tissues from *Arabidopsis* embryos to oil palm mesocarp. PLoS ONE 8(7):68887
Manimekalai R, Deeshma KP, Manju KP, Soumya VP, Sunaiba M, Nair S, Ananda KS (2012) Molecular marker-based genetic variability among Yellow Leaf Disease (YLD) resistant and susceptible arecanut (*Areca catechu* L.) genotypes. Indian J Hortic 69 (4):455–461
Manimekalai R, Nair S, Naganeeswaran A, Karun A, Malhotra S, Hubbali V (2018) Transcriptome sequencing and *de novo* assembly in arecanut, *Areca catechu* L elucidates the secondary metabolite pathway genes. Biotechnol Rep 17:63–69
Manju KP, Manimekalai R, Arunachalam V (2011) Microsatellites in palm (Arecaceae) sequences. Bioinformation 7(7):347
Manju KP, Manimekalai R, Naganeeswaran SA, Arunachalam V, Karun A (2016) Microsatellites mining in date palm ('*Phoenix dactylifera*' L.) and their cross transferability across 'Arecaceae' family. Plant Omics 9(3):191
Marondedze C, Gehring C, Thomas L (2014) Dynamic changes in the date palm fruit proteome during development and ripening. Hortic Res 1:14039
Mathew LS, Spannagl M, Al-Malki A, George B, Torres MF, Al-Dous EK, Al-Azwani EK, Hussein E, Mathew S, Mayer KF, Mohamoud YA (2014) A first genetic map of date palm (*Phoenix dactylifera*) reveals long-range genome structure conservation in the palms. BMC Genom 15(1):285
Meerow AW, Krueger RR, Singh R, Low ETL, Ithnin M, Ooi LCL (2012) Coconut, date, and oil palm genomics. In: Schnell R, Priyadarshan P (eds) Genomics of tree crops. Springer, New York, NY, pp 299–351
Moonrungsee N, Peamaroon N, Boonmee A, Suwancharoen S, Jakmunee J (2018) Evaluation of tyrosinase inhibitory activity in Salak (*Salacca zalacca*) extracts using the digital image-based colorimetric method. Chem Pap 72(11):2729–2736
Moore D, Alexander L (1990) Resistance of coconuts in St. Lucia to attack by the coconut mite *Eriophyes guerreronis* Keifer. Trop Agr 67(1):33–36
Mujer CV, Ramirez DA, Mendoza EMT (1984) α-d-galactosidase deficiency in coconut endosperm: its possible pleiotropic effects in makapuno. Phytochemistry 23(4):893–894
Murugesan P, Aswathy GM, Sunil Kumar K, Masilamani P, Kumar V, Ravi V (2017) Oil palm (*Elaeis guineensis*) genetic resources for abiotic stress tolerance: a review. Indian J Agr Sci 171:12–17
Navia D, Gondim MGC, Aratchige NS, de Moraes GJ (2013) A review of the status of the coconut mite, *Aceria guerreronis* (Acari: Eriophyidae), a major tropical mite pest. Exp Appl Acarol 59(1–2):67–94
Nejat N, Cahill DM, Vadamalai G, Ziemann M, Rookes J, Naderali N (2015) Transcriptomics-based analysis using RNA-Seq of the coconut (*Cocos nucifera*) leaf in response to yellow decline phytoplasma infection. Mol Genet Genomics 290(5):1899–1910

Neoh BK, Teh HF, Wong YC, Ooi TEK, Cheah SS, Appleton DR (2017) Study of oil palm photosynthesis using omics technologies. In: Abdullah S, Chai-Ling H, Wagstaff C (eds) Crop improvement. Springer, Cham, pp 27–46

Nguyen QT, Bandupriya HD, Foale M, Adkins SW (2016) Biology, propagation and utilization of elite coconut varieties (makapuno and aromatics). Plant Physiol Biochem 109:579–589

Paul B, Rajesh MK, Kumar N (2010a) Bioinformatics approach towards mining of candidate genes for water stress in coconut. In: Thomas GV, Krishnakumar V, Jerard BA (eds) Improving productivity and profitability in coconut farming proceedings international conference on coconut biodiversity for prosperity, Central Plantation Crops Research Institute, Kasaragod, Kerala, India, pp 109–115

Paul B, Rajesh MK, Kiran AG, Thomas GV, Kumar N (2010b) Differential expression of genes regulated in response to water stress in coconut. In: Thomas GV, Krishnakumar V, Jerard BA (eds) Improving productivity and profitability in coconut farming proceedings international conference on coconut biodiversity for prosperity, Central Plantation Crops Research Institute, Kasaragod, Kerala, India. pp 116–122

Pereira S, Gravendeel B, Wijntjes P, Vos RA (2016) OrchID: a generalized framework for taxonomic classification of images using evolved artificial neural networks. BioRxiv, 070904

Phoeurk C, Somana J, Sornwatana T, Udompaisarn S, Traewachiwiphak S, Sirichaiyakul P, Phongsak T, Arthan D (2018) Three novel mutations in *α-galactosidase* gene involving in galactomannan degradation in endosperm of curd coconut. Phytochemistry 156:33–42

Pootakham W, Jomchai N, Ruang-areerate P, Shearman JR, Sonthirod C, Sangsrakru D, Tragoonrung S, Tangphatsornruang S (2015) Genome-wide SNP discovery and identification of QTL associated with agronomic traits in oil palm using genotyping-by-sequencing (GBS). Genomics 105(5):288–295

Premkrishnan BV, Arunachalam V (2012) In silico RAPD priming sites in expressed sequences and iSCAR markers for oil palm. Comp Funct Genomics. https://doi.org/10.1155/2012/913709

Purwoko D, Cartealy IC, Tajuddin T, Dinarti D, Sudarsono S (2019) SSR identification and marker development for sago palm based on NGS genome data. Breed Sci 69(1):1–10

Rahul CU, Babu M, Hemalatha N, Rajesh MK (2015) Computational prediction of the secretome of *Ganoderma lucidum*. Int J Innovative Res Computer Commun Eng 3(7):275–280

Rajesh MK, Jerard BA, Preethi P, Thomas RJ, Fayas TP, Rachana KE, Karun A (2013) Development of a RAPD-derived SCAR marker associated with tall-type palm trait in coconut. Sci Hortic 150:312–316

Rajesh MK, Rachana KE, Naganeeswaran SA, Shafeeq R, Thomas RJ, Shareefa M, Merin B, Anitha K (2015) Identification of expressed resistance gene analog sequences in coconut leaf transcriptome and their evolutionary analysis. Turk J Agr For 39(3):489–502

Rajesh MK, Fayas TP, Naganeeswaran S, Rachana KE, Bhavyashree U, Sajini KK, Karun A (2016a) *De novo* assembly and characterization of global transcriptome of coconut palm (*Cocos nucifera* L.) embryogenic calli using Illumina paired-end sequencing. Protoplasma 253(3):913–928.

Rajesh MK, Sabana AA, Rachana KE, Rahman S, Ananda KS, Karun A (2016b) Development of a SCoT-derived SCAR marker associated with tall-type palm trait in arecanut and its utilization in hybrid (dwarf x tall) authentication. Indian J Genet Plant Breed 76(1):119–122

Rajesh MK, Rachana KE, Kulkarni K, Sahu BB, Thomas RJ, Karun A (2018) Comparative transcriptome profiling of healthy and diseased Chowghat Green Dwarf coconut palms from root (wilt) disease hot spots. European J Plant Pathol 151(1):173–193

Rajesh MK, Gangaraj KP, Prabhudas SK, Prasad TK (2020a) The complete chloroplast genome data of *Areca catechu* (Arecaceae). Data Brief 33: https://doi.org/10.1016/j.dib.2020.106444

Rajesh MK, Chowdappa P, Behera SK, Kasaragod S, Gangaraj KP, Kotimoole CN, Nekrakalaya B, Mohanty V, Sampgod RB, Banerjee G, Das AJ, Niral V, Karun A, Mahato AK, Gaikwad K, Singh NK, Keshava Prasad TS (2020b) Assembly and annotation of the nuclear and organellar genomes of a dwarf coconut (Chowghat Green Dwarf) possessing enhanced disease resistance. OMICS: J Integr Biol 24(12):726–742

Ramesh SV, Hebbar KB, Rajesh MK, Abhin SP, Gangaraj KP, Athul Bobby (2020). Transcriptome analysis of *Cocos nucifera* L., seedlings having contrasting water use efficiency (WUE) under water-deficit stress: molecular insights and genetic markers for drought tolerance. In: Proceedings of The 1st International Electronic Conference on Plant Science. https://doi.org/10.3390/IECPS2020-08853

Rance KA, Mayes S, Price Z, Jack PL, Corley RHV (2001) Quantitative trait loci for yield components in oil palm (*Elaeis guineensis* Jacq.). Theor Appl Genet 103(8):1302–1310

Rasool K, Khan M, Aldawood A, Tufail M, Mukhtar M, Takeda M (2015) Identification of proteins modulated in the date palm stem infested with red palm weevil (*Rhynchophorus ferrugineus* Oliv.) using two dimensional differential gel electrophoresis and mass spectrometry. Int J Mol Sci 16(8):19326–19346

Rasool KG, KhanMA Tufail M, Husain M, Mehmood K, Mukhtar M, Takeda M, Aldawood AS (2018) Differential proteomic analysis of date palm leaves infested with the red palm weevil (Coleoptera: Curculionidae). Florida Entomologist 101(2):290–299

Riedel M, Riederer M, Becker D, Herran A, Kullaya A, Arana-López G, Peña-Rodríguez L, Billotte N, Sniady V, Rohde W, Ritter E (2009) Cuticular wax composition in *Cocos nucifera* L.: physicochemical analysis of wax components and mapping of their

QTLs onto the coconut molecular linkage map. Tree Genet Genomes 5(1):53

Riju A, Arunachalam V (2009) Data mining for simple sequence repeats in oil palm expressed sequence tags. Nature Precedings http://precedings.nature.com/documents/3581/version/1

Riju A, Chandrasekar A, Arunachalam V (2007) Mining for single nucleotide polymorphisms and insertions/deletions in expressed sequence tag libraries of oil palm. Bioinformation 2(4):128–131

Rohman A, Che Man YB (2009) Monitoring of virgin coconut oil (VCO) adulteration with palm oil using Fourier transform infrared spectroscopy. J Food Lipids 16(4):618–628

Röser M, Johnson MAT, Hanson L (1997) Nuclear DNA amounts in palms (Arecaceae). Botanica Acta 110 (1):79–89

Ruiz M, Sempéré G, Hamelin C (2017) Using TropGeneDB: a database containing data on molecular markers, QTLs, maps, genotypes, and phenotypes for tropical crops. In: van Dijk A (ed) Plant genomics databases. Methods in molecular biology, vol 1533. Humana Press, New York, NY

Sabir JS, Arasappan D, Bahieldin A, Abo-Aba S, Bafeel S, Zari TA, Edris S, Shokry AM, Gadalla NO, Ramadan AM, Atef A (2014) Whole mitochondrial and plastid genome SNP analysis of nine date palm cultivars reveals plastid heteroplasmy and close phylogenetic relationships among cultivars. PLoS ONE 9(4):94158

Saensuk C, Wanchana S, Choowongkomon K, Wongpornchai S, Kraithong T, Imsabai W, Chaichoompu E, Ruanjaichon V, Toojinda T, Vanavichit A, Arikit S (2016) *De novo* transcriptome assembly and identification of the gene conferring a "pandan-like" aroma in coconut (*Cocos nucifera* L.). Plant Sci 252:324–334

Sahu SK, Singh R, Kathiresan K (2016) Multi-gene phylogenetic analysis reveals the multiple origin and evolution of mangrove physiological traits through exaptation. Estuarine Coastal Shelf Sci 183:41–51

Sakulsathaporn A, Wonnapinij P, Vuttipongchaikij S, Apisitwanich S (2017) The complete chloroplast genome sequence of Asian Palmyra palm (*Borassus flabellifer*). BMC Res Notes 10(1):740

Sanusi NS, Rosli R, Halim MA, Chan KL, Nagappan J, Azizi N, Amiruddin N, Tatarinova TV, Low ET (2018) PalmXplore: oil palm gene database. Database 2018:1–9. https://doi.org/10.1093/pcp/pcz237

Shalini KV, Manjunatha S, Lebrun P, Berger A, Baudouin L, Pirany N, Ranganath RM, Prasad DT (2007) Identification of molecular markers associated with mite resistance in coconut (*Cocos nucifera* L.). Genome 50(1):35–42

Sharma P, Kothari SL, Rathore M, Gour V (2018) Properties, variations, roles, and potential applications of epicuticular wax: a review. Turk J Bot 42(2):135–149

Shelomi M, Lin SS, Liu LY. (2019) Transcriptome and microbiome of coconut rhinoceros beetle (*Oryctes rhinoceros*) larvae. BMC Genomics 20. https://doi.org/10.1186/s12864-019-6352-3

Singab AN, El-Taher EMM, Elgindi MR, Kassem MES (2015) *Phoenix roebelenii* O'Brien DNA profiling, bioactive constituents, antioxidant and hepatoprotective activities. Asian Pacific J Trop Dis 5(7):552–558

Singh R, Low ETL, Ooi LCL, Ong-Abdullah M, Ting NC, Nagappan J, Nookiah R, Amiruddin MD, Rosli R, Manaf MAA, Chan KL (2013a) The oil palm *SHELL* gene controls oil yield and encodes a homologue of SEEDSTICK. Nature 500(7462):340–344

Singh R, Ong-Abdullah M, Low ETL, Manaf MAA, Rosli R, Nookiah R, Ooi LCL, Ooi SE, Chan KL, Halim MA, Azizi N (2013b) Oil palm genome sequence reveals divergence of interfertile species in old and new worlds. Nature 500(7462):335–339

Singh R, Low ETL, Ooi LCL, Ong-Abdullah M, Nookiah R, Ting NC, Marjuni M, Chan PL, Ithnin M, Manaf MAA, Nagappan J (2014) The oil palm *VIRESCENS* gene controls fruit colour and encodes a R2R3-MYB. Nature Commun 5:4106

Smith SA, Donoghue MJ (2008) Rates of molecular evolution are linked to life history in flowering plants. Sciences 322(5898):86–89

Somyong S, Poopear S, Jomchai N, Uthaipaisanwong P, Ruang-Areerate P, Sangsrakru D, Sonthirod C, Ukoskit K, Tragoonrung S, Tangphatsornruang S (2015) The AKR gene family and modifying sex ratios in palms through abiotic stress responsiveness. Funct Integr Genomics 15(3):349–362

Somyong S, Poopear S, Sunner SK, Wanlayaporn K, Jomchai N, Yoocha T, Ukoskit K, Tangphatsornruang S, Tragoonrung S (2016) ACC oxidase and miRNA 159a, and their involvement in fresh fruit bunch yield (FFB) via sex ratio determination in oil palm. Mol Genet Genomics 291(3):1243–1257

Somyong S, Walayaporn K, Jomchai N, Hassan SH, Yodyingyong T, Phumichai C, Limsrivilai A, Saklang A, Suvanalert S, Sonthirod C, Anggradita LD (2019) Identifying a DELLA gene as a height controlling gene in oil palm. Chiang Mai J Sci 46 (1):32–45

Suresh K, Nagamani C, Ramachandrudu K, Mathur RK (2010) Gas-exchange characteristics, leaf water potential and chlorophyll a fluorescence in oil palm (*Elaeis guineensis* Jacq.) seedlings under water stress and recovery. Photosynthetica 48(3):430–436

Tan YA, Low KW, Lee CK, Low KS (2010) Imaging technique for quantification of oil palm fruit ripeness and oil content. Eur J Lipid Sci Tech 112(8):838–843

Tan YC, Yeoh KA, Wong MY, Ho CL (2013) Expression profiles of putative defence-related proteins in oil palm (*Elaeis guineensis*) colonized by *Ganoderma boninense*. J Plant Physiol 170:1455–1460

Tautenhahn R, Patti GJ, Rinehart D, Siuzdak G (2012) XCMS Online: a web-based platform to process untargeted metabolomic data. Anal Chem 84 (11):5035–5039

Tee SS, Tan YC, Abdullah F, Ong-Abdullah M, Ho CL (2013) Transcriptome of oil palm (*Elaeis guineensis* Jacq.) roots treated with *Ganoderma boninense*. Tree Genetics & Genomes 9(2):377–386

Teh HF, Neoh BK, Hong MPL, Low JYS, Ng TLM, Ithnin N, Thang YM, Mohamed M, Chew FT, Yusof HM, Kulaveerasingam H (2013) Differential metabolite profiles during fruit development in high-yielding oil palm mesocarp. PLoS ONE 8(4):61344

Teh HF, Neoh BK, Wong YC, Kwong QB, Ooi TEK, Ng TLM, Tiong SH, Low JYS, Danial AD, Ersad MA, Kulaveerasingam H (2014) Hormones, polyamines, and cell wall metabolism during oil palm fruit mesocarp development and ripening. J Agr Food Chem 62(32):8143–8152

Teh CK, Ong AL, Kwong QB, Apparow S, Chew FT, Mayes S, Mohamed M, Appleton D, Kulaveerasingam H (2016) Genome-wide association study identifies three key loci for high mesocarp oil content in perennial crop oil palm. Sci Rep 6:19075

Theerawitaya C, Samphumphaung T, Cha-um S, Yamada N, Takabe T (2014) Responses of Nipa palm (*Nypa fruticans*) seedlings, a mangrove species, to salt stress in pot culture. Flora 209(10):597–603

Thomas T, Kumar SN, Cherian VK, Kasturi Bai KV, Rajagopal V (2006) Role of certain biochemical compounds in adaptation of coconut to different weather conditions—a study in two agro-climatic regions of India. Indian J Hortic 63(1):1–7

Ting NC, Jansen J, Mayes S, Massawe F, Sambanthamurthi R, Ooi LCL, Chin CW, Arulandoo X, Seng TY, Alwee SSRS, Ithnin M (2014) High density SNP and SSR-based genetic maps of two independent oil palm hybrids. BMC Genom 15(1):309

Tranbarger TJ, Dussert S, Joët T, Argout X, Summo M, Champion A, Cros D, Omore A, Nouy B, Morcillo F (2011) Regulatory mechanisms underlying oil palm fruit mesocarp maturation, ripening, and functional specialization in lipid and carotenoid metabolism. Plant Physiol 156(2):564–584

Uthaipaisanwong P, Chanprasert J, Shearman JR, Sangsrakru D, Yoocha T, Jomchai N, Jantasuriyarat C, Tragoonrung S, Tangphatsornruang S (2012) Characterization of the chloroplast genome sequence of oil palm (*Elaeis guineensis* Jacq.). Gene 500(2):172–180

Uthaipaisanwong P, Somyong S, Tangphatsornruang S, Yoocha T, Jantasuriyarat C (2017) Development and characterization of simple sequence repeats derived from mitochondrial genome of oil palm using next generation sequencing. Thai J Sci Tech 6(3):288–300

Verma SK, Jasrotia RS, Iquebal MA, Jaiswal S, Angadi UB, Rai A, Kumar D (2017) Deciphering genes associated with root wilt disease of coconut and development of its transcriptomic database (CnTDB). Physiol Mol Plant P 100:255–263

Wilson MA, Gaut B, Clegg MT (1990) Chloroplast DNA evolves slowly in the palm family (Arecaceae). Mol Biol Eevol 7(4):303–314

Xiao W, Mason AS, Xiao Y, Liu Z, Yang Y, Lei X, Wu X, Ma Z, Peng M (2014a) Analysis of multiple transcriptomes of the African oil palm (*Elaeis guineensis*) to identify reference genes for RT-qPCR. J Biotechnol 184:63–73

Xiao Y, Zhou L, Xia W, Mason AS, Yang Y, Ma Z, Peng M (2014b) Exploiting transcriptome data for the development and characterization of gene-based SSR markers related to cold tolerance in oil palm (*Elaeis guineensis*). BMC Plant Biol 14:384

Xiao Y, Xia W, Mason AS, Cao Z, Fan H, Zhang B, Zhang J, Ma Z, Peng M, Huang D (2019) Genetic control of fatty acid composition in coconut (*Cocos nucifera*), African oil palm (*Elaeis guineensis*), and date palm (*Phoenix dactylifera*). Planta 249(2):333–350

Yaish MW (2015) Proline accumulation is a general response to abiotic stress in the date palm tree (*Phoenix dactylifera* L.). Genet Mol Res 14 (3):9943–9950

Yaish MW, Kumar PP (2015) Salt tolerance research in date palm tree (*Phoenix dactylifera* L.), past, present, and future perspectives. Front Plant Sci 6:348

Yamada N, Cha-Um S, Kageyama H, Promden W, Tanaka Y, Kirdmanee C, Takabe T (2011) Isolation and characterization of proline/betaine transporter gene from oil palm. Tree Physiol 31(4):462–468

Yang H, Gan SM, Yin GT, Xu HC (2005) Identification of random amplified polymorphic DNA markers linked to sex determination in *Calamus simplicifolius* C. F. Wei. J Integr Plant Biol 47:1249–1253

Yang M, Zhang X, Liu G, Yin Y, Chen K, Yun Q, Zhao D, Al-Mssallem IS, Yu J (2010) The complete chloroplast genome sequence of date palm (*Phoenix dactylifera* L.). PLoS ONE 5(9):12762. https://doi.org/10.1371/journal.pone.0012762

Yeap WC, Ooi TEK, Namasivayam P, Kulaveerasingam H, Ho CL (2012) *EgRBP42* encoding an hnRNP-like RNA-binding protein from *Elaeis guineensis* Jacq. is responsive to abiotic stresses. Plant cell Rep 31(10): 1829–1843

Yin Y, Zhang X, Fang Y, Pan L, Sun G, Xin C, Abdullah MMB, Yu X, Hu S, Al-Mssallem IS, Yu J (2012) High-throughput sequencing-based gene profiling on multi-staged fruit development of date palm (*Phoenix dactylifera* L.). Plant Mol Biol 78(6): 617–626

Zhao H, Sun H, Li L, Lou Y, Li R, Qi L, Gao Z (2017) Transcriptome-based investigation of cirrus development and identifying microsatellite markers in rattan (*Daemonorops jenkinsiana*). Sci Rep 7:46107

Zuniga LC (1953) The probable inheritance of the makapuno character of the coconut. Philipp Agr 36:402–413

Genome Sequencing, Transcriptomics, Proteomics and Metabolomics

8

M. K. Rajesh, S. V. Ramesh, Anitha Karun, and P. Chowdappa

Abstract

The availability of enormous amount of sequence data, driven by next generation sequencing (NGS) techniques and multi-omics-integrated strategies, has contributed considerably to expanding and intensifying our understanding on the molecular make-up of plants. The use of 'omics'-based technologies for molecular breeding in palms, especially coconut, has, in general, lagged behind in comparison with other monocots such as cereals. However, remarkable progress has been made in the application of 'omics' tool box in coconut palm in the last six years. The availability of three high-quality genome assemblies provides immense opportunities for crop improvement in this perennial species to accelerate the strategies of marker-assisted selection and development of genomic selection (GS) models. In addition, the genome blueprint can provide novel insights into the molecular basis underpinning key agronomic traits, which are mainly quantitative in nature. To comprehend the application of basic omics in coconut vis-à-vis traits of agronomic importance, somatic embryogenesis, as well as responses to varied biotic and abiotic stresses, it is imperative to delve into the genome structure, functional integrity and collinearity of coconut with oil palm and date palm. This chapter covers the present status of achievements in the field of coconut omics via genome sequencing initiatives in coconut palm, gene expressing and small RNA profiling, metabolomics and proteomics. These multi-omics approaches can definitely provide a much needed fillip not only to decipher the evolution of the complex coconut genome but also to the extant crop improvement programs in coconut.

M. K. Rajesh (✉) · S. V. Ramesh · A. Karun · P. Chowdappa
ICAR-Central Plantation Crops Research Institute, Kasaragod, Kerala, India
e-mail: rajesh.mk@icar.gov.in

S. V. Ramesh
e-mail: ramesh.sv@icar.gov.in

A. Karun
e-mail: anitha.karun@icar.gov.in

P. Chowdappa
e-mail: pallem22@gmail.com

8.1 Introduction

'Omics' studies have the capacity to unravel the gamut of evolutionary history of an organism. The coconut palm is one of the most economically useful trees of the tropics, which has been in cultivation from time immemorial. A diploid species ($2n = 2x = 36$), the mean genome size has been estimated to be 5.57 pg/2C, with 1C = 2723.73 Mb (Freitas Neto et al. 2016), which is relatively large than the evolutionarily

M. K. Rajesh et al. (eds.), *The Coconut Genome*, Compendium of Plant Genomes,
https://doi.org/10.1007/978-3-030-76649-8_8

related palms, viz. date palm (*Phoenix dactylifera*; 671 Mb; Al-Mssallem et al. 2013) and oil palm (*Elaeis guineensis* and *E. oleifera*; 1.8 Gb; Singh et al. 2013). In spite of its global importance, multifarious uses and emerging potential as a nutraceutical in the recent times, genomics and transcriptomics resources in coconut were very scarce till 2014.

The genome sequencing initiatives from China, the Philippines and India have provided comprehensive information of the genes underlying important traits. A remarkable observation emerging from these studies has been that more than 70% of coconut genome is composed of repetitive elements. Concurrently, high-throughput RNA-Sequencing (RNA-Seq) assays have resulted in identification of novel transcripts and regulatory genetic components and the quantification of gene expression in coconut. A high-density genetic linkage map has been generated recently using genotyping-by-sequencing (GBS) approach. These achievements will enable fine mapping of major loci, genome-wide association studies (GWAS), development of genomic selection (GS) models and provide the much needed acceleration to breeding efforts in this perennial palm. Complementing the data generated from the transcriptomics, proteomics and metabolomics studies in coconut, the genome data will immensely aid the coconut researchers in linking genotype to phenotype. In addition, these studies would greatly help in comprehending the complex molecular regulatory mechanisms underlying key traits, viz. copra and oil yield, oil quality and tolerance to adverse environments, pests and diseases, and provide attractive opportunities in coconut breeding programs, mainly reduction in the length and number of breeding cycles to develop a new variety/hybrid.

8.2 Genomics Studies in Coconut

Whole genome sequencing of coconut has been undertaken through a few independent efforts. While Hainan Tall cultivar was used by researchers from China, the sequencing initiatives from the Philippines and India relied on indigenous Green Dwarfs of the respective countries.

8.2.1 Sequencing of Hainan Tall

The first draft of the whole genome sequence (WGS) of *Cocos nucifera* L. was released by a consortium, led by the Chinese Academy of Tropical Agricultural Sciences, who sequenced the Hainan Tall variety (Xiao et al. 2017). The Hainan Tall, which occupies the major area under coconut cultivation in Hainan province, China, is endowed with tolerance to drought, salinity and cold stressors. Using a combination of paired-end (PE) and mate-pair (MP) libraries on an Illumina HiSeq 2000 sequencer, 419.67 Gb raw reads were generated to achieve a 173.32× read depth coverage of the predicted genome of 2.42 Gb. The final genome assembly of scaffolds provided a total length of 2.20 Gb (N50 = 418 Kb), which corresponds to 90.91% of the estimated genome size. The genome was predicted to encode a total of 28,039 protein-coding genes, with benchmarking universal single-copy orthologs (BUSCO) assessment indicating 74.1% genome completeness. Around 72.75% of the genome comprises repeat regions and transposable elements (TEs), with around 92% of TEs represented by long-terminal repeat retrotransposons elements (LTRs). The expansion of gene families which impart salt tolerance, viz. Na^+/H^+ antiporters and ion channels, was noteworthy, suggesting a role for these genes in enabling adaptation of Hainan Tall to saline environments. Moreover, evolutionary analysis revealed the divergence time of coconut from *Elaeis guineensis* (the African oil palm), the closest relative whose genome has been sequenced, to be approximately 46 mya (Xiao et al. 2017).

8.2.2 Sequencing of Catigan Green Dwarf

The assembly and annotation of WGS of the 'Catigan Green Dwarf' cultivar (CATD) were

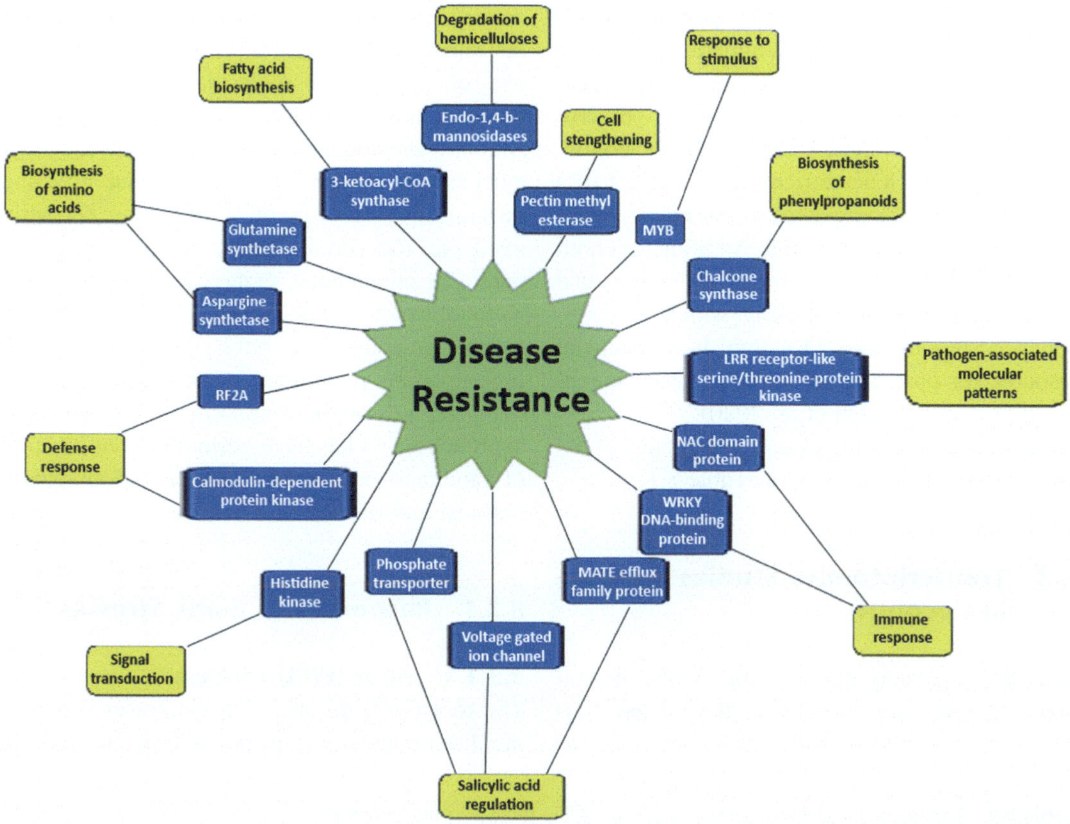

Fig. 8.1 A model of the defence response in coconut to root (wilt) disease. Figure adapted from Rajesh et al. (2018)

carried out at the University of Philippines utilizing a combination of Pacific Biosciences (PacBio), Illumina and Chicago sequencing (Lantican et al. 2019). The final genome assembly of 7998 scaffolds (N50 = 570.487 Kb) indicated a total length of 2.1 Gb (97.6% of 2.15 Gb estimated genome size). The authors could find 58,503 SNP variants by comparing the CATD genome with the previously sequenced Hainan Tall genome. A total of 34,958 gene models could be predicted with high confidence. In addition to 7139 unique genomic SSRs, genic SSRs were mined from genes with putative roles in drought response, biotic stress resistance and oil biosynthesis; these genetic markers possess immense potential for use in molecular breeding approaches for varietal improvement in coconut. Interestingly, based on comparison of synteny of exons and variations in the genome sequences of members of Arecaceae, it was inferred that palms might have undergone at least three rounds of whole genomic duplications (WGD) during the course of evolution (Lantican et al. 2019).

8.2.3 Sequencing of Chowghat Green Dwarf

Recently, Rajesh et al. (2020), in a collaborative effort led by ICAR-Central Plantation Crops Research Institute (ICAR-CPCRI), India, have undertaken the assembly and annotation of the nuclear and organellar genomes of Chowghat Green Dwarf cultivar, which possessed enhanced tolerance to the root (wilt) disease. A hybrid assembly, comprising short and long reads Illumina HiSeq 4000 and Pacific Biosciences RSII platforms, respectively, resulted in a draft genome of 1.93 Gb, distributed over 26,855 scaffolds. After application of stringent filtering, a total of 13,707 genes were predicted, which encode 11,181 proteins. Validation of these predicted genes was undertaken through transcriptome sequences available from diverse coconut tissues. Hundred and twelve nucleotide-binding-site and leucine-rich repeat (NBS-LRR) loci, belonging to six classes of resistance genes, were mined from the genome. Additionally, both the organellar genomes of CGD were also described. The genome of CGD, which possesses enhanced disease resistance trait, is a beneficial resource for hastening genomics-assisted breeding in coconut (Rajesh et al. 2020).

A comparison of the assembly statistics of the three genomes is presented in Table 8.1.

8.3 Transcriptomics Studies in Coconut

Over the past many decades, traditional coconut breeding programs worldwide have been concentrating on selection for attributes such as copra yield, quality of tender nut water, disease resistance and drought tolerance. Nevertheless, use of appropriate molecular techniques can enable a much better comprehension of organization of the coconut genome and, subsequently, enhance the effectiveness of coconut breeding programs. Transcriptomics data are currently available for certain genes entailed in plant–pathogen interactions, embryogenesis, fatty acid metabolism, aroma and DNA methylation. In the following sections, we have concentrated on the investigations pertaining to gene expression utilizing high-throughput transcriptome sequencing approaches, which have permitted identification of candidate genes that influence various physiological and biochemical processes.

8.3.1 Responses to Biotic Stresses

8.3.1.1 Root (Wilt) Disease

The root (wilt) disease is the foremost debilitating disease and a serious production constraint in

Table 8.1 A comparison of the assembly statistics of the three coconut genomes

Parameters	Hainan Tall (Xiao et al. 2017)	Catigan Green Dwarf (Lantican et al. 2019)	Chowghat Green Dwarf (Rajesh et al. 2020)
Total sequences	1,11,366	7998	26,885
Total bases	2,20,24,55,121	2,10,24,17,611	1,93,00,87,115
Min sequence length	499	1912	102
Max sequence length	11,99,63,539	87,79,653	12,61,010
Average sequence length	19,776.73	2,62,867.92	71,790.48
Median sequence length	1139	1,20,849	41,589
N25 length	6,24,58,273	15,74,781	2,40,194
N50 length	12,17,559	5,70,487	1,28,735
N75 length	1,56,844	2,29,523	64,118
N90 length	32,679	1,11,544	32,613
N95 length	3169	73,798	21,678
As	30.57%	31.06%	30.70%
Ts	30.56%	31.04%	30.69%
Gs	18.18%	18.81%	18.50%
Cs	18.18%	18.81%	18.49%
(A + T)s	61.13%	62.09%	61.40%
(G + C)s	36.36%	37.62%	36.98%
Ns	2.51%	0.28%	1.62%

India, especially in the states of Kerala and Tamil Nadu. Most of the coconut varieties or hybrids screened are susceptible to the root (wilt) disease; however, CGD has been reported to possess enhanced field tolerance to this disease (Nair et al. 2004). To gain molecular insights into the resistance mechanisms, a study of global gene expression pattern of diseases and healthy CGD palms was undertaken by comparative RNA-Seq on an Illumina platform by Rajesh et al. (2015, 2018). Differential gene expression (DGE) analyses revealed a total of 2718 transcripts to be differentially expressed between the healthy and diseased palms; also, 136 transcripts are specifically expressed in diseased palms, whereas 454 transcripts are unique to healthy palms. The molecular gene regulatory modulations, that define the healthy and diseased palms, indicated significant enrichment of gene ontology (GO) and Kyoto Encyclopaedia of Genes and Genomes (KEGG) pathways involved in binding of metal ions, fixation of carbon, secondary metabolites biosynthesis, etc. The upregulation of certain proteins, which are typical of plant–pathogen interactions, like calmodulin-like 41 and WRKY DNA binding proteins, is noteworthy. The differential abundances of key genes were corroborated by quantitative real-time PCR (qRT-PCR), validating the RNA-seq technique for quantitative analysis of the coconut transcriptome.

Based on the results obtained from this study, a model for coconut-pathogen interaction was put forth (Fig. 8.1). The primary trigger of the coconut defence responses is affected by the interaction between the host membrane receptors and pathogen-associated molecular patterns (PAMPs) like effectors. Receptor and PAMP interaction commences the signal transduction cascades. Protein kinases are key players in these regulatory cascades; they influence Ca^{2+} influxes and, in turn, calcium signalling. Salicylic acid is also enlisted, which ensues activation of genes involved in defence responses (e.g. chalcone synthase) in healthy palms with the assistance of different classes of transcription factors (TFs) containing WRKY and NAC domains (Rajesh et al. 2018) (Fig. 8.1).

8.3.1.2 Yellow Decline Disease

Similar to the root (wilt) disease in India, the coconut yellow decline (CYD) is a debilitating phytoplasmal disease in Malaysia, affecting coconut and ornamental palms. A screening study of Malaysian coconut germplasm by Nejat et al. (2009) indicated the high tolerance of Malayan Red Dwarf (MRD) to CYD disease in comparison to Malayan Talls and Yellow Dwarfs. A comparative transcriptome analyses of leaves of healthy and phytoplasma-infected MRD palms by Nejat et al. (2015) generated 72,019,264 reads from healthy leaves and 70,935,896 reads from diseased leaves, which were then assembled into 108,994 and 148,264 contigs, respectively. Totally, 39, 873 transcripts were differentially expressed; while 18,013 transcripts were upregulated, 21,860 transcripts were down-regulated in the leaves of diseased palms when compared to the leaves of healthy palms. Typically, genes involved in defence responses against pathogens, especially defence signalling, were upregulated in diseased palms. The upregulated transcripts include those coding for pathogenesis-related proteins (PRs) [viz. PR-5 (thaumatin-like proteins; TLPs), PR-9 (peroxidases), PR-2 (endo-1,3-beta-glucosidases), PR-3 (chitinases) and PR-10 (Bet v 1-type proteins)], 1-aminocyclopropane-1-carboxylate oxidase (ACO), polyphenol oxidases, amino acid permeases, ABC transporters, cytochrome P450, flavonol synthases, WRKY transcription factors, etc. On the other hand, transcripts which were down-regulated mainly include those involved in cytokinin biosynthesis and photosynthesis. The study, therefore, provided direct evidences for the activation of defence mechanisms in the wake of phytoplasma attack in the MRD cultivar (Nejat et al. 2015).

8.3.2 Fatty Acid Biosynthesis and Metabolism

Transcriptome analysis of juvenile leaves and endosperm of coconut cultivar Hainan Tall yielded a total of 57,304 unigenes (Fan et al. 2013). Mapping of a sizable proportion of unigenes

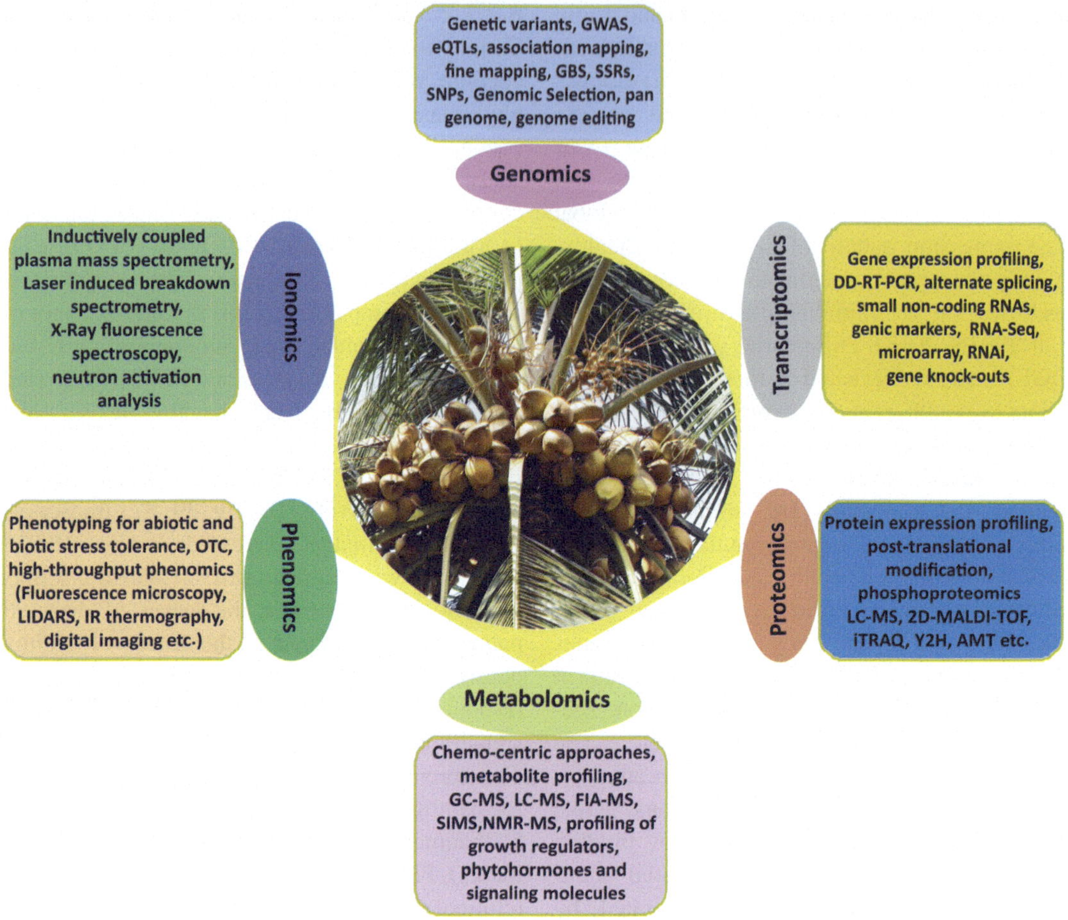

Fig. 8.2 Framework components of multi-omics approaches with an immense potential to usher in an era of coconut omics-driven advances (*GWAS* Genome-wide association study; *eQTLs* Expression quantitative trait loci; *GBS* Genotyping-by-sequencing; *SSR* Simple sequence repeats; *SNPs* Single nucleotide polymorphisms; *DD-RT-PCR* Differential-display reverse transcription-PCR; *RNAi* RNA interference; *LC-MS* Liquid chromatography with mass spectrometry; *2D-GE* Two-dimensional gel electrophoresis; *MALDI-TOF* Matrix-assisted laser desorption/ionization-time of flight; *iTRAQ* Isobaric tag for relative and absolute quantitation; *Y2H* Yeast two-hybrid; *AMT* Accurate mass and time tag; *FIA-MS* Flow injection analysis tandem mass spectrometry; *SIMS* Secondary ion mass spectrometry; *NMR-MS* Nuclear magnetic resonance–Mass spectrometry; *OTC* Open-top-chamber; *LIDAR* Light detection and ranging; *ICP-MS* Inductively coupled plasma mass spectrometry; *LIBS* Laser-induced breakdown spectrometry; *XRF* X-ray fluorescence spectroscopy; *NAA* Neutron activation analysis)

(23,168 unigenes) to KEGG pathways identified 215 pathways which are involved in diverse aspects such as galactose metabolism, plant–pathogen interaction and signal transduction mediated by phytohormone. In addition, 347 unigenes representing fatty acid biosynthesis and lipid metabolism were also documented. These unigenes represented various steps involved in lipid biosynthesis namely fatty acid biosynthesis, biosynthesis of unsaturated fatty acid, elongation of fatty acids, other lipid metabolism and citrate cycle. Of particular interest among these unigenes are those (20 in number) involved in the production of fatty acyl-ACP thioesterase, which is a crucial enzyme involved in the termination of carbon chain elongation and hence determines the chain length of fatty acids. Thus, it was proposed that upregulation of fatty acyl-ACP thioesterase is a key factor in the development of coconut endosperm which causes relatively high accumulation of medium chain fatty acids (e.g. lauric acid). Thus, this transcriptome study has greatly assisted in understanding the molecular features that underlie the fatty acid biosynthesis in coconut.

8.3.3 Somatic and Zygotic Embryogenesis

To obtain molecular insights into the phenomenon of in vitro recalcitrance in coconut, Rajesh et al. (2016) analysed the global transcriptome of coconut plumular tissue-derived embryogenic calli. From the assembled reads, a total of 14 genes possessing key roles in somatic embryogenesis (SE) were identified. These genes could be categorized into transcription factors, extracellular proteins, receptor-like kinases, auxin responsive genes, metabolic enzymes and kinases (Table 8.2). Experimental validation of the expression pattern of these 14 genes, as deduced from the sequencing data, was undertaken in different stages of in vitro development via qRT-PCR. The stage-specific differential expression pattern of these 14 genes, as revealed by the qRT-PCR analysis, provided insights into their probable functions in vitro regeneration pathway in coconut. The potential of use of these genes as markers of coconut SE was also highlighted in the study.

The data derived through transcriptome analyses of different tissues, viz. non-photosynthetic embryos (immature, mature and microspore-derived embryo) and photosynthetic mature leaves, were assembled into 32621 unigenes, and were utilized by Bandupriya et al. (2016) to catalogue genes involved in coconut embryogenesis. This study has led to the identification of transcripts having putative roles in embryogenesis encoding various proteins such as chitinase, β-1,3-glucanase, ATP synthase CF0 subunit, thaumatin-like protein and metallothionein-like protein.

8.3.4 RNA-Directed DNA Methylation

Epigenetic modifications play key roles during development of embryo, endosperm and seed (Köhler and Makarevich 2006). In addition to various small RNAs, such epigenetic modifications often include the involvement of RNA-directed DNA methylation (RdDM) machinery (Matzke and Mosher 2014). Transcriptome profiling of coconut leaves and seeds (embryo and endosperm), with an objective of identifying genetic factors involved in RdDM, revealed that the transcripts of enzymes methyl transferase (MET1) and chromomethylase-3 (CMT) were differentially expressed in endosperm compared to embryo or leaves (Huang et al. 2014). Further, gene expression analysis during various stages of endosperm development revealed that four genes, viz. *DRM*, *NRPD1*, *NRPE1* and *MET1*, showed increased expression, which eventually was diminished when the endosperm reached a solid thick layer. This study, therefore, provided insights into the molecular interplay of RdDM factors and small RNAs in the developmental stages of coconut endosperm (Huang et al. 2014).

8.3.5 Aroma

The Aromatic Green Dwarf is a unique accession, native to Thailand. The liquid endosperm of this accession has characteristic 'pandan-like' aroma, and hence, transcriptome sequencing and analysis were conducted to decipher the molecular basis for this phenomenon (Saensuk et al. 2016). The biochemical compound 2-acetyl-1-pyrroline (2AP) is responsible for imparting the characteristic aroma not only in coconut but also in aromatic rice among others. Transcriptome assembly and analysis of liquid endosperm of Aromatic Green Dwarf characterized the rice orthologous gene(s) that are involved in the biosynthesis of 2AP. Comparative transcriptome analysis of aromatic and non-aromatic Green Dwarf genotypes revealed the differences in the length of transcripts encoding 2AP (Saensuk et al. 2016).

Table 8.2 Catalogue of 14 transcripts related to somatic embryogenesis obtained from RNA-Seq data of coconut embryogenic calli

Sl. No.	Transcript id	Transcript length (bp)	E-value	Gene	Source
A. Transcription factors					
1.	530086	500	1E-22	AP2/ERF domain-containing transcription factor (*AP2*)	*Populus trichocarpa*
2.	439278	231	5E-15	WUSCHEL (*WUS*)	*Medicago truncatula*
3.	581560	4694	0	PICKLE (*PKL*)	*Vitis vinifera*
4.	553982	765	3.00E-25	AINTEGUMENTA (*ANT*)	*Populus canescens*
5.	560808	921	0	WRKY (*WRKY*)	*Phoenix dactylifera*
B. Extra-cellular proteins					
6.	556118	804	6E-25	Embryogenic cell protein (*ECP*)	*Daucus carota*
7.	555772	797	0	Germin-like protein (*GLP*)	*Phoenix dactylifera*
8.	562916	986	2.00E-63	Arabinogalactan protein (*AGP*)	*Vitis vinifera*
9.	533908	527	3.00E-179	Late embryogenesis-abundant protein (*LEA*)	*Phoenix dactylifera*
C. Receptor-like kinases					
10.	578404	2255	0	Somatic embryogenesis receptor kinase (*SERK*)	*Oryza sativa*
11.	577164	2021	0	CLAVATA1 (*CLV*)	*Phoenix dactylifera*
D. Auxin responsive genes					
12.	539062	570	0	Small auxin upregulated RNA (SAUR)	*Phoenix dactylifera*
E. Metabolic enzymes					
13.	567344	1170	0	Glutathione S-transferase (*GST*)	*Phoenix dactylifera*
F. Kinases					
14.	564618	1047	0	Mitogen-activated protein kinase (*MAPK*)	*Phoenix dactylifera*

Source Rajesh et al. (2016)

8.3.6 Water-Deficit Stress

In order to study the molecular response of coconut to water-deficit stress, two-year-old coconut seedlings of the varieties Kalpasree and Kalpatharu subjected to soil water-deficit regimes (25% of available of soil moisture and control) were analysed by performing whole transcriptome sequencing. Biochemical, physiological and growth parameters while the seedlings were under water-deficit stress showed differential enzymatic antioxidants, lipid peroxidation status and water use efficiency trait between the coconut genotypes (Ramesh et al. 2019). The whole plant water use efficiency at control condition was significantly low in Kalpatharu (4.06) compared to Kalpasree (4.74). Nevertheless, under severe stress (25% ASM) Kalpatharu exhibited highest WUE (5.68) as against dwarf variety Kalpasree (3.84). The leaf transcriptome profiles of the control and water-deficit stressed seedlings were examined utilizing paired-end RNA-Seq. In total, ~7300 differentially expressed genes have been identified between the seedlings under water-deficit stress and control. Analysis of control and stressed Kalpasree leaf transcriptome showed significant upregulation of PHLOEM PROTEIN 2-LIKE A1-like, WRKY transcription factor 40 isoform X1 and down-regulation of glycerol-3-phosphate acyltransferase 3 transcripts. On the other hand, upregulation of transcripts encoding polyamine oxidase, arabinose 5-phosphate isomerase among others and down-regulation of aquaporin PIP1-2 transcript was documented in Kalpatharu leaves. Besides, long non-coding RNA and genic SSRs were also identified from the transcriptome data to further enrich the genomic resources of coconut palm which would pave way for its utilization in developing climate-smart coconut crop (Ramesh et al. 2020).

8.3.7 Small RNA Transcriptomics

Endogenous small RNAs are class of genetic elements with profound influence on the growth and development of an organism. Among them microRNAs (miRNAs), characterized by 20–24nt length non-coding RNAs have been recognized with gene regulatory roles. Initial exploration of coconut-derived miRNAs utilized miRNA microarray to profile the expression status of miRNAs during endosperm development (Li et al. 2009).

Comparative analysis of miRNA expression profile in the mature and immature endosperm of coconut suggests that 23 miRNAs are upregulated whereas nine miRNAs are down-regulated specific to mature endosperm (Li et al. 2009). Furthermore, the role of expression of miR397, which targets plant laccases thereby ensuring the maintenance of endosperm cells in meristematic state, could be deciphered (Li et al. 2009). Advancements and the availability of coconut transcriptome sequences further spurred the search for novel miRNAs in coconut. Naganeeswaran et al. (2015) utilized the coconut leaf transcriptome profile and identified 16 miRNAs belonging to 11 miRNA families. Interestingly, most of the target genes predicted for the coconut miRNAs are transcription factors (TFs) suggesting the centrality of miRNAs in endogenous gene regulation. Sabana et al. (2018) explored the coconut miRNAs from the transcriptome data of embryogenic calli yielding a total of 27 novel mature miRNA sequences belonging to 15 diverse miRNA families. In addition, the conserved miRNAs identified have been shown to potentially play significant role in the processes of phytohormone-mediated signalling and somatic embryogenesis pathways suggesting the miRNA-mediated gene regulatory nuances in overcoming in vitro recalcitrance in coconut (Sabana et al. 2018). Building on their earlier findings, Sabana et al. (2020) profiled small RNAs in embryogenic and non-embryogenic calli and identified 110 conserved miRNAs and 48 miRNAs specific for embryogenic calli and 21 miRNAs confined to non-embryogenic calli. It has uncovered the role of conserved and species-specific miRNAs in controlling the process of somatic embryogenesis (SE) in coconut and provides critical resource for manipulating the SE turnover in the crop. The finding that miRNAs in the diet could be absorbed by the human digestive system and potentially affects the human gene expression led to the study of extracellular miRNAs in the coconut water (Zhao et al., 2018). This study revealed the presence of relatively high number of miRNAs in mature water than in immature coconut water implying the functional diversity of mature water. Notably the concentration of miR528-5p, rose by 112 times suggesting the vigorous metabolism of germinating seeds (Zhao et al. 2018). Also, this study warrants further investigation to analyse the role of cross-kingdom or cross-species miRNA interaction.

8.4 Proteomics Studies in Coconut

Proteomics studies allow for high-throughput analysis of protein components in a genome-wide scale. Further, proteomic analysis generates wealth of information regarding the biological systems by not only providing the protein expression patterns or concentrations but also helps in deciphering the interacting protein partners in the cellular milieu. Application of proteomics techniques is indispensable now more than ever in coconut. Since last decade or so, an apparent escalation in the use multiple-omics technologies, including 2D-gel electrophoresis and iTRAQ-based protein profiling techniques, to decipher the molecular responses of coconut to various stresses, has been witnessed. The following section describes the salient achievements made in the field of proteomics in coconut.

8.4.1 Nutritional Quality

Coconut proteome was explored to comprehend its nutritional quality and identifying suitable candidate antigens for immunological studies utilizing two-dimensional electrophoresis (2-DE) technique coupled with MALDI-TOF/TOF-MS analysis (Huang et al. 2016). 2-DE of proteins obtained from mature endosperm yielded approximately 200 spots. Based on their molecular mass, the proteins were categorized into seven Taxa (a to g) and MALDI-TOF/TOF-MS characterization further identified that proteins of Taxa a, b and g belonged to 7S globulins, whereas Taxa d and e were found as glutelin and receptor-like protein kinase, respectively (Huang et al. 2016). Investigations on thermal stability of coconut proteins revealed that heating (100 °C) for 40 min caused a slight decrease in the stability. However, autoclaving the protein extracts resulted in considerable changes in the protein profile due to protein degradation (Huang et al. 2016). This study not only provided the information regarding thermal stability of protein fractions but also suggested that taxa a and e could be applied as a potential food ingredients because of their thermal stability (Huang et al. 2016).

8.4.2 Cold Stress

The iTRAQ (isobaric tags for relative and absolute quantitation) approach was resorted to analyse the proteomic response of two coconut genotypes Hainan Tall [BenDi (BD)] and Aromatic coconut [XiangShui (XS)] when subjected to cold stress (8 °C for 2 and 5 days) (Yang et al. 2020). Cold stress treatment (8 °C) for 2 days caused perturbations of 193 upregulated and 134 down-regulated proteins in BD, whereas respective differentially expressed proteins in XS were 140 and 155 (Yang et al. 2020). Similarly, cold stress for 5 days has also evinced differential response in the genotypes with abundance of upregulated proteins in the cultivar BD compared to XS (Yang et al. 2020). Classification of proteins up- and down-regulated during cold stress in both the cultivars identified that multiple stress responses such as abiotic, biotic and oxidative stresses are involved in response to cold stress. Proteome responses of both the cultivars have divulged that enhanced expression of proteins involved in metabolism and respiratory-related proteins (Yang et al. 2020). Thus, cold stress tolerance in BD (Hainan Tall) genotype could be attributed to the accumulation of stress-responsive proteins and by effective scavenging of ROS. It was also suggested that post-translational modifications, role of protein turnover and chaperones could play a major role in providing cold stress adaptation in coconut (Yang et al. 2020)

8.4.3 Somatic and Zygotic Embryogenesis

Proteins expressed in the different developmental stages of somatic and zygotic embryogenesis in coconut were identified by combining the one-dimensional SDS-PAGE gel electrophoresis with matrix-assisted laser desorption/ionization-time-of-flight mass spectrometry (MALDI-TOF/TOF-MS). The SDS-PAGE analysis showed that a number of protein bands resolved in the different stages studied were higher in somatic (27–29 bands from 15 to 100 kDa) than in zygotic (19–23 bands from 15 to 120 kDa). Seven major protein bands were observed common in somatic embryo (SE) and zygotic embryo (ZE), and four were identified as 6-phosphogluconate dehydrogenase (54.9 kDa), LIM-domain-containing protein (24.49 kDa), thioredoxin (15.6 kDa), glycine rich cell wall structural protein (12.78 kDa) and the other three as uncharacterized proteins. The unique proteins identified in ZE represent Zeta-carotene desaturase (64 kDa), Adenylate kinase (26.67) and AT5G11810-like protein (25.77). Zeta-carotene desaturase is the highly expressed protein, present throughout the zygotic stages and absent in somatic stages. The important proteins identified in the different stages of somatic embryogenesis were eukaryotic initiation factor 4A (eIF4A) and aldo-keto reductase. The former was present in the stages of somatic embryo, meristemoid and embryonic shoot meristem, whereas the latter was present throughout the stages of somatic embryogenesis. The accumulated evidence of eukaryotic initiation factor 4A during the somatic embryo development indicates that protein metabolism is a key factor in SE, especially during the late stages of embryo development (Lakshmi Jayaraj 2019).

8.5 Metabolomics Studies in Coconut

Large-scale plant metabolite profiling and development of appropriate metabolomic assays have helped the investigators in unravelling metabolites with biochemical potential and in understanding the associated metabolic pathways. Thus, metabolite profiling assists in the characterization of chemical footprints of a species under various stages of growth and development. In addition, metabolomics coupled with ionomics have contributed in identification of plant bioactives with nutraceutical potential. Integration of metabolomics approaches with other genome-wide analysis techniques such as transcriptomics, and quantitative genetics have a greater relevance for crop improvement programs than ever. Given this, the following section briefly highlights the major accomplishments of metabolomics in coconut and its relevance in the breeding efforts of coconut.

Nutritional and metabolomics characterization of coconut water at various developmental stages was carried out by Kumar et al. (2021). Considering the nutritional and commerce potential, the nutritional profile of coconut water of two varieties namely Chowghat Orange Dwarf (COD) and Malayan Yellow Dwarf (MYD) was determined at four different nut stages (Kumar et al. 2021) using gas chromatography–mass spectrometry (GC–MS) and ultra-performance liquid chromatography (UPLC)-based metabolomics approaches. In total, 17 amino acids were documented in both the varieties. No significant differences in the content of amino acids between the cultivars were observed during the various developmental stages. All the amino acids except alanine showed an increase in concentration from the maturity stage 1 (5–6 months after pollination-MAP) to maturity stage 4 (11–12 MAP). However, ascorbic acid content of the nut water decreased with an increase in maturity in both the varieties. Further GC-MS-based profiling of metabolites in the coconut water of varieties MYD and COD identified 18 and 19 metabolites, respectively. The metabolites included fatty acids, flavonoids, organic acids, phenolics, sugars and sugar alcohol. Besides, the metabolite biomarkers identified from the nut water of these varieties could discriminate the cultivars COD—which has two additional biomarkers namely caffeic—and myristic acids; MYD—fumaric acid and stearic acid were confined to it. Further, the metabolite profiling helped in identifying the alterations in biochemical pathways notable among them are increased fatty acids metabolism; biosynthesis of phenylpropanoids and phenolic acids; etc. (Kumar et al. 2021). Similarly, metabolic finger printing of matured coconut water collected from Wenchang City (Hainan, China), following harvest operation, was analysed (Chen et al. 2018). Non-targeted UPLC-MS/MS analysis of the nut water from stored coconut samples (for 0–5 months) identified 1405 precursor molecules in the positive ion mode and 1275 precursor molecules in the negative ion mode. Further, orthogonal partial least squares (OPLS) data analysis offered better discrimination of samples groups and divided them into two clusters (Cluster I: A, B, C, D corresponding to 0, 1, 2, 3 months of storage and Cluster II: E, F referring 4 and 5 months of storage). Nine metabolite biomarkers, viz. dibutylphthalate, L-leucine, (S)-malate, L-valine, deethylatrazine, L-threonine, (R)-pantothenic acid, taurine and malic acid, are significantly upregulated, even up to sixfolds change, in nut water stored for 0–3 months of storage. However, down-regulation of gamma-aminobutyric acid zwitter ion, acetoacetate and keto-D-fructose was documented. Considering the metabolic profiling of nut water during storage it is suggested that coconut water should not be stored beyond 4 months (Chen et al. 2018).

Hainan-native high-growing coconuts were analysed for the changes in the metabolites utilizing UPLC-MS/MS datasets following the nut harvest and storage under refrigerated conditions (4 °C) (Zhang et al. 2020). Additionally, differentially expressed metabolites and their fold changes are also investigated for hierarchical cluster analysis and association analysis (Zhang et al. 2020). Differential expression analysis of metabolites divided the samples into two categories: group 1 (0 week) and group 2 (1–6 weeks of storage) suggesting immense metabolic changes in the nut water composition even when stored under 4 °C. Further, among the differentially expressed metabolites 52 were found to be significantly upregulated, whereas 14 metabolites are down-regulated. Between these two categories of nut water, it was found that Yohimbic acid was the most influential upregulated metabolite and vinpocetine was the most influential down-regulated metabolite between groups 1 and 2 (Zhang et al. 2020). Though previous studies have correlated the changes in the content of aldehydes and lipids of tender nut water as reason for spoilage, this study has delineated that amino acid metabolism is predominantly affected (Zhang et al. 2020).

8.6 Future Perspectives

It is anticipated that the relatively sluggish rate of application of genomics-assisted breeding efforts in coconut would receive a greater fillip due to the availability of high-quality whole genome sequences derived from the diverse genotypes. Development of high-density genetic linkage map and assembling of large coconut genome sequences into 16 pseudomolecules by Yang et al. (2021) are one such improvement required for the coconut breeders' tool box. Though euchromatin region of coconut genome is well-represented, low recombination frequency in heterochromatin region restricted the quality and accurate assembly which warrants use of novel mapping technologies such as Hi-C and optical mapping approaches. Further, large-scale identification of QTLs linked to the phenotypic trait of importance is a mammoth task which has implications for improving the quantity and quality of oil considering the premium price coconut oil fetches in the international market owing to its perceived and proven nutraceutical potential. Also, identification of synteny blocks between the genomes of coconut and related genera (oil palm to be specific) would accelerate the application of marker-assisted breeding efforts in coconut.

More effective strategies are necessary for breeding coconut for key traits like enhanced yield and climate resilience. Implementation of genomic selection (GS) can result in a paradigm shift in coconut breeding by expediting the breeding process and enhancing the selection intensity in comparison with the conventional techniques. In addition, application of multi-omics approaches investigating functional elements such as transcripts, proteins, metabolites and ions of coconut provides a holistic picture to tackle the environment-induced stressors and to analyse the traits of quality importance (Fig. 8.2).

References

Al-Mssallem IS, Hu S, Zhang X, Lin Q, Liu W, Tan J, Yu X, Liu J, Pan L, Zhang T, Yin Y (2013) Genome sequence of the date palm *Phoenix dactylifera* L. Nature Commun 4(1):1–9

Bandupriya HD, Fernando SC, Vidhanaarachchil YR (2016) Micropropagation and androgenesis in coconut: an assessment of Sri Lankan implication. Cocos 22(1):31–47

Chen W, Zhang G, Chen W, Zhong Q, Chen H (2018) Metabolomic profiling of matured coconut water during post-harvest storage revealed discrimination and distinct changes in metabolites. RSC Adv 8 (55):31396–31405

Fan H, Xiao Y, Yang Y, Xia W, Mason AS, Xia Z, Qiao F, Zhao S, Tang H (2013) RNA-Seq analysis of *Cocos nucifera*: transcriptome sequencing and *de novo* assembly for subsequent functional genomics approaches. PLoS ONE 8(3):

Freitas Neto M, Pereira TN, Geronimo IG, Azevedo AO, Ramos SR, Pereira MG (2016) Coconut genome size determined by flow cytometry: tall versus dwarf types. Embrapa Tabuleiros Costeiros-Artigo em periódico indexado (ALICE)

Huang YY, Lee CP, Fu JL, Chang BC, Matzke AJ, Matzke M (2014) De novo transcriptome sequence assembly from coconut leaves and seeds with a focus on factors involved in RNA-directed DNA methylation. G3: Genes Genomes Genet 4(11):2147–2157

Huang J, Liu X, Lan Q, Lai X, Luo Z, Yang G (2016) Proteomic profile of coconuts. Eur Food Res Technol 242(3):449–455

Köhler C, Makarevich G (2006) Epigenetic mechanisms governing seed development in plants. EMBO Rep 7 (12):1223–1227

Kumar M, Saini SS, Agrawal PK, Roy P, Sircar D (2021) Nutritional and metabolomics characterization of the coconut water at different nut developmental stages. J Food Compos Anal 96:

Lakshmi Jayaraj K (2019) Comparative proteomic analysis of somatic and zygotic embryo in coconut (*Cocos nucifera* L.). Ph.D Dissertation submitted to Mangalore University, p 304

Lantican DV, Strickler SR, Canama AO, Gardoce RR, Mueller LA, Galvez HF (2019) De novo genome sequence assembly of dwarf coconut (*Cocos nucifera* L. 'Catigan Green Dwarf') provides insights into genomic variation between coconut types and related palm species. G3: Genes Genomes Genet 9(8):2377–2393

Li D, Zheng Y, Wan L, Zhu X, Wang Z (2009) Differentially expressed microRNAs during solid endosperm development in coconut (*Cocos nucifera* L.). Sci Hortic 122(4):666–669

Matzke MA, Mosher RA (2014) RNA-directed DNA methylation: an epigenetic pathway of increasing complexity. Nature Rev Genet 15(6):394–408

Naganeeswaran S, Fayas TP, Rachana KE, Rajesh MK (2015) Computational prediction and characterization of miRNA from coconut leaf transcriptome. J Appl Hortic 17(1):12–17

Nair RV, Jacob PM, Ajithkumar R (2004) Screening of coconut varieties against root (wilt) disease. J Plantn Crops 32:50–51

Nejat N, Sijam K, Abdullah SN, Vadamalai G, Dickinson M (2009) Phytoplasmas associated with disease of coconut in Malaysia: phylogenetic groups and host plant species. Plant Pathol 58(6):1152–1160

Nejat N, Cahill DM, Vadamalai G, Ziemann M, Rookes J, Naderali N (2015) Transcriptomics-based analysis using RNA-Seq of the coconut (*Cocos nucifera*) leaf in response to yellow decline phytoplasma infection. Mol Genet Genom 290(5):1899–1910

Rajesh MK, Rachana KE, Naganeeswaran SA, Shafeeq R, Thomas RJ, Shareefa M, Merin B, Anitha K (2015) Identification of expressed resistance gene analog sequences in coconut leaf transcriptome and their evolutionary analysis. Turk J Agric For 39(3): 489–502

Rajesh MK, Fayas TP, Naganeeswaran S, Rachana KE, Bhavyashree U, Sajini KK, Karun A (2016) *De novo* assembly and characterization of global transcriptome of coconut palm (*Cocos nucifera* L.) embryogenic calli using Illumina paired-end sequencing. Protoplasma 253(3):913–928

Rajesh MK, Rachana KE, Kulkarni K, Sahu BB, Thomas RJ, Karun A (2018) Comparative transcriptome profiling of healthy and diseased Chowghat Green Dwarf coconut palms from root (wilt) disease hot spots. Eur J Plant Pathol 151(1):173–193

Rajesh MK, Chowdappa P, Behera SK, Kasaragod S, Gangaraj KP, Kotimoole CN, Nekrakalaya B, Mohanty V, Sampgod RB, Banerjee G, Das AJ, Niral V, Karun A, Mahato AK, Gaikwad K, Singh NK, Keshava Prasad TS (2020) Assembly and annotation of the nuclear and organellar genomes of a dwarf coconut (Chowghat Green Dwarf) possessing enhanced disease resistance. OMICS: J Integrative Biol 24(12):726–742

Ramesh SV, Hebbar KB, Rajesh MK, Archana P (2019) Soil water-deficit differentially modulates the expression of stress associated genes (SAGs) of *Cocos nucifera* L. seedlings with contrasting water-use efficiency (WUE). In: Proceedings of 23rd biennial symposium on plantation crops (PLACROSYM-XXIII) held from 6–8 March, 2019, at central coffee research institute, coffee research station, Chikkamagaluru, Karnataka, pp 93. ISBN No.: 978-81-932263-8-4

Ramesh SV, Hebbar KB, Rajesh MK, Abhin SP, Gangaraj KP, Athul Bobby (2020). Transcriptome analysis of *Cocos nucifera* L., seedlings having contrasting water use efficiency (WUE) under water-deficit stress: molecular insights and genetic markers for drought tolerance. In: Proceedings of the 1st International Electronic Conference on Plant Science. https://doi.org/10.3390/iecps2020-08853

Sabana AA, Antony G, Rahul CU, Rajesh MK (2018) *In silico* identification of microRNAs and their targets associated with coconut embryogenic calli. Agri Gene 7:59–65

Sabana AA, Rajesh MK, Antony G (2020) Dynamic changes in the expression pattern of miRNAs and associated target genes during coconut somatic embryogenesis. Planta 251(4):1–8

Saensuk C, Wanchana S, Choowongkomon K, Wongpornchai S, Kraithong T, Imsabai W, Chaichoompu E, Ruanjaichon V, Toojinda T, Vanavichit A, Arikit S (2016) *De novo* transcriptome assembly and identification of the gene conferring a "pandan-like" aroma in coconut (*Cocos nucifera* L.). Plant Sci 252:324–334

Singh R, Ong-Abdullah M, Low ET, Manaf MA, Rosli R, Nookiah R, Ooi LC, Ooi SE, Chan KL, Halim MA, Azizi N (2013) Oil palm genome sequence reveals divergence of interfertile species in old and new worlds. Nature 500(7462):335–339

Xiao Y, Xu P, Fan H, Baudouin L, Xia W, Bocs S, Xu J, Li Q, Guo A, Zhou L, Li J (2017) The genome draft of coconut (*Cocos nucifera*). Giga Sci 6(11):gix095

Yang Y, Saand MA, Abdelaal WB, Zhang J, Wu Y, Li J, Fan H, Wang F (2020) iTRAQ-based comparative proteomic analysis of two coconut varieties reveals aromatic coconut cold-sensitive in response to low temperature. J Proteom 220: https://doi.org/10.1016/j.jprot.2020.103766

Yang Y, Bocs S, Fan H, Armero A, Baudouin L, Xu P, Xu J, This D, Hamelin C, Iqbal A, Qadri R (2021) Coconut genome assembly enables evolutionary analysis of palms and highlights signaling pathways involved in salt tolerance. Commun Biol 4(1):1–4

Zhang Y, Chen W, Chen H, Zhong Q, Yun Y, Chen W (2020) Metabolomics analysis of the deterioration mechanism and storage time limit of tender coconut water during storage. Foods 9(1):46. https://doi.org/10.3390/foods9010046

Zhao Z, Yu S, Li M, Gui X, Li P (2018) Isolation of exosome-like nanoparticles and analysis of microRNAs derived from coconut water based on small RNA high-throughput sequencing. J Agric Food Chem 66(11):2749–2757

9 Mitochondrial and Chloroplast Genomes

S. V. Ramesh, M. K. Rajesh, Ajeet Singh, and K. B. Hebbar

Abstract

Coconut (*Cocos nucifera* L.), which belongs to family Arecaceae, is an important economic crop of the tropics. Despite its economic significance, applications of genomics science in the coconut crop improvement programs has gained momentum only in recent times. Besides the whole genome and transcriptome sequencing approaches, genomics of plastids and mitochondria of coconut are worthwhile objects of examination for delineating the molecular evolutionary genomics of lands plants in general and palms in particular. Plant mitochondria and chloroplasts have semi-autonomous genetic systems. The genome sequences of both the mitochondria and chloroplast of coconut were made publicly available prior to the sequencing of its nuclear genome. In this chapter, we present the comprehensive structural and functional organization of organellar genomes of coconut. Also, the outcomes of the molecular phylogeny and comparative genomics of chloroplast and mitochondria among the monocots are discussed.

9.1 Introduction

Cocos nucifera L., a member of the palm family (Arecaceae), is one of the most economically important crops in tropics and subtropics, where it is the principal source of food, drink, medicine, construction material and several industrial applications (Gunn et al. 2011). Notwithstanding the morphological diversity in coconut, it is taxonomically defined as a single species under the genus *Cocos*. Based on the stature and breeding, coconut cultivars are divided into talls—that show outcrossing and dwarfs—are usually self-crossing in nature (Perera et al. 2003; Gunn et al. 2011). Genomics-based crop improvement techniques have been sparsely employed in coconut when compared to economically important palms like oil palm or date palm (Al-Dous et al. 2011; Al-Mssallem et al. 2013; Singh et al. 2009; Dussert et al. 2013). The true potential of genomics assisted-plant breeding techniques have not been unleashed in coconut crop improvement programs due to various reasons such as its long life cycle, complex *in vitro* regeneration capability, poor fertility, and procrastination in the use of genome sequencing

S. V. Ramesh (✉) · M. K. Rajesh · A. Singh · K. B. Hebbar
ICAR-Central Plantation Crops Research Institute, Kasaragod, Kerala, India
e-mail: ramesh.sv@icar.gov.in

M. K. Rajesh
e-mail: rajesh.mk@icar.gov.in

A. Singh
e-mail: ajeet.singh1@icar.gov.in

K. B. Hebbar
e-mail: hebbar.kb@icar.gov.in

M. K. Rajesh et al. (eds.), *The Coconut Genome*, Compendium of Plant Genomes,
https://doi.org/10.1007/978-3-030-76649-8_9

technologies. Alsaihati et al. (2014) reported a draft coconut genome based on seven libraries of pair-end and mate-pair genome sequencing utilizing Illumina's HiSeq platform with an overall coverage of 94.5% of 2.6 GB genome. It was followed by the whole genome of tall cultivar "Hainan Tall" by Xiao et al. (2017) that generated a scaffold length of 2.2 Gb (with a scaffold N50 of 418 kb) representing over 91% of the estimated coconut genome size. It was also predicted that the genome encodes 28,039 proteins which was less compared to the protein-coding sequences of *Phoenix dactylifera* (41,660) and *Elaeis guineensis* (34,802). Also, 72.75% of the coconut genome comprises of transposable elements (Xiao et al. 2017). Whole genome sequence (WGS) assembly and annotation of a dwarf coconut variety, 'Catigan Green Dwarf' (CATD), was made available (Lantican et al. 2019). Also, a total of 34,958 high-confidence gene models associated with the agronomically important traits were predicted. Rajesh et al. (2020) sequenced a green dwarf cultivar-Chowghat Green Dwarf-from India and this discovery laid special emphasis on characterizing genes conferring resistance to diseases. Rapid strides in the development of novel sequencing platforms such as next-generation sequencing (NGS) techniques have helped in investigating the organellar genomes of coconut (Huang et al. 2013; Aljohi et al. 2016) long before the exploration of nuclear genome of the crop. As of August, 2021, 12,527 mitochondria (mt) genomes, 5660 chloroplast genomes, and genomes of 62 apicoplast, 3 kinetoplast, 2 chromatophore, 1 cyanelle, 1064 other plastids are available in the NCBI organelle database opening up unparalleled possibilities (https://www.ncbi.nlm.nih.gov/genome/browse#!/organelles/). Despite the variation in the structure and organization of plastidial and mitochondrial genomes, a greater degree of conservation is observed in the gene sets among the lineages of terrestrial plants. It makes the exploration of organellar genomes an exciting branch of genomic science, especially from the perspective of evolutionary biology.

9.2 Mitochondrial Genome

Mitochondria are the sub-cellular organelles involved in the process of energy-production wherein chemical energy is converted into ATP through the process of oxidative phosphorylation. Mitochondria are involved in various cellular functions such as homeostasis, programmed cell death, and biosynthesis of various metabolites (McBride et al. 1995). Mitochondria contains DNA as its genetic material in its matrix and have been considered as a remnant of an ancestral α-proteobacterium—a symbiont of eukaryotes (Lang et al. 1999). Plant mitochondrial genomes have unique features such as highly variable gene content across plant taxa due to frequent genetic exchanges between nuclear and plastidial genomes; rapid evolution of genome structure compared to modifications in its primary sequence; low mutation rates compared to plastidial or nuclear genomes; large number of intron-containing genes; high frequency of RNA editing thereby conserving the functional attributes of mitochondrial proteins etc. (Wu et al. 2015). Hence, these unique features of mitochondrial genomes make it very cumbersome to sequence, annotate, and analyze them.

Mitochondrial (mt) genome of a coconut variety (Oman Local Tall) was obtained from Roche/454 GS FLX sequence reads followed by the homopolymer correction using the Illumina reads (Aljohi et al. 2016). The mitochondrial genome is of 678,653 bp in length and codes for the 72 proteins among the predicted 87 protein-coding genes (8.62% of mt genome) (Fig. 9.1). Also, nine truncated proteins, 23 tRNAs, and three rRNAs coding genes were also documented constituting a gene content of 11.43%. Around 5.07% of the genome sequence, encoding 15 protein-coding genes (*rpl14*, *rpl33* and *rps14*), and some rRNA and tRNA genes account for the genomic regions that are derived from the chloroplasts. The GC contents of protein-coding genes, pseudogenes, tRNAs, rRNAs, and the remaining non-coding sequences are 44.5% (58,895 bp), 47.7% (5294 bp), 41.1% (3092 bp),

53.5% (10,261 bp), and 45.5% (601,111 bp), respectively. The mt genome harbors 0.49% tandem (3310 bp) and 17.26% long repeats ($\geq$100 bp). Among the protein-coding genes, 50 genes encode known functional proteins, whereas remaining 22 loci encode hypothetical proteins. Most of the known functional proteins are components involved in electron transport chain including various subunits of ATP synthase F1 complex, nicotinamide adenine dinucleotide dehydrogenase, and cytochrome c. Recently, mitochondrial genome (744,799 bp) of a dwarf coconut cultivar Chowghat Green Dwarf (CGD) was published (Rajesh et al. 2020). The mitochondrial genome of CGD has a GC content of 41.86% and encodes 123 genes. The CGD mitochondrial genome exhibited characteristic features similar to that of Oman Local Tall cultivar already reported (Aljohi et al. 2016). Molecular phylogenetic analysis of mitochondrial proteins with that of other plant species revealed that coconut mt proteins clustered with those of monocots *P. dactylifera* and *B. umbellatus* (Fig. 9.2). Comparative genomic analysis of coconut mt genome and other plant species have identified that *sdh* gene is unique to the coconut among the monocots. However, mitochondrial genomes of monocots have *rps3* gene which is distinctly absent in coconut. Only 12 among the tRNA genes have introns of which three tRNA genes are of chloroplastic origin. These findings lend credence to the theory that in due course of time mt tRNA genes are being replaced by those of cp-derived tRNAs (Fang et al. 2012; Tian et al. 2007).

9.2.1 Genomic Regions of Chloroplastic Origin in Mitochondria

Mitochondrial genome of coconut harbors 33 regions of chloroplastic origin with a length range of 64 to 3365 bp and are characterized with a GC content of 42%. Further, 38% of this region codes for proteins. Considering the similarities between coconut and date palm in the plastidial origin of mt genomic regions, it was suggested that chloroplast-derived DNA sequences in mitochondria have evolved to match the recipient genome soon after the transfer (Fang et al. 2012; Yang et al. 2010).

Mitochondrial genome of coconut has 28 intron-containing genes of which 16 encode proteins and 12 code for tRNA genes. Interestingly, comparisons of mitochondrial tRNA genes of various plant species including *Marchantia polymorpha*, *Cycas taitungensis*, *Butomus umbellatus*, *Phoenix dactylifera*, *Zea mays*, and *Vitis vinifera* with coconut tRNA genes revealed that the latter possesses largest number of intron-containing genes. Percentage of tandem repeat DNA sequences of coconut mt genome (0.49%) is comparable to that of *P. dactylifera*; however, the coconut mt has greater proportion (>17%) of long repeats (>=100 bp). Nonetheless, the long repeats of coconut mt genome are similar to that of other monocot species.

9.2.2 Mitochondrial Genome Sequence Variation

Sequence variation analysis revealed that mt genome has at least 100 single nucleotide polymorphisms (SNPs) and two insertions. Also, the greater proportion of sequence (48 of the total) variation arises from the regions of chloroplastic origin. Comparative analysis of mt genomes of *P. dactylifera* and *C. nucifera* identified 2,442 SNPs and 1,122 InDels. Also, five SNPs are found in protein-coding genes include three synonymous SNPs of *rps1*, *rps2*, and *rpl14* and two non-synonymous SNPs of *orf247-ct*. There are 23 transitions (Ti) and 77 transversions (Tv), and excluding the chloroplast-derived sequences, the Ti/Tv ratio is 0.06 which is very low compared to the Ti/Tv ratio of nuclear genome ($\sim$2.0). These critical differences once again underscore the functional divergence in DNA repair mechanism between mitochondrial and nuclear genomes.

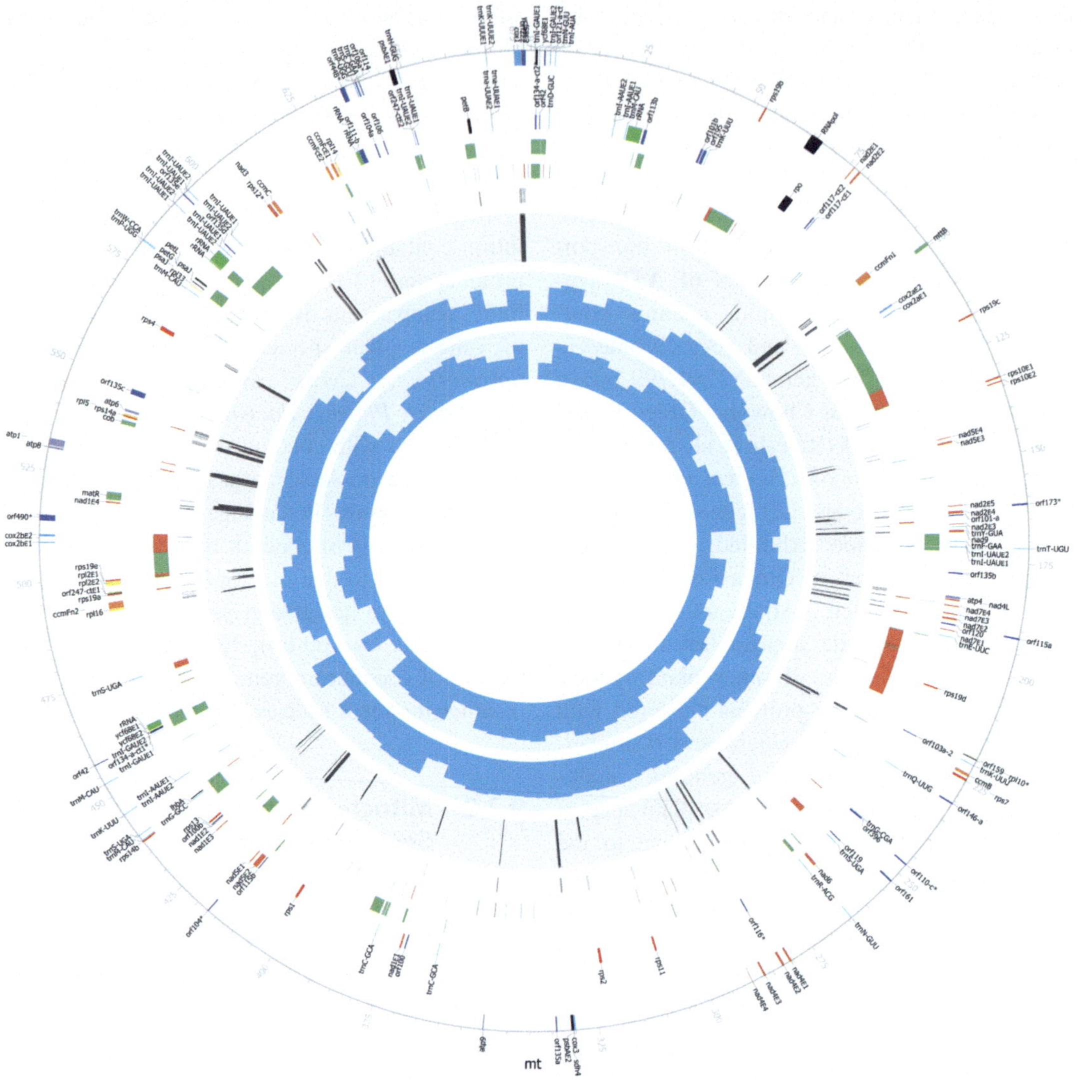

Fig. 9.1 Circular display of mitochondrial genome of coconut (*Cocos nucifera* L.) *Source* Aljohi et al. (2016). https://doi.org/10.1371/journal.pone.0163990

9.2.3 RNA Editing

RNA editing is an important mechanism that accounts for genetic plasticity of an organism as it allows production of different protein products from a single gene by modifying the nucleotide sequence features of the gene (Cuenca et al. 2013). RNA-seq data from coconut, analyzed for RNA editing, predicted around 845 RNA editing sites (92 synonymous and 753 non-synonymous) in protein-coding genes, and 36 sites in regions of chloroplastic origin. Between Tall and Dwarf cultivars, 665 RNA editing sites are shared, whereas 675 and 835 sites are unique to the cultivars, respectively.

9.3 Chloroplast Genome

Chloroplasts are sub-cellular organelles that are involved in the conversion of solar energy into carbohydrates, concomitantly evolving oxygen in the process called photosynthesis. Undoubtedly, photosynthesis is an important function of

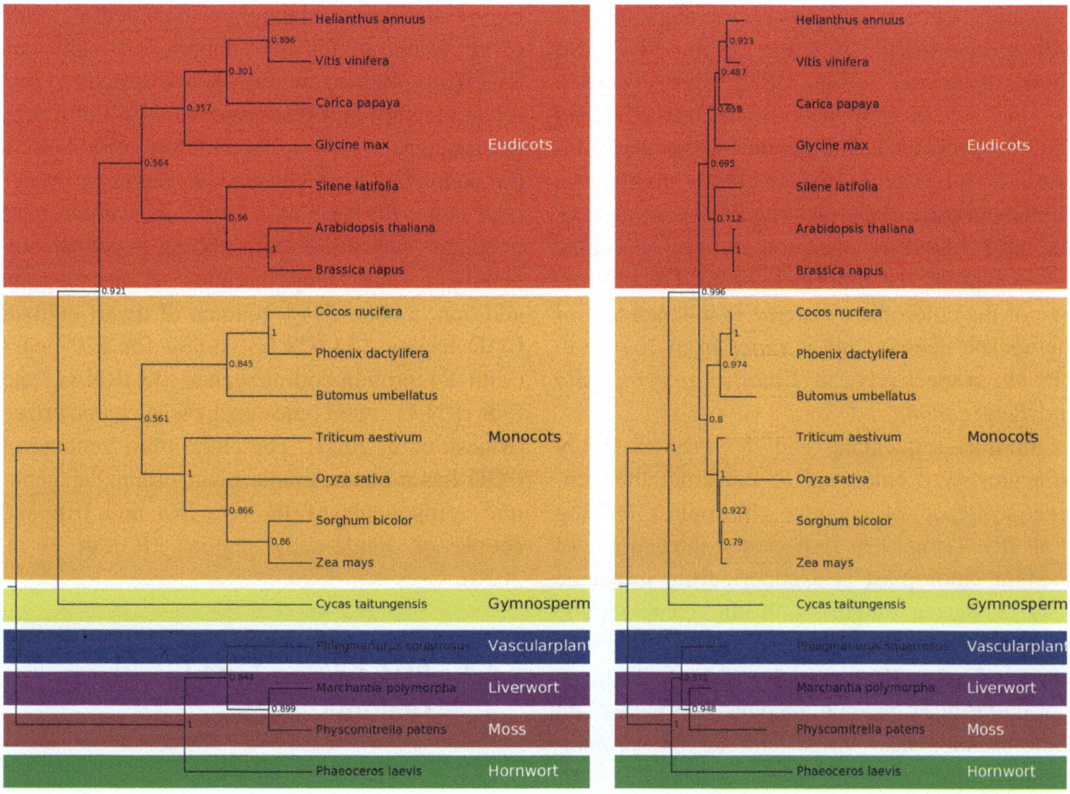

Fig. 9.2 Molecular phylogenetic analysis of mitochondrial proteins of coconut and 19 other plant species. The phylogenetic trees depicted are generated by the maximum parsimony (left) and maximum likelihood (right) methodologies in MEGA 6.06 *Source* Aljohi et al. (2016). https://doi.org/10.1371/journal.pone.0163990

chloroplasts. However, plastids have a greater role to play in plant physiology and in growth and development by involving in biosynthesis of metabolites, amino acids, fatty acids, hormones, vitamins, and also in the assimilation of minerals. Hence, chloroplasts are rightly called as metabolic centers reacting to variations in environmental stress-related signals (Daniell et al. 2016).

9.3.1 Significance of Chloroplast Genome

The sequence variations of chloroplast genomes have wider implications in comprehending the factors that contribute to the adaptation mechanisms of economically important crop plants so as to facilitate breeding and conservation of invaluable traits (Wambugu et al. 2015; Brozynska et al. 2016). Also, genome sequences of chloroplasts have helped in resolving the phylogenetic clades or evolutionary relationships among the plants. Further, the pattern of genetic exchanges among the chloroplast, mitochondria, and nuclear genomes of a particular species is an exciting area of research that would help understand the organization and functioning of genetic materials. On the translational research and applications front, understanding the chloroplast genomics has yielded important tools to produce stress-tolerant plants, vaccines and bio-based pharmaceuticals.

9.3.2 Organization and Structure of Chloroplast Genome

Structure and organization of chloroplast genomes of plants are relatively conserved. Genome is a single circular DNA molecule characterized with a quadripartite structure composed of two copies of inverted repeats (IR) region separating the small and large single-copy regions (SSC and LSC, respectively). The genome generally comprises 120–130 genes that are chiefly involved in photosynthesis. Important regulatory sequences are also found in the non-coding intergenic spacer regions (Daniell et al. 2006). Overall, the size of the chloroplast genome is independent of the nuclear genome, but it varies from 107 kb to 218 kb, respectively, in *Cathaya argyrophylla* and *Pelargonium*.

Illumina sequencing and *de novo* assembly strategies were employed to construct the genome sequence of coconut chloroplast (Huang et al. 2013) (Fig. 9.3). Expectedly, the genome of coconut chloroplast is a quadripartite molecule (154,731 bp in length) with the characteristic SSC (17,391 bp) and LSC (84,230 bp) which are separated by an IR region. It was also predicted that chloroplast genome encodes 130 genes and four pseudogenes. LSC harbors 84 protein-coding genes, 38 tRNA genes, and eight rRNA genes, whereas SSC has a copy of *ycf1, rps19*, and couple of copies of *ycf15*. In general, genes of chloroplast have only one intron with exceptions being *ycf3* and *clpP* that have two introns each. Most of the genes encoding proteins have AUG as start codon; however, ACG, GUG start codons are also not uncommon. The codon usage frequency of coconut chloroplast genome is comparable to that of many flowering plants (Yang et al. 2010; Uthaipaisanwong et al. 2012; Qian et al. 2013). Similarly, a strong bias toward A or T in the third position of synonymous codons is also documented. The prevalently encoded amino acid is leucine, whereas the least preferred one is sulfur-containing cysteine.

Identification of repetitive sequences in the genome of chloroplast revealed eight tandem repeats, three direct repeats, and two inverted repeats. IR regions of genome have three repeats, whereas ten repeats are present in LSC region. The nature and content of repetitive sequences of coconut chloroplast genome is similar to that found in oil palm and date palm. The repeat region of chloroplast genome sequences is found to be shared among the coconut, oil palm, and data palm plastid genomes. It was suggested that recombination and rearrangements in the repetitive sequences of chloroplast DNA could help in the stabilization of chloroplast genomes (Marechal et al. 2009). Palm family has chloroplast genomes with fewer and shorter repeats compared to other crop families (Table 9.1). In addition, a chloroplast genome of dwarf cultivar CGD having 154,628 bp coding for 129 genes (with 84 protein-coding genes, 38 tRNAs, and four rRNAs (two copies each) was characterized (Rajesh et al. 2020). The chloroplast genome of CGD has a characteristic plant organellar genomic components of IR viz., IRa and IRb and couple of single-copy regions (Rajesh et al. 2020).

9.3.3 RNA Editing in Chloroplast Genome

RNA transcripts of protein coding genes of chloroplast and mitochondrial genomes are altered post synthesis in a phenomenon called RNA editing (Gray and Covello 1993). Bioinformatics analysis of chloroplast genome of coconut identified around 83 potential RNA editing sites spread over 27 genes (Huang et al. 2013). RT-PCR-further confirmed the presence of 64 sites along with six more genes (*acc*D, *mat*K, *ndh*B, *ndh*G, *ndh*H, and *rpo*A) that are prone to RNA editing. In coconut, the base changes were primarily from C-to-U that result in non-silent changes, and majority of the amino acids conversions are from hydrophilic to hydrophobic (63) followed by hydrophobic to hydrophobic amino acids (11) and very rare hydrophobic to hydrophilic amino acids (1). Even though evolutionary analysis of RNA editing in land plants revealed that a number of edited sites are declining with time and predominant change is silent, chloroplast exhibits extraordinarily high number of RNA editing loci, and many gene deletions such as *acc*D, *ycf*1, *ycf*2, *ycf*4, *psa*I, and *rpo*A.

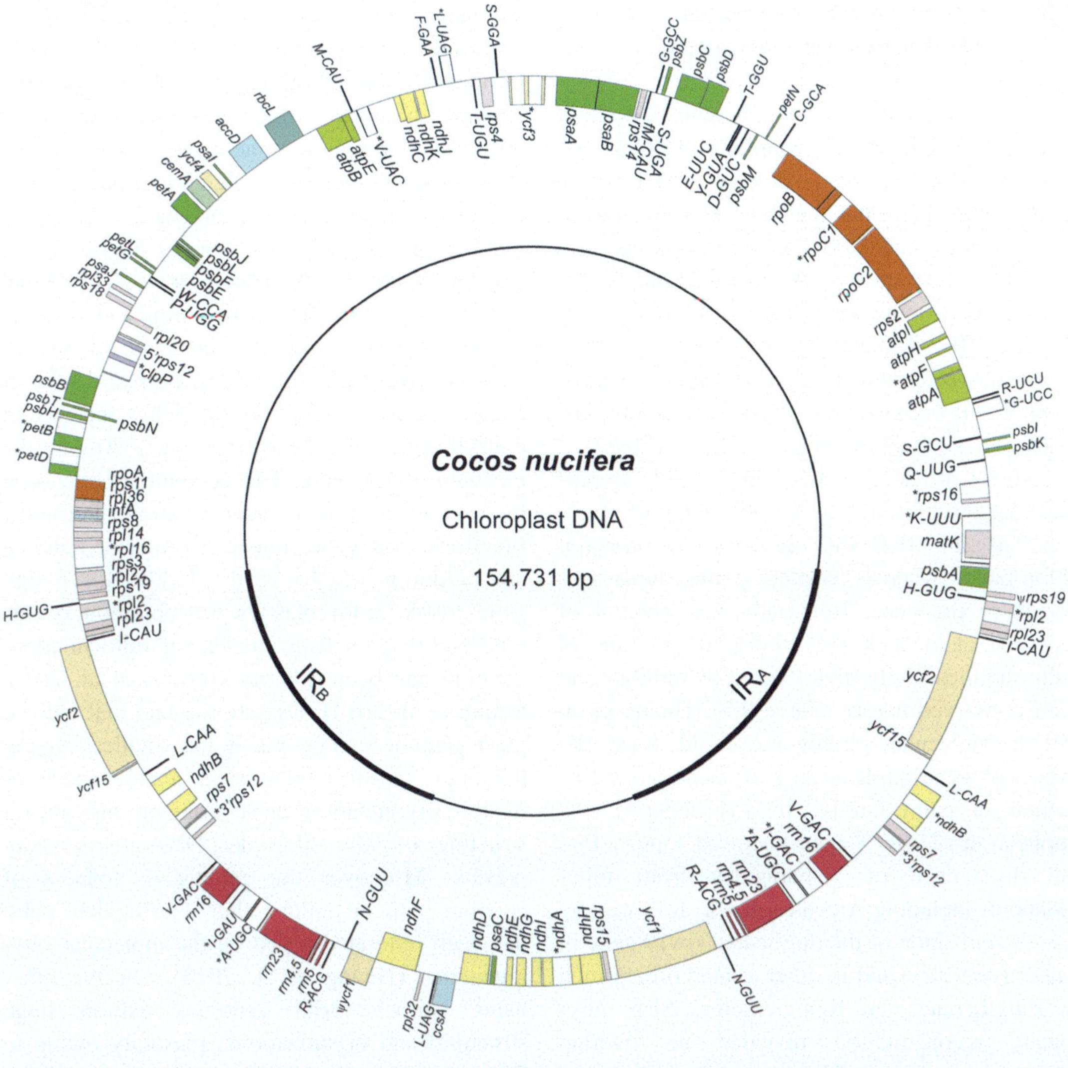

Fig. 9.3 Chloroplast genome map of coconut (*Cocos nucifera* L.) *Source* Huang et al. (2013). https://doi.org/10.1371/journal.pone.0074736

Table 9.1 Characteristic features of repetitive sequences of chloroplast genomes of coconut and other crops

Sl. no.	Family	No. of repeats	Longest repeat length (bp)	Reference
1	Coconut	13	30	Huang et al. (2013)
2	Oil palm	11	39	Uthaipaisanwong et al. (2012)
3	Poaceae	>20	132	Zhang et al. (2011), Saski et al. (2007)
4	Cymbidium	232	30–61	Yang et al. (2013)
5	Citrus	29	30–59	Bausher et al. (2006)
6	Solanaceae	42	56	Daniell et al. (2006)
7	*Gossypium*	54	64	Lee et al. (2006)

9.3.4 Comparative Genomics of Chloroplast Genomes

Among the six palm species, coconut has one of the smallest chloroplast genomes (154,731 bp) and the IRs are also shortest (53,110 bp), whereas date palm has the largest of the chloroplast genome and IR of 158,462 bp and 54,552 bp, respectively. As in other plants, the chloroplast genomes of palms are AT rich (Yang et al. 2010; Uthaipaisanwong et al. 2012). Comparative genomic analysis revealed that gene content and organization of coconut chloroplast DNA are similar to other palms (Table 9.2) (Huang et al. 2013). However, notable exceptions are the presence of a single copy of *rps19* gene at the IRB/SSC junction in coconut chloroplast, whereas related palms have two copies of the gene. Relatively low content of repetitive elements in the chloroplast genomes of palm families imply high degree of stability and their conserved nature (Table 9.2) (Huang et al. 2013). Molecular phylogeny could trace the events of gene duplication and pseudogene formation in palm family. It was deduced that duplication of *rps19* gene occurred long before the divergence of Asparagales from other monocots including Arecales. It was followed by pseudogenization of the duplicated *rps*19 gene in *Cocos* (Arecales) and in other related orders such as Zingiberales and Ranunculales. Also, phylogeny reconstruction revealed an overlap between genes viz., *ndh*F and pseudo *ycf*1 among the Arecales and Dasypogonaceae. Similar overlap was also observed between Canellales and Chloranthales of Magnoliids suggesting the common ancestry and delineating the fact that atleast three independent events have contributed to this phenomenon.

9.4 Perspectives

Photosynthetic and metabolic pathways are pivotal to the survival of plants, the plastidial genome is under high degree of selection pressure, and hence, it demonstrates utmost functional and structural conservation (Wicke et al. 2011). Further, the structural features of the plastidial genomes have been conserved over a long evolutionary time scales. Nonetheless, loss of one IR copy often associated with changes in single-copy regions, gene losses, or changes in repeat numbers are observed in lineages such as Fabaceae or Geraniance (Guisinger et al. 2011; Weng et al. 2014). Thus, IR has been implicated in the maintenance of structural integrity of plastid genomes over time. Interestingly, great degree of plastidial genome reconfiguration has occurred in plant lineages that are heterotrophic. Among the mechanisms postulated to account for the gene losses, transcriptional gene regulation, alternative functions, and gene transfer to nuclear genome are predominan (Wicke et al. 2013). Though gene losses from plastids are common occurrence, gene gain from nuclear or mitochondrial genome has been a rarity (Iorizzo et al. 2012; Straub et al. 2014). Despite the fact that chloroplast genome of coconut is the smallest among the palm families, conserved features such as overall organization, gene content, and repeat structures of palm chloroplast DNA are markedly present. Moreover the genetic relatedness of coconut with oil palm rather than to date palm has been reiterated based on the molecular phylogenetics (Huang et al. 2013). On the other hand, mitochondrial genomes exhibit huge structural and organizational plasticity owing to DNA repair and replication mechanism (Christensen 2013). Unlike the plastidial genomes, mitochondrial genomes are prone to foreign DNA insertion derived either from nuclear genomes, plastids, or gene copies from host plants (Sloan and Wu 2014; Barkman et al. 2007). Also, compared to mitochondrial genomes of animals, chondromes of plants encode more genes and also harbor more introns in their genes. Phylogenetic studies by Aljohi et al. (2016), based on the genome sequences of mitochondria, revealed that Arecaceae has diverged from the rest of the monocots earlier in the evolution of flowering plants.

Table 9.2 Comparative analysis of chloroplast genomes of palm species *Source* Huang et al. (2013). https://doi.org/10.1371/journal.pone.0074736

Characteristic features	*Phoenix*	*Elaeis*	*Cocos*
Size (bp)	158,462	156,973	154,731
LSC	86,198	85,192	84,230
SSC	17,712	17,639	17,391
IR	54,552	54,142	53,110
GC content (%)	37.23	37.40	37.44
Number of genes	131	131	129
Protein-coding genes	85	85	84
Bases (bp)	192,511	10,782	90,130
rRNAs	8	8	8
tRNAs	38	38	38
Pseudogenes	1	1	2
Genes with intron(s)	22	22	22
Protein-coding genes	14	14	14
tRNAs	8	8	8

9.5 Concluding Remarks and Prospectives

In summary, genomes of organelles are most valuable resources for comprehending the evolution of plant kingdom. Availability of plastid genomes could open up large-scale comparative genomes in the field of systematic botany similar to the application of mitochondrial genomics in animal science. Genome sequences of chloroplast and mitochondria of coconut are such valuable tools to study the palm biology. Technical challenges in this endeavor are enormous as sequencing is prone to error and mitochondrial genomics is still lagging compared to the advancements in plastid genomics. However, advent of cost-effective, rapid next-generation sequencing technologies could effectively bring down the error rate and uplift the utility of mitochondrial genomics in plant systematics. Another exciting area of research is genetic engineering of plastid genomes for obtaining superior plant phenotypes to confer stress tolerance, production of therapeutic vaccines, enzymes, biopharmaceuticals, etc. The advantages of this strategy are that it circumvents 'gene escape' via pollen as plastids are maternally inherited and high concentration of protein production. Therefore, it is anticipated that the technological advances such as genome editing and plastid engineering could create improved crop phenotypes.

References

Al-Dous EK, George B, Al-Mahmoud ME, Al-Jaber MY, Wang H, Salameh YM, Al-Azwani EK, Chaluvadi S, Pontaroli AC, DeBarry J, Arondel V, Ohlrogge J, Saie IJ, Suliman-Elmeer KM, Bennetzen JL, Krueg-ger RR, Malek JA (2011) *De novo* genome sequencing and comparative genomics of date palm (*Phoenix dactylifera*). Nat Biotech 29(6):521–527

Aljohi HA, Liu W, Lin Q, Zhao Y, Zeng J, Alamer A, Alanazi IO, Alawad AO, Al-Sadi AM, Hu S, Yu J (2016) Complete sequence and analysis of coconut palm (*Cocos nucifera* L.) mitochondrial genome. PLoS One 11(10):e163990. https://doi.org/10.1371/journal.pone.0163990

Al-Mssallem IS, Hu S, Zhang X, Lin Q, Liu W, Tan J, Yu X, Liu J, Pan L, Zhang T, Yin Y (2013) Genome sequence of the date palm *Phoenix dactylifera* L. Nature Commun 4(1):1–9. https://doi.org/10.1038/ncomms3274

Alsaihati B, Liu W, Lin Q, Al-Mssallem IS (2014) Coconut genome *de novo* sequencing. In: Plant and animal genome XXII conference; poster presented at plant and animal genome (PAG) meeting XXII, San Diego

Barkman TJ, McNeal JR, Lim SH, Coat G, Croom HB, Young ND, dePamphilis CW (2007) Mitochondrial DNA suggests at least 11 origins of parasitism in angiosperms and reveals genomic chimerism in parasitic plants. BMC Evol Biol 7:248

Bausher M, Singh N, Lee SB, Jansen R, Daniell H (2006) The complete chloroplast genome sequence of *Citrus sinensis* (L.) Osbeck var 'Ridge Pineapple': organization and phylogenetic relationships to other angiosperms. BMC Plant Biol 6:21

Brozynska M, Furtado A, Henry RJ (2016) Genomics of crop wild relatives: expanding the gene pool for crop improvement. Plant Biotechnol J 14:1070–1085

Christensen AC (2013) Plant mitochondrial genome evolution can be explained by DNA repair mechanisms. Genome Biol Evol 5:1079–1086

Cuenca A, Petersen G, Seberg O (2013) The complete sequence of the mitochondrial genome of *Butomus umbellatus*—a member of an early branching lineage of Monocotyledons. PLoS ONE 8(4):

Daniell H, Lee S-B, Grevich J, Saski C, Quesada-Vargas T, Guda C, Tomkins J, Jansen RK (2006) Complete chloroplast genome sequences of *Solanum bulbocastanum, Solanum lycopersicum* and comparative analyses with other Solanaceae genomes. Theor Appl Genet 112:1503–1518

Daniell H, Lin CS, Yu M, Chang WJ (2016) Chloroplast genomes: diversity, evolution, and applications in genetic engineering. Genome Biol 17(1):1–29. https://doi.org/10.1186/s13059-016-1004-2

Dussert S, Guerin C, Andersson M, Joët T, Tranbarger TJ, Pizot M, Sarah G, Omore A, Durand-Gasselin T, Morcillo F (2013) Comparative transcriptome analysis of three oil palm fruit and seed tissues that differ in oil content and fatty acid composition. Plant Physiol 162 (3):1337–1358

Fang Y, Wu H, Zhang T, Yang M, Yin Y, Pan L, Yu X, Zhang X, Hu S, Al-Mssallem IS, Yu J (2012) A complete sequence and transcriptomic analyses of date palm (*Phoenix dactylifera* L.) mitochondrial genome. PloS One 7(5):e37164. https://doi.org/10.1371/journal.pone.0037164

Gray MW, Covello PS (1993) RNA editing in plant mitochondria and chloroplasts. The FASEB J 7(1):64–71. https://doi.org/10.1096/fasebj.7.1.8422976

Guisinger MM, Kuehl JV, Boore JL, Jansen RK (2011) Extreme reconfiguration of plastid genomes in the angiosperm family Geraniaceae: rearrangements, repeats, and codon usage. Mol Biol Evol 28:583–600

Gunn BF, Baudouin L, Olsen KM (2011) Independent origins of cultivated coconut (*Cocos nucifera* L.) in the Old World tropics. PLoS ONE 6:e21143

Huang YY, Matzke AJM, Matzke M (2013) Complete sequence and comparative analysis of the chloroplast genome of coconut palm (*Cocos nucifera*). PLoS ONE 8: https://doi.org/10.1371/journal.pone.0074736

Iorizzo M, Senalik D, Szklarczyk M, Grzebelus D, Spooner D, Simon P (2012) *De novo* assembly of the carrot mitochondrial genome using next generation sequencing of whole genomic DNA provides first evidence of DNA transfer into an angiosperm plastid genome. BMC Plant Biol 12(1):61

Lang BF, Gray MW, Burger G (1999) Mitochondrial genome evolution and the origin of eukaryotes. Ann Rev Genet 33(1):351–397

Lantican DV, Strickler SR, Canama AO, Gardoce RR, Mueller LA and Galvez HF (2019) *De novo* genome sequence assembly of dwarf coconut (*Cocos nucifera* L.'Catigan Green Dwarf') provides insights into genomic variation between coconut types and related palm species. G3: Genes Genomes Genet 9(8):2377–2393

Lee S-B, Kaittanis C, Jansen R, Hostetler J, Tallon L, Town CD, Daniell H (2006) The complete chloroplast genome sequence of *Gossypium hirsutum*: organization and phylogenetic relationships to other angiosperms. BMC Genom 7(1):61. https://doi.org/10.1186/1471-2164-7-61

Marechal A, Parent JS, Veronneau-Lafortune F, Joyeux A, Lang BF, Brisson N (2009) Whirly proteins maintain plastid genome stability in *Arabidopsis*. Proc Natl Acad Sci U S A 106:14693–14698

McBride KE, Svab Z, Schaaf DJ, Hogan PS, Stalker DM, Maliga P (1995) Amplification of a chimeric *Bacillus* gene in chloroplasts leads to an extraordinary level of an insecticidal protein in tobacco. Biotechnol (N Y) 13:362–365

Perera L, Russell JR, Provan J, Powell W (2003) Studying genetic relationships among coconut varieties/populations using microsatellite markers. Euphytica 132:121–128

Qian J, Song J, Gao H, Zhu Y, Xu J, Pang X, Yao H, Sun C, Li C, Liu J, Xu H (2013) The complete chloroplast genome sequence of the medicinal plant *Salvia miltiorrhiza*. PLoS ONE 8: https://doi.org/10.1371/journal.pone.0057607

Rajesh MK, Chowdappa P, Behera SK, Kasaragod S, Gangaraj KP, Kotimoole CN, Nekrakalaya B, Mohanty V, Sampgod RB, Banerjee G, Das AJ, Niral V, Karun A, Mahato AK, Gaikwad K, Singh NK, Keshava Prasad TS (2020) Assembly and annotation of the nuclear and organellar genomes of a dwarf coconut (Chowghat Green Dwarf) possessing enhanced disease resistance. OMICS: J Integr Biol 24(12):726–742

Saski C, Lee SB, Fjellheim S, Guda C, Jansen RK, Luo H, Tomkins J, Rognli OA, Daniell H, Clarke JL (2007) Complete chloroplast genome sequences of *Hordeum vulgare, Sorghum bicolor* and *Agrostis stolonifera*, and comparative analyses with other grass genomes. Theor Appl Genet 115:571–590

Singh R, Tan SG, Panandam JM, Rahman RA, Ooi LC, Low ET, Sharma M, Jansen J, Cheah SC (2009) Mapping quantitative trait loci (QTLs) for fatty acid composition in an interspecific cross of oil palm. BMC Plant Biol 9:114–132

Sloan DB, Wu Z (2014) History of plastid DNA insertions reveals weak deletion and AT mutation biases in angiosperm mitochondrial genomes. Genome Biol Evol 6(12):3210–3221

Straub SC, Moore MJ, Soltis PS, Soltis DE, Liston A, Livshultz T (2014) Phylogenetic signal detection from an ancient rapid radiation: effects of noise reduction, long-branch attraction, and model selection in crown clade Apocynaceae. Mol Phylogenet Evol 80:169–185

Tian X, Zheng J, Hu S, Yu J (2007) The discriminatory transfer routes of tRNA genes among organellar and nuclear genomes in flowering plants: a genome-wide investigation of indica rice. J Mol Evol 64(3):299–307

Uthaipaisanwong P, Chanprasert J, Shearman JR, Sangsrakru D, Yoocha T, Jomchai N, Jantasuriyarat C, Tragoonrung S, Tangphatsornruang S (2012) Characterization of the chloroplast genome sequence of oil palm (*Elaeis guineensis* Jacq.). Gene 500:172–180

Wambugu P, Brozynska M, Furtado A, Waters D, Henry R (2015) Relationships of wild and domesticated rices (*Oryza* AA genome species) based upon whole chloroplast genome sequences. Sci Rep 5:13957. https://doi.org/10.1038/srep13957

Weng ML, Blazier JC, Govindu M, Jansen RK (2014) Reconstruction of the ancestral plastid genome in Geraniaceae reveals a correlation between genome rearrangements, repeats and nucleotide substitution rates. Mol Biol Evol 31:645–659

Wicke S, Schneeweiss GM, dePamphilis CW, Müller KF, Quandt D (2011) The evolution of the plastid chromosome in land plants: gene content, gene order, gene function. Plant Mol Biol 76:273–297. https://doi.org/10.1007/s11103-011-9762-4

Wicke S, Müller KF, dePamphilis CW, Quandt D, Wickett NJ, Zhang Y, Renner SS, Schneeweiss GM (2013) Mechanisms of functional and physical genome reduction in photosynthetic and non-photosynthetic parasitic plants of the broomrape family. Plant Cell 25(10):3711–3725. https://doi.org/10.1105/tpc.113.113373

Wu Z, Cuthbert JM, Taylor DR, Sloan DB (2015) The massive mitochondrial genome of the angiosperm *Silene noctiflora* is evolving by gain or loss of entire chromosomes. Proc Natl Acad Sci 112(33):10185–10191

Xiao Y, Xu P, Fan H, Baudouin L, Xia W, Bocs S, Xu J, Li Q, Guo A, Zhou L, Li J (2017) The genome draft of coconut (*Cocos nucifera*). GigaSci 6(11):p.gix095. https://doi.org/10.1093/gigascience/gix095

Yang M, Zhang X, Liu G, Yin Y, Chen K, Yun Q, Zhao D, Al-Mssallem IS, Yu J (2010) The complete chloroplast genome sequence of date palm (*Phoenix dactylifera* L.). PloS ONE 5(9):p.e12762. https://doi.org/10.1371/journal.pone.0012762

Yang JB, Tang M, Li HT, Zhang ZR, Li DZ (2013) Complete chloroplast genome of the genus *Cymbidium*: lights into the species identification, phylogenetic implications and population genetic analyses. BMC Evol Biol 13(1):1–12

Zhang YJ, Ma PF, Li DZ (2011) High-throughput sequencing of six bamboo chloroplast genomes: phylogenetic implications for temperate woody bamboos (Poaceae: Bambusoideae). PLoS ONE 6(5): https://doi.org/10.1371/journal.pone.0020596

10 Endosperm Oil Biosynthesis: A Case Study for Trait Related Gene Evolution in Coconut

V. Arunachalam, S. V. Ramesh, S. Paulraj, B. Kalyana Babu, K. S. Muralikrishna, and M. K. Rajesh

Abstract

Coconut palm is one of the important vegetable oil crops of the tropics and sub-tropics. Studies delineating the oil biosynthesis in coconut palm are of paramount importance due to its market potential and quality [high proportion of medium-chain fatty acids (MCFAs), especially lauric acid]. The availability of coconut transcriptome data, coconut genome assemblies, and functional gene expression profiles has greatly aided in systematically identifying *Arabidopsis* orthologs involved in lipid and carbohydrate metabolic pathways. These findings could provide valuable information into the evolutionary relationships between the lipid and carbohydrate metabolism pathways, identifying genes involved in regulating quantitative and qualitative variation in fatty acid and lipid components. This chapter reviews the genes involved in coconut endosperm fatty acid and triacylglycerol (TAG) synthesis and the findings of comparative genomics into the fruit composition, oil biosynthesis, fatty acid composition, and oil quality in coconut. A roadmap for future work on the subject is highlighted for exploiting the oil biosynthesis genes in the coconut palm.

V. Arunachalam (✉)
ICAR-Central Coastal Agricultural Research Institute, Ela, Old Goa, India
e-mail: v.arunachalam@icar.gov.in

S. V. Ramesh · S. Paulraj · K. S. Muralikrishna · M. K. Rajesh
ICAR-Central Plantation Crops Research Institute, Kasaragod, Kerala, India
e-mail: ramesh.sv@icar.gov.in

S. Paulraj
e-mail: s.paulraj@icar.gov.in

K. S. Muralikrishna
e-mail: muralikrishna.ks@icar.gov.in

M. K. Rajesh
e-mail: rajesh.mk@icar.gov.in

B. K. Babu
ICAR-Indian Institute Of Oil Palm Research, Pedavegi, Andhra Pradesh, India
e-mail: B.Babu@icar.gov.in

10.1 Introduction

Cocos nucifera L. (Palmae: Arecaceae) is an economically important large seed oil crop of tropical and subtropical regions. Botanically, the coconut fruit is a fibrous drupe with three main layers viz., external exocarp layer (having characteristic color green, orange or yellow, etc.), the middle mesocarp layer (fibrous husk region), and the endocarp (internal hard layer or shell). The endocarp encloses the edible or oil containing economic part called endosperm or kernel. A thin, brown layer of seed coat that lies between the endocarp and the endosperm is called testa as it closely covers the endosperm. Pollination and fertilization results in a diploid embryo (2n) and a triploid endosperm (3n); the latter persists in

M. K. Rajesh et al. (eds.), *The Coconut Genome*, Compendium of Plant Genomes,
https://doi.org/10.1007/978-3-030-76649-8_10

the mature seed and serves as a nutrient resource for the germinating embryo. It is well known that coconut endosperm accumulates and store oil of approximately 65% of its weight (Laureles et al. 2002; Kumar 2011). Coconut endosperm oil is rich in saturated fatty acids (Over 90%) and over 80% of these belong to medium-chain fatty acids (MCFAs). Thus, a meta-analysis of the composition of economically important and commonly consumed vegetable oils reveals that the MCFAs content of coconut oil is quite high (Table 10.1). A comprehensive compositional analysis of fatty acids of coconut oil reveals that the oil is predominantly rich in lauric acid (48%; C12:0) (Naresh Kumar et al. 2000, 2004). However, other major fatty acid components of coconut oil are myristic acid (19% C14:0), palmitic acid (8.5%; C16:0). Further, it has been demonstrated that the MCFAs derived from coconut oil show various health benefits such as rich antioxidant status, antimicrobial activity, and cardioprotective effects (Ramesh et al. 2019).

The oil biosynthesis is an energy-intensive biochemical process involving different subcellular location and a multitude of enzymes. Fatty acid biosynthesis occurs in plastids, and assembly of fatty acids occurs due to acyl carrier proteins (ACP) in a cycle of reactions adding 2C every cycle. The fatty acids are further processed by thioesterases such as FatB enzymes in hydrolysis reactions or elongated further by Kas II enzymes (Li-Beisson et al. 2010). Once the fatty acids are formed or elongated, they form substrates for the triacylglycerol (TAG) synthesis. The TAG assembly occurs in the endoplasmic reticulum, and the process is catalyzed by lysophosphatidic acid acyltransferase (LPAAT). Genetic components of oil biosynthesis in coconut are a subject matter of interest because of the energy, economic, and nutritional potential of the oil and its processed variants (Ramesh et al. 2019, 2020).

10.2 Coconut Endosperm Development and Compositional Changes in Fatty Acid

Developmental stages of the coconut fruit and its endosperm were described by Child (1974). During the initial six months of the fruit

Table 10.1 Comparative percentage composition of fatty acids of coconut oil and other economically important saturated and unsaturated vegetable oils (Compiled from Naresh Kumar 2007; Kostik et al. 2013; Dorni et al. 2018)

Fatty acids	Edible oils rich in saturated MCFAs		Edible oils rich in unsaturated LCFAs			
	Coconut	Oil palm kernel	Sunflower	Olive	Soybean	Flax seed
C6:0 (caproic)	0.083	–	0.26	–	–	–
C8:0 (caprylic)	5.19	3.72	0.13	–	–	–
C10:0 (capric)	5.84	4.12	0.34	–	–	–
C12:0 (lauric)	48.04	43.57	0.07	–	–	–
C14:0 (myristic)	19.04	16.09	5.57	0.35	0.21	–
C16:0 (palmitic)	9.42	8.33	0.13	8.34	10.05	5.5
C16:1 (palmitoleic)	–	–	0.05	–	–	–
C17:0 (margaric)	–	–	3.714	0.11	0.11	–
C18:0 (stearic)	3.06	2.14	29.56	2.83	4.04	3.5
C18:1 (oleic)	7.92	22.5	*59.55*	78.4	26.63	22.1
C18:2 (linoleic)	1.38	1.25	0.24	7	51.83	20.5
C18:3 (linolenic)	0.067	–	0.64	–	6.58	47.5
C20:0 (arachidonic)	0.14	0.15	0.65	0.29	0.38	0.65
C22:0 (behenic)	–	–	0.6	0.13	0.58	–
C22:1 (erucic)	–	–	0.21	–	–	–
C24 (lignoceric)	0.03	0.3	–	0.03	0.23	–

development, the volume of the fruit increases since the central cavity remains filled with liquid endosperm (tender nut water). The liquid endosperm gets converted into a solid kernel layer in the next 3–4 months, starting from a thin jelly-like layer. Oil biosynthesis in coconut kernel or endosperm starts from the 4th month to the 7th month, coinciding with solid endosperm/kernel development. The rate of oil biosynthesis and saturation of oil with medium chain fatty acids (MCFAs) rapidly increase from the 7th month until the 12th month. Biochemically, oil synthesis in the coconut endosperm begins in the intracellular compartment of plastids and continues in the cytoplasm in the acyl-CoA dependent or independent pathways (Kartha 1964). The oil produced as a result of biochemical pathways is stored in the endosperm or embryo in discrete bodies called oleosomes.

The qualitative and quantitative changes in the fatty acids of endosperm vary with its developmental stages. In the developing endosperm, fatty acids of C_8–C_{18} are produced from the carbon donors acetyl-CoA and malonyl-CoA. Fatty acid composition analysis during the endosperm development reveal that the concentration of MCFAs (C6, C8, C10 and C12) increase during the developmental phases of liquid and solid endosperm. In contrast, longer chain fatty acids (LCFAs) (C18:3 and C22) are found when the endosperm is in the liquid state, and it tends to decrease severely with the maturation of endosperm into a solid kernel. It was documented that since the 6th month, the endosperm biosynthesizes MCFAs and continues to do so until 12 months, thereby producing about 50% of C12 (lauric) and C14 (myristic) acids (Child 1974). Hence in the mature endosperm of coconut, MCFAs predominate and the proportion of LCFAs decreases (Ceniza et al. 1974). Later it was proved that fatty acid and triacylglycerol content of oils derived from the coconut hybrids showed a range of 47–51% (Rodriguez et al. 1998). Besides, environmental conditions (growing conditions) and genotype influence the fatty acid profile of coconut oil (Balleza and Sierra 1972). Further, genotypic variations in the fatty acid profile of cultivars and hybrids and their suitability for particular industrial and food-grade purposes have been reported (Naresh Kumar et al. 2000, 2004).

10.3 Genetic Elements and Molecular Workhorses of Fatty Acid Biosynthesis

Genes, enzyme complexes, transcription factors (TFs), and endogenous small RNAs involved in plant fatty acid biosynthesis are presented in Table 10.2. Liang et al. (2014) utilized the suppression subtractive hybridization technique to identify differentially expressed genes during various stages of endosperm development. This study delineated that the coconut MADS-box transcription factors (TFs) are upregulated during all the stages of endosperm development and suggested that functional analysis of this class of TFs could shed light on endosperm metabolism (Liang et al. 2014). Among the ACP-thioesterases, *CnFatB3* alone was found in SSH libraries, and expression analysis of FatB enzyme encoding genes further revealed continuous downregulation of *CnFatB1* during endosperm ripening. In contrast, expression of *CnFatB2* and *CnFatB3* increased 1.4 and 11.5-fold, respectively. Enhanced expression of *CnFatB3* during endosperm ripening explains the predominance of lauric acid ($C_{12:0}$) content in coconut oil as CnFatB3 enzyme is specific for $C_{12:0}$ and $C_{14:1}$ fatty acids. Further enhanced expression of genes encoding enzymes pyruvate kinase (PK), especially cytosolic PK (PK_c), during the early stages and significant decline during the late stages of the endosperm warrants further investigation to link the carbohydrate and lipid metabolism both occurs simultaneously during the maturation of coconut endosperm. This study thus laid the foundation for exploring the various genetic and biochemical components of coconut endosperm development (Liang et al. 2014).

Analysis of gene expression in the leaves and endosperm of coconut divulged a host of enzymes in the catalytic reactions that convert C3 molecule pyruvate into various fatty acids. In coconut, out of 62 enzymes involved in fatty acid biosynthesis, 32 enzymes have their coding

Table 10.2 Genes and enzyme complexes involved in fatty acid biosynthesis

Sl. No.	Enzymes/enzyme complex/genes	Features	Reference(s)
1	Pyruvate dehydrogenase complex (PDHC)	• Supplies acetyl-CoA precursor for the fatty acid synthesis • Relatively high expression in oil-producing tissues of coconut and oil palm • Enzyme with four subunits, namely *E*1-α and -β, *E*2, and *E*3 • Varied gene copy number in Arecaceae • Coconut has four copies of *E*2, but only three of them were found to be highly expressed	Wong et al. (2017), Dussert et al. (2013), Xiao et al. (2019)
2	Acetyl–CoA carboxylase (ACCase) complex	• Comprises enzymes such as Acyl carrier protein (ACP), ketoacyl-ACP reductase (KAR), and hydroxyacyl-ACP dehydratase (HAD) • Involved in the initial carboxylation step of conversion of acetyl-CoA to malonyl-CoA • Regulated by light and redox status involving phosphorylation and interaction with a PII protein	Bourrellier et al. (2010)
3	Acyl-CoA-binding proteins (ACBPs)	• Acyl-CoA-binding proteins (ACBPs) are involved in binding and trafficking acyl-CoA esters and ensure continues carbon supply for fatty acid synthesis in plastids • Seven *ACBPs (EgACBP1* to *EgACBP7*) in the genome of *E. guineensis* • Expression of *EgACBP1* was high in mesocarp and endosperm. Role in seed development and lipid biosynthesis	Xiao and Chye (2011), Du et al. (2016)
4	Ketoacyl-ACP synthase (*KASI*, -KASII, and KAS III)	• KAS enzymes are crucial for fatty acid carbon chain condensation and elongation (from C4 to C18 fatty acids) • KAS III—is involved in the initial condensation step of malonyl-CoA acyl carrier protein (ACP) and acetyl-CoA to produce a C4 compound • KASI: in the conversion of C4:0–C14:0 ACP • KASII: final condensation of high C-chain palmitoyl-ACPs to stearoyl-ACPs	Yuan et al. (2015a, b), Schuch et al. (1997)
5	Acyl-ACP thioesterase A (FATA) and Acyl-ACP thioesterase B (FATB)	• FatAs show substrate preference toward 18:1-ACP, whereas FatBs primarily hydrolyze saturated acyl-ACPs with 8 to 18 carbons • Expression of *CnFatB3* from coconut (*Cocos nucifera*) endosperm mainly contributed to enhanced production of C12:0	Jones et al. (1995), Yuan et al. (2014)

(continued)

Table 10.2 (continued)

Sl. No.	Enzymes/enzyme complex/genes	Features	Reference(s)
		• *CnFatB3* and *EgFatB3* exhibit high expression in coconut and oil palm endosperms, respectively • Five *FatB* genes in coconut (*CnFatB1*, *CnFatB2-1*, *CnFatB2-2*, *CnFatB3*, and *CnFatB4*) reported	
6	Long-chain acyl-CoA synthetases (LACS)	• Catalyze the esterification of free fatty acids to CoA • 11–13 *LACS* genes are identified in Arecaceae (coconut palm, oil palm, and date palm)	Dussert et al. (2013), Bourgis et al. (2011), Troncoso-Ponce et al. (2011)
7	Stearoyl-acyl carrier protein desaturase (*SAD*)	• Three Arabidopsis orthologs (At2g43710, At1g43800, and At3g02630) identified in Arecaceae • At2g43710 was highly expressed in coconut/oil palm oil-producing tissues • Stearoyl-acyl carrier protein desaturase designated *CocoFAD* gene improves the C16:1 and C18:1 components and reduce the C18:0 component	Gao et al. (2014)
8	Acyl transferases [diacylglycerol acyltransferase (DGAT), glycerol-3-phosphate acyltransferase (GPAT) and lysophosphatidic acid acyltransferase (LPAAT)	• Involved in TAG synthesis • *CnDGAT2* gene characterized in coconut endosperm was found to express a maximum from seven months of pollination • *CnGPAT9* gene is characterized in coconut palm and was shown to involve in C16:0 assembly and plays a major role in the accumulation of medium-chain fatty acid • *CnLPAAT* gene has shown to play a role in reducing the C12:0 and C14:0 contents based on heterologous gene expression studies	Zheng et al. (2019), Yuan et al. (2015a, b)
9	Fructose-1, 6-bisphosphate aldolase (FBA)–cytosolic	• *CoFBA1* gene was characterized in coconut • Heterologous expression showed changes in fatty acid content and lipid composition	Feng et al. (2020)
10	Wrinkled1 /WRI1 Transcription factor	• 'Master regulator' for transcriptional control of TAG biosynthesis in plants • Seed developmental regulators, LEAFY COTYLEDON1 (LEC1), and LEAFY COTYLEDON2 (LEC2) are identified as transcriptional regulators of WRI • FUSCA3 (FUS3) is another TF that is speculated to mediate expression of *AtWRI1* similar to LEC2 • WRI1 is also known to be regulated by the activity of a transcriptional repressor MYB89	Xiao et al. (2017), Fei et al. (2020), Sun et al. (2017), Kong et al. (2020)

(continued)

Table 10.2 (continued)

Sl. No.	Enzymes/enzyme complex/genes	Features	Reference(s)
		• Analysis of the oil palm genome showed that 207 out of the 338 genes governing lipid synthesis possess the AW [CnTnG](n)$_7$[CG] motif that binds with WRI • It belongs to the AP2/EREBP group of transcription factors. CnWRI is cloned and sequenced and shows high expression of immature endosperm compared to other tissues • The activity of WRI1 activity is influenced by post-translational modifications and gene interactions	
11	Coconut miRNAs involved in endosperm development	• Differentially expressed miRNAs in the mature and immature endosperm • Transcription factors are the targets of miRNAs	Li et al. (2009)

regions located in the homologous segments of the genomes of African oil palm or date palm. Conspicuous enzymes such as pyruvate dehydrogenase complex (PDHC), acyl carrier protein (ACP), stearoyl-ACP desaturases (SAD), and acyl-ACP thioesterase A (FATA) have been shown to possess the conserved gene copy numbers among the palms suggesting their common evolutionary lineage. Moreover, the proteins or enzymes of plastidial origin involved in the biochemical conversion of pyruvate to fatty acids exhibited relatively high expression in oil-producing tissues viz., endosperm (than in embryo or leaf) and mesocarp (than in leaf) of coconut and oil palm, respectively. The oil biosynthesis genes in coconut endosperm are acyl-ACP thioesterase A and B (*FatA* and *FatB*); β-ketoacyl-ACP synthases I, II, and III (*KasI*, *KasII*, and *KasIII*); and *LPAAT* genes (Bates et al. 2013). *LPAAT* is a major gene responsible for the high lauric acid accumulation in coconut (Reynolds et al. 2019; Lantican et al. 2019). Expression profile of endogenous small RNAs, i.e., miRNAs during the developmental stages of coconut endosperm utilizing microRNAs microarray technique identified a set of differentially expressed miRNAs, and target transcripts for some of these miRNAs were also predicted (Li et al. 2009). This study identified 176 coconut miRNAs from mature and immature tissues, of which 23 miRNAs were upregulated, and nine were downregulated in the mature endosperm (Li et al. 2009). Also, TFs such as GRL3, homeobox-leucine zipper family protein (HB8), DEAD/DEAH box, and S locus F-box-related domain were identified as targets for miRNAs (Li et al. 2009).

10.4 Molecular Breeding Approaches in the Exploration of Oil Yield Trait

Fruit and its components determine the oil yield in coconut and oil palm as the oil is expelled from the kernel of solid coconut endosperm and mesocarp and seed kernel of oil palm. Currently, palm oil (along with palm kernel oil) constitutes one of the most important and chiefly consumed vegetable oils, with ~75.2 million tons traded globally (Kushairi et al. 2018). The high fresh fruit bunch (FFB) yields and mesocarp content are the major palm oil yield determinants—the prime focus of the oil palm breeders. Along with this short height increment, best oil quality parameters, biotic, and abiotic stress are also major traits that influence oil yield and FFB production in oil palm. The widely grown high-yielding oil palm variety, the thin-shelled *tenera* having oil: bunch of more than 20% is generally obtained by crossing the female parent thick-shelled *dura* (O/B approximately 17%) with the male parent shell-less *pisifera*. The *pisifera* is used as the pollen parent since it is generally female sterile (Babu et al. 2017). There is considerable genetic variation in photosynthetic rates in the modern oil palm breeding population (Smith 1991). In Indian germplasm, sufficient variability is available to exploit them to produce palms with a high bunch index.

In coconut, quantitative trait loci (QTL) analysis for fruit component traits was conducted by Baudouin et al. (2006). The authors identified four major QTLs linked to endosperm humidity. One of them was co-located in linkage group 12 flanked by mEgCIR2433 and CnCirK1 microsatellite markers (Baudouin et al. 2006). Earlier, fruit component analysis was conducted to identify the major trait linked to DNA markers in oil palm (Rance et al. 2001). On the other hand, in oil palm, about 16 putative QTLs distributed in seven linkage groups were linked to the yield of fruit bunches, oil per bunch, and other component traits (Jeennor and Volkaert 2014). Genes involved in sucrose metabolism are downregulated in high oil yielding oil palm (Wong et al. 2017). Tenera, characterized with a medium-thick shell, showed heterosis for oil yield and is commercially exploited to enhance oil palm productivity. Tenera hybrid is derived by crossing two different genotypes controlled by a single gene, thick-shelled dura form with thin-shelled pisifera form. Genomic investigations identified the shell thickness (SHELL) gene as a *SEEDSTICK* (STK) transcription factor regulated by the formation of heterodimers (Singh et al. 2013). Oil yield per dry matter was governed by three loci in oil palm (Teh et al. 2016). The authors also mapped two major QTLs on chromosome 6 and one on chromosome 5 governing the trait of oil per mesocarp in oil palm. The third QTL on chromosome 5 was found closer to the gene pyruvate kinase with a known role in fatty acid biosynthesis (Teh et al. 2016).

A high-density linkage map was constructed in oil palm using 1357 SNPs and 123 SSR markers. They covered a length of 1527 cM, at an average length of 1.03 cM across the 16 chromosomes (Bai et al. 2017). This study has identified QTLs linked to oil to bunch and oil to dry mesocarp in the LG1, LG8, and LG10 in an F1 mapping population. These loci explained a phenotypic variance of 7.6–13.3%. The beneficial palms having an O/B of 30.97% found to have two QTLs at a high p-value of 0.01, higher than that of trees without any beneficial QTL genotypes (average O/B of 28.24%) (Bai et al. 2017).

Also, in oil palm, the genotyping-by-sequencing (GBS) method was used by Osoria-Guarin et al. (2019) in 378 interspecific (O × G) hybrid population for 10 agronomic traits using 3776 single nucleotide polymorphisms for genome-wide association analysis (GWAS). The genomic regions harboring candidate genes govern multiple functions such as the growth of tissue, physiological processes, and cellular trafficking. However, it is pertinent to acknowledge the fact that GWAS in perennial crops with outcrossing behaviors have serious limitations due to the insufficiency of marker density which is required for population diversity and the effects of population structure causing high false positive rates. Teh et al. (2016) studied GWAS on *tenera* palms which were selected based on the linkage disequilibrium (LD) distance, using 200 K SNPs for oil to dry mesocarp content (O/DM). They found eighty SNP loci associated with *O/DM,* and three key signals were identified. Ithnin et al. (2017) performed genotyping analysis of 422 oil palms obtained from the Malaysian Palm Oil Board (MPOB), Angola germplasm collection and also documented the traits of economic importance from these palms. Multi-locus GWAS was conducted utilizing genome-wide efficient and least absolute shrinkage and selection operator (LASSO) models. This study had led to the identification of 19 QTLs governing eight traits. Among these, four QTLs identified in Angola germplasm known to govern oil palm bunch components were found on the chromosomes 4, 8, and 11 suggesting the potentiality of these QTLs for introgression in other advanced breeding populations to improve the bunch and oil characters.

10.5 Revelations from Comparative Genomics

Fruits of 144 species of palms are a potential source of vegetable oils (Guerin et al. 2020). Many species of palms are devoid of oil; hence, the variation offers scope for comparative omics analysis to understand the oil biosynthesis genes. With the availability of genome sequences of three major palms, namely date palm (*Phoenix dactylifera*) (Al-Dous et al. 2011), African oil palm (*Elaeis guineensis*) (Singh et al. 2013), and coconut (*Cocos nucifera*) (Xiao et al. 2017) considerable interests have been generated to compare the genetic components of oil biosynthesis pathways of these palms (Xiao et al. 2019). The underlying reason for the interests in comparison among these species emerges from their inherent differences in the oil production profiles. As stated above, coconut endosperm (copra) is rich in oil ~63%, of which 50% is constituted by lauric acid. In contrast, the mesocarp of African oil palm is characterized by 90% of oil, while date palm mesocarp is devoid of any oil. Interestingly the kernel of African oil palm also possesses oil which is predominant in lauric acid, similar to that of coconut. In contrast, seeds of date palm have a little oil predominantly of oleic acid (Xiao et al. 2019).

Comparative transcriptome and metabolome profiling of oil palm and date palm mesocarp suggest the importance of the supply of pyruvate in plastid tissue rather than TAG assembly (Bourgis et al. 2011). Also, transcriptional regulation of lipid and carotenoid metabolism of oil palm mesocarp was deciphered applying the RNA-seq approach (Tranbarger et al. 2011). Transcriptome and fatty acid profiles of three different tissues, namely mesocarp, endosperm, and embryo of African oil palm revealed marked differences in the contribution from cytosolic or plastidial glycolytic routes between the mesocarp and endosperm; however, no transcriptional differences in the genes involved in the conversion of sucrose to pyruvate were observed (Dussert et al. 2013). Comparative fatty acid composition and genomic analysis in coconut tissues (leaf, endosperm and embryo) and oil palm tissues (leaf, mesocarp), date palm tissues (leaf, mesocarp) revealed 806, 840, and 849 genes from their respective genomes. Around 73% and 62% of the coconut-derived lipid biosynthesis-related genes are available in the homologous genome segments of *Elaeis guineensis* and *Phoenix dactylifera*, respectively (Xiao et al. 2019). Despite the evolutionary divergence between the oil palm and coconut and consequently the appreciable differences in their gene expression pattern of fatty acid biosynthesis and TAG conversion, both the species utilize the conserved oil biosynthesis pathways and store oils in their respective tissues (Xiao et al. 2019). Comparative analysis of lipid-related genes of three palm species to that of Arabidopsis revealed that ~25% of the genes are single copy, and 50–55% of the genes showed two or three orthologous genes copies for every gene in Arabidopsis (Xiao et al. 2019). Despite the conserved structural features of *cis*-acting elements of genes among the three palms, the lipid-related genes showed little or no expression. For multiple gene families (*WRI1* and *FatB*), it was inferred that sub-functionalization of these genes occurred well in advance before speciation (Xiao et al. 2019).

Notwithstanding the functional differences in the tissue types of endosperm and embryo, the transcriptional response of plastidial fatty acid synthesis genes are comparable (Xiao et al. 2019). Nevertheless, gene expression bias was observed between coconut endosperm and oil palm mesocarp (Xiao et al. 2019) and between the fruits of oil palm and date palm (Bourgis et al. 2011). Comparative expression analysis of *CnFat* genes correlated the high expression levels of *CnFatB* to increased accumulation of MCFA, i.e., lauric acid (Liang et al. 2014; Yuan et al. 2017; Xiao et al. 2019). Analysis of homologous segments of *FatB* genes (5 nos.) of palm species displays a high degree of conservation. However, the divergence between CnFatB2-2 and PdFatB2-2 suggests the role of this isoform in endosperm lipid biosynthesis in coconut (Xiao et al. 2019). Three ortholog groups of *WRI1* are identified in the palm species based on structural gene analysis. The ortholog group 1 and ortholog group II showed high expression in oil palm mesocarp and coconut and oil palm endosperm, respectively, implying the differential role for WRI orthologs in tissue-specific oil accumulation. Expression analysis of genes involved in carbon partitioning between glycolysis and oil biosynthesis pathways and the involvement of various sub-cellular locations identify that coconut endosperm shows high expression levels of phosphoenolpyruvate transporter 1 (PPT1) and plastid-fructose bisphosphate aldolase (FBA) and plastid pyruvate kinase (PK) relative to their cytosolic counterparts (Xiao et al. 2019).

10.6 Mining of Coconut Genome Assembly

The availability of relatively good quality coconut genome assemblies has spurred the interest in exploring oil biosynthesis genes of coconut (Xiao et al. 2017; Lantican et al. 2018; Rajesh et al. 2020). Since the genes involved in endosperm oil biosynthesis was not investigated much in coconut, the complete genome sequence of coconut genotype, Catigan Green Dwarf (CATD) (Lantican et al. 2018), was mined for six candidate oil biosynthesis genes involved in MCFA production utilizing suitable *in silico* tools (Manohar et al. 2019). The candidate genes were mapped in the genomic scaffolds using alignment tools that work on the principle of sequence homology. Further, the gene sequences of coconut *Kas* II and *Kas* III were characterized. Eco-TILLING (Ecotype Targeting Induced Local Lesions IN Genomes) approach was utilized in validating the sequence features in a panel of 48 coconut varieties. An SNP was detected in *LPAAT* of varieties 'West African Tall' (WAT) and 'Aguinaldo Tall' (AGDT), with an immense potential for genomics assisted breeding in coconut for improved or enhanced oil production (Manohar et al. 2019). It is important to validate the SNP so as to correlate the oil content or other phenotypic data using a suitable mapping population.

10.7 Association Mapping Studies in Coconut

Association mapping analysis was performed in search for elite alleles linked to fatty acid content using SSR markers in a total of 80 coconut accessions representing six different populations, namely Red Dwarf, Yellow Dwarf, Hainan Red, Hainan Tall, Malayan Yellow Dwarf × West African Tall (MAWA), and Aromatica Green Dwarf (Zhou et al. 2020). The accessions were grouped into the higher-fatty acid-containing germplasm subgroup (HFA) and lower-fatty acid-containing germplasm subgroup (LFA). The study identified 10 microsatellite loci; most of them at chromosome 11 was associated with fatty acid content (Zhou et al. 2020). Further exploration of elite alleles and their donor or carrier material is imperative for breeding speciality coconut cultivars. It revealed that the Aromatica Green Dwarf was found to be a donor cultivar for fatty acid content with an allele CnFAtB3-359 with a major positive effect (Zhou et al. 2020). Thus, this association analysis identified elite alleles with potential use in marker-aided selection and identification of coconut accessions as suitable parents to incorporate fatty acid-related traits.

10.8 Insights from the Endosperm Trait Mutations

Aromatic, makapuno (Nguyen et al. 2016), and sweet kernel are three popular and premium endosperm mutants of coconut and extensively explored for genetics, genomics, and biochemical investigations (Arunachalam and Rajesh, 2017).

DNA methylation events affect the endosperm development in the mutant (makapuno/gelly like) soft endosperm (Angeles et al. 2018). Aromatic mutant differed from normal coconut only in one base substitution (Dumhai et al. 2019) in the aldehyde dehydrogenase gene (*BADH2*). Galactosidase activity is very low in the soft endosperm genotypes (Phoeurk et al. 2018) compared to normal coconut endosperm. Unsaturated fatty acids, linoleic, and oleic acids were high in sweet kernel endosperm genotype (*Mohachao Narel*) compared to normal coconut endosperm (Samsudeen et al. 2013; Kumar and Samsudeen 2018). These mutants, being controlled by a single recessive gene each with pleiotropic effects, offer lot of scope in understanding the genes oil biosynthesis of coconut palms.

10.9 Concluding Remarks

Palm seeds generally do not store considerable oil, with a notable exception of the tribe Cocoseae implying the potentialities of coconut and oil palm in supplying edible oils and meeting the vegetable oil requirements. Ecologically it suggests the adaptive strategies of oil rich seeds to enhance their seed germination and seedling growth during the conditions of adversity. In this backdrop, despite the substantial improvements in the understanding of oil biosynthesis pathways in coconut much remains to be explored especially mapping QTLs linked to high oil yield or desirable quality and application of genomics assisted breeding for oil trait. However, it is anticipated that the availability of quality genome assemblies and cost-effective genome resequencing efforts would offer a way forward. Also, it is obligatory to explore the oil biosynthesis trait of coconut employing diverse molecular breeding techniques such as GWAS, GBS, and association mapping studies, etc., after considering the successful accomplishments in the related genera African oil palm. Further, the insights gained from endosperm variants such as makapuno, sweet kernel, and aroma liquid endosperm offer an alternative perspective of the coconut endosperm.

References

Al-Dous EK, George B, Al-Mahmoud ME, Al-Jaber MY, Wang H, Salameh YM, Al-Azwani EK, Chaluvadi S, Pontaroli AC, DeBarry J, Arondel V (2011) *De novo* genome sequencing and comparative genomics of date palm (*Phoenix dactylifera*). Nat Biotechnol 29 (6):521–527

Angeles JG, Lado JP, Pascual ED, Cueto CA, Laurena AC, Laude RP (2018) Towards the understanding of important coconut endosperm phenotypes: is there an epigenetic control? Agronomy 8(10):225

Arunachalam V, Rajesh MK (2017) Coconut genetic diversity, conservation and utilization. In: Ahuja MR, Jain SM (eds). Biodiversity and conservation of woody plants, pp 3–36. Springer International Publishing AG

Babu BK, Mathur RK, Kumar PN, Ramajayam D, Ravichandran G, Venu MVB (2017) Development, identification and validation of CAPS marker for SHELL trait which governs *dura, pisifera* and *tenera* fruit forms in oil palm (*Elaeis guineensis* Jacq.). PLoS One 12(2):e0171933. https://doi.org/10.1371/journal.pone.0171933

Bai B, Wang L, Lee M, Zhang Y, Alfiko Y, Ye BQ, Wan ZY, Lim CH, Suwanto A, Chua NH, Yue GH (2017) Genome-wide identification of markers for selecting higher oil content in oil palm. BMC Plant Biol 17:93

Balleza CF, Sierra ZN (1972) Proximate analysis of the coconut endosperm in progressive stages of development. Phil J Coco Stud 16:37–43

Bates PD, Stymne S, Ohlrogge J (2013) Biochemical pathways in seed oil synthesis. Curr Opin Plant Biol 16:358–364

Baudouin L, Lebrun P, Konan JL, Ritter E, Berger A, Billotte N (2006) QTL analysis of fruit components in the progeny of a Rennell Island Tall coconut (*Cocos nucifera* L.) individual. Theor Appl Genet 112 (2):258–268

Bourgis F, Kilaru A, Cao X, Ngando-Ebongue GF, Drira N, Ohlrogge JB, Arondel V (2011) Comparative transcriptome and metabolite analysis of oil palm and date palm mesocarp that differ dramatically in carbon partitioning. Proc Natl Aca Sci USA 108(30):12527–12532

Bourrellier AB, Valot B, Guillot A, Ambard-Bretteville F, Vidal J, Hodges M (2010) Chloroplast acetyl-CoA carboxylase activity is 2-oxoglutarate-regulated by interaction of PII with the biotin carboxyl carrier subunit. Proc Natl Acad Sci USA 107:502–507

Ceniza MS, Ueda S, Sugimura Y (1974) *In vitro* culture of coconut endosperm: Callus induction and its fatty acids. Phil J Coco Stud 18(2):16–19

Child R (1974) Coconuts, 2nd ed. Longman Group Ltd., London, UK. ISBN0-582-46675-X

Dorni C, Sharma P, Saikia G, Longvah T (2018) Fatty acid profile of edible oils and fats consumed in India. Food Chem 238:9–15

Du ZY, Arias T, Meng W, Chye ML (2016) Plant acyl-CoA-binding proteins: an emerging family involved in plant development and stress responses. Prog Lipid Res 63:165–181

Dumhai R, Wanchana S, Saensuk C, Choowongkomon K, Mahatheeranont S, Kraithong T, Toojinda T, Vanavichit A, Arikit S (2019) Discovery of a novel *CnAMADH2* allele associated with higher levels of 2-acetyl-1-pyrroline (2AP) in yellow dwarf coconut (*Cocos nucifera* L.). Sci Hortic 243:490–497

Dussert S, Guerin C, Andersson M, Joët T, Tranbarger TJ, Pizot M, Sarah G, Omore A, Durand-Gasselin T, Morcillo F (2013) Comparative transcriptome analysis of three oil palm fruit and seed tissues that differ in oil content and fatty acid composition. Plant Physiol 162 (3):1337–1358

Fei W, Yang S, Hu J, Yang F, Qu G, Peng D, Zhou B (2020) Research advances of WRINKLED1 (WRI1) in plants. Funct Plant Biol 47(3):185–194

Feng D, Gao L, Zheng Y, Li D, Zhou P (2020) Molecular cloning and function characterisation of a cytoplasmic fructose-1, 6-bisphosphate aldolase gene from coconut (*Cocos nucifera* L.). The J Hortic Sci Biotechnol 95 (6):703–711

Gao L, Sun R, Liang Y, Zhang M, Zheng Y, Li D (2014) Cloning and functional expression of a cDNA encoding stearoyl-ACP Δ9-desaturase from the endosperm of coconut (*Cocos nucifera* L.). Gene 549(1):70–76

Guerin C, Serret J, Montúfar R, Vaissayre V, Bastos-Siqueira A, Durand-Gasselin T, Tregear J, Morcillo F, Dussert S (2020) Palm seed and fruit lipid composition: phylogenetic and ecological perspectives. Ann Bot 125(1):157–172

Ithnin M, Xu Y, Marjuni M, Serdari NM, Amiruddin MD, Low ET, Tan YC, Yap SJ, Ooi LC, Nookiah R, Singh R (2017) Multiple locus genome-wide association studies for important economic traits of oil palm. Tree Genet Genom 13:103

Jeennor S, Volkaert H (2014) Mapping of quantitative trait loci (QTLs) for oil yield using SSRs and gene-based markers in African oil palm (*Elaeis guineensis* Jacq.). Tree Genet Genomes 10:1–14. https://doi.org/10.1007/s11295-013-0655-3

Jones A, Davies HM, Voelker TA (1995) Palmitoyl-acyl carrier protein (ACP) thioesterase and the evolutionary origin of plant acyl-ACP thioesterases. Plant Cell 7:359–371

Kartha ARS (1964) Biosynthesis of fat in ripening coconut: the *in vivo* ‘quantum’ synthesis of triglycerides. J Sci Food Agric 15(5):299–302

Kong Q, Yang Y, Guo L, Yuan L, Ma W (2020) Molecular basis of plant oil biosynthesis: insights gained from studying the WRINKLED1 transcription factor. Front Plant Sci 11:24. https://doi.org/10.3389/fpls.2020.00024

Kostik V, Memeti S, Bauer B (2013) Fatty acid composition of edible oils and fats. JHED 4:112–116

Kumar SN (2011) Variability in coconut (*Cocos nucifera* L.) germplasm and hybrids for fatty acid profile of oil. J Agric Food Chem 59:13050–13058

Kumar A, Samsudeen K (2018) Seasonal variation and developmental changes in the biochemical composition of coconut kernel in *Mohachao Narel*, a sweet endosperm coconut (*Cocos nucifera* L.) population from Maharashtra. Trends Biosci 11:1599–1603

Kushairi A, Loh SK, Azman I, Hishamuddin E, Ong-Abdullah M, Izuddin ZB, Razmah G, Sundram S, Parveez GK (2018) Oil palm economic performance in Malaysia and R&D progress in 2017. J Oil Palm Res 30:163–195. https://doi.org/10.21894/jopr.2018.0030

Lantican D, Strickler S, Canama A, Gardoce R, Mueller L, Galvez H (2018) The coconut genome: providing a reference sequence towards coconut varietal improvement. In: Proceedings of the plant and animal genome XXVI conference, San Diego, CA, USA, pp 13–17

Lantican DV, Strickler SR, Canama AO, Gardoce RR, Mueller LA, Galvez HF (2019) *De novo* genome sequence assembly of dwarf coconut (*Cocos nucifera* L. ‘Catigan Green Dwarf’) provides insights into genomic variation between coconut types and related palm species. G3: Genes Genomes Genet 9:6

Laureles LR, Rodriguez FM, Rean˜o CE, Santos GA, Laurena AC, Mendoza EMT (2002) Variability in fatty acid and triacylglycerol composition of the oil of coconut (*Cocos nucifera* L.) hybrids and their parentals. J Agric Food Chem 50:1581–1586

Li D, Zheng Y, Wan L, Zhu X, Wang Z (2009) Differentially expressed microRNAs during solid endosperm development in coconut (*Cocos nucifera* L.). Sci Hortic 122(4):666–669

Liang Y, Yuan Y, Liu T, Mao W, Zheng Y, Li D (2014) Identification and computational annotation of genes differentially expressed in pulp development of *Cocos nucifera* L. by suppression subtractive hybridization. BMC Plant Biol 14(1):205

Li-Beisson Y, Shorrosh B, Beisson F, Andersson MX, Arondel V, Bates PD, Baud S, Bird D, DeBono A, Durrett TP, Franke RB (2010) Acyl-lipid metabolism. Arabidopsis Book 8:e0133. Published online 2010 Jun 11. https://doi.org/10.1199/tab.0133

Manohar AN, Lantican DV, Dancel MP, Gardoce RR, Galvez HF (2019) Genome-guided molecular characterization of oil genes in coconut (*Cocos nucifera* L.). Philipp J Sci 148(S1):183–191

Naresh Kumar S (2007) Capillary gas chromatography method for fatty acid analysis of coconut oil. J Plantn Crops 35:23–27

Naresh Kumar S, Champakam B, Rajagopal V (2000) Fatty acid composition of coconut oil among the cultivars—an insight into industrial application. Ind Coconut J 31:25–28

Naresh Kumar S, Champakam B, Rajagopal V (2004) Variability in coconut cultivars for lipid and fatty acid composition of oil. Trop Agric (trinidad and Tobago) 81(1):34–40

Nguyen QT, Bandupriya HD, Foale M, Adkins SW (2016) Biology, propagation and utilization of elite coconut varieties (makapuno and aromatics). Plant Physiol Biochem 109:579–589

Osorio-Guarín JA, Garzón-Martínez GA, Delgadillo-Duran P, Bastidas S, Moreno LP, Enciso-Rodríguez FE, Cornejo OE, Barrero LS (2019) Genome-wide association study (GWAS) for morphological and yield-related traits in an oil palm hybrid (*Elaeis oleifera x Elaeis guineensis*) population. BMC Plant Biol 19:533. https://doi.org/10.1186/s12870-019-2153-8

Phoeurk C, Somana J, Sornwatana T, Udompaisarn S, Traewachiwiphak S, Sirichaiyakul P, Phongsak T, Arthan D (2018) Three novel mutations in α-galactosidase gene involving in galactomannan degradation in endosperm of curd coconut. Phytochem 156:33–42

Rajesh MK, Chowdappa P, Behera SK, Kasaragod S, Gangaraj KP, Kotimoole CN, Nekrakalaya B, Mohanty V, Sampgod RB, Banerjee G, Das AJ (2020) Assembly and annotation of the nuclear and organellar genomes of a dwarf coconut (Chowghat Green Dwarf) Possessing enhanced disease resistance. OMICS A J Integ Biol 24(12):726–742

Ramesh SV, Krishnan V, Praveen S, Hebbar KB (2019) Coconut oil–scientific facts. Curr Sci 117(4):564–565

Ramesh SV, Pandiselvam R, Thushara R, Manikantan MR, Hebbar KB, Beegum S, Mathew AC, Neenu S, Shil S (2020) Engineering intervention for production of virgin coconut oil by hot process and multivariate analysis of quality attributes of virgin coconut oil extracted by various methods. J Food Process Eng 43(6):e13395

Rance KA, Mayes S, Price Z, Jack PL, Corley RH (2001) Quantitative trait loci for yield components in oil palm (*Elaeis guineensis* Jacq.). Theor Appl Genet 103: 1302–1310

Reynolds KB, Cullerne DP, El Tahchy A, Rolland V, Blanchard CL, Wood CC, Singh SP, Petrie JR (2019) Identification of genes involved in lipid biosynthesis through *de novo* transcriptome assembly from *Cocos nucifera* developing endosperm. Plant Cell Physiol 60 (5):945–960

Rodriguez FM, Laurena AC, Laureles LR, Santos GA, Mendoza EM (1998) Determination of fatty acid and triglyceride composition of different coconut (*Cocos nucifera* L.) cultivars and hybrids. Phil J Crop Sci 23:84

Samsudeen K, Rajesh MK, Nagwaker DD, Reshmi R, Kumar PA, Devadas K, Anitha K (2013) Diversity in *Mohachao Narel*, a sweet endosperm coconut (*Cocos nucifera* L.) population from Maharashtra, India. Natl Acad Sci Lett 36(3):319-330

Schuch R, Winter E, Bruck FM, Brummel M, Spener F (1997) *β*-Ketoacyl-acyl carrier protein synthases in the regulation of fatty acid synthase activity in higher plants-an overview. Fett-Lipid 99(8):278–281

Singh R, Ong-Abdullah M, Low ET, Manaf MA, Rosli R, Nookiah R, Ooi LC, Ooi SE, Chan KL, Halim MA, Azizi N (2013) Oil palm genome sequence reveals divergence of inter fertile species in old and new worlds. Nature 500(7462):335–339

Smith BG (1991) Oil palm breeding: the potential for using physiological selection criteria. Paper presented at International oil palm conference, Kuala Lumpur

Sun R, Ye R, Gao L, Zhang L, Wang R, Mao T, Zheng Y, Li D, Lin Y (2017) Characterization and ectopic expression of CoWRI1, an AP2/EREBP domain-containing transcription factor from coconut (*Cocos nucifera* L.) endosperm, changes the seeds oil content in transgenic *Arabidopsis thaliana* and rice (*Oryza sativa* L.). Front Plant Sci 8:63

Teh CK, Ong AL, Kwong QB, Apparow S, Chew FT, Mayes S, Mohamed M, Appleton D, Kulaveerasingam H (2016) Genome-wide association study identifies three key loci for high mesocarp oil content in perennial crop oil palm. Sci Rep 6:19075

Tranbarger TJ, Dussert S, Joët T, Argout X, Summo M, Champion A, Cros D, Omore A, Nouy B, Morcillo F (2011) Regulatory mechanisms underlying oil palm fruit mesocarp maturation, ripening, and functional specialization in lipid and carotenoid metabolism. Plant Physiol 156(2):564–584

Troncoso-Ponce MA, Kilaru A, Cao X, Durrett TP, Fan J, Jensen JK, Thrower NA, Pauly M, Wilkerson C, Ohlrogge JB (2011) Comparative deep transcriptional profiling of four developing oilseeds. Plant J 68:1014–1027

Wong YC, Teh HF, Mebus K, Ooi TE, Kwong QB, Koo KL, Ong CK, Mayes S, Chew FT, Appleton DR, Kulaveerasingam H (2017) Differential gene expression at different stages of mesocarp development in high- and low-yielding oil palm. BMC Genomics 18:470. https://doi.org/10.1186/s12864-017-3855-7

Xiao S, Chye ML (2011) New roles for acyl-CoA-binding proteins (ACBPs) in plant development, stress responses and lipid metabolism. Prog Lipid Res 50:141–151

Xiao Y, Xu P, Fan H, Baudouin L, Xia W, Bocs S, Xu J, Li Q, Guo A, Zhou L, Li J (2017) The genome draft of coconut (*Cocos nucifera*). Giga Sci 6(11):p.gix095

Xiao Y, Xia W, Mason AS, Cao Z, Fan H, Zhang B, Zhang J, Ma Z, Peng M, Huang D (2019) Genetic control of fatty acid composition in coconut (*Cocos nucifera*), African oil palm (*Elaeis guineensis*), and date palm (*Phoenix dactylifera*). Planta 249(2):333–350

Yuan Y, Chen Y, Yan S, Liang Y, Zheng Y, Dongdong L (2014) Molecular cloning and characterisation of an acyl carrier protein thioesterase gene (CocoFatB1) expressed in the endosperm of coconut (*Cocos nucifera*) and its heterologous expression in *Nicotiana tabacum* to engineer the accumulation of different fatty acids. Funct Plant Biol 41(1):80–86

Yuan Y, Liang Y, Gao L, Sun R, Zheng Y, Li D (2015) Functional heterologous expression of a lysophosphatidic acid acyltransferase from coconut (*Cocos nucifera* L.) endosperm in *Saccharomyces cerevisiae* and *Nicotiana tabacum*. Sci Hortic 192:224–230

Yuan Y, Liang Y, Li B, Zheng Y, Luo X, Dongdong L (2015) Cloning and function characterization of a β-Ketoacyl-acyl-ACP Synthase I from Coconut (*Cocos nucifera* L.) endosperm. Plant Mol Biol Rep 33:1131–1140. https://doi.org/10.1007/s11105-014-0816-z

Yuan Y, Gao L, Sun R, Yu T, Liang Y, Li D, Zheng Y (2017) Seed-specific expression of an acyl-acyl carrier protein thioesterase CnFatB3 from coconut (*Cocos nucifera* L.) increases the accumulation of medium-chain fatty acids in transgenic Arabidopsis seeds. Sci Hortic 223:5–9

Zheng Y, Jin Y, Yuan Y, Feng D, Chen L, Li D, Zhou P (2019) Identification and function analysis of a type 2 diacylglycerol acyltransferase (DGAT2) from the endosperm of coconut (*Cocos nucifera* L.). Gene 702:75–82

Zhou L, Yarra R, Cao H (2020) SSR based association mapping analysis for fatty acid content in coconut flesh and exploration of the elite alleles in *Cocos nucifera* L. Curr Plant Biol 21:100141

11 Aroma and Fragrance: A Case Study for Trait-Related Gene Evolution in Coconut

V. Arunachalam, S. V. Ramesh, M. K. Rajesh, and K. S. Muralikrishna

Abstract

High-value coconut mutants offer competitive advantage of growing coconut palm vis-a-vis other oil seed crops including oil palm. Endosperms of Makapuno and aromatic coconut mutants generally fetch a premium price in the export market and thereby have a huge potential to earn valuable foreign exchange. In this chapter, the developments in the field of biochemistry, genetics, genomics and bioinformatics of aromatic mutants of plants, with special reference to the coconut palms, are discussed. Pandan-like aroma in liquid endosperm of Nam Hom Green Dwarf and Yellow Dwarf from Thailand has been attributed to the accumulation of the aromatic compound 2-acetyl-1-pyrroline (2-AP). The mutation in coconut is caused by single non-synonymous base substitution at two different locations in a single gene, viz., *BADH2* (betaine aldehyde dehydrogenase). Homology modelling studies have indicated that these amino acid changes increase the activity of aminoaldehyde dehydrognease, which leads to accumulation of aroma compounds. Mutations causing aroma trait have been extensively investigated in rice, soybean, cucurbits and other crops. The comparative analysis of the same with that of coconut palm is presented herein. A roadmap for the future breeding work in exploiting the gene conferring aroma trait in coconut improvement is also highlighted.

11.1 Introduction

The production of pandan-like fragrance or aroma in crops and cultivars of many species namely rice (Buttery et al. 1982), soybean (Fushimi and Masuda 2001), sorghum (Kottur 1919), cucumber (Pramnoi et al. 2013) and coconut (Chomcalow 1999) are known (Somta et al. 2019). Economically, the fragrant genotypes of these crops are valued high compared to their non-fragrant counterparts in the international markets. Aromatic and Makapuno coconuts are endosperm mutants of very high commercial significance (Nguyen et al. 2016). Lactones are known to be important flavour imparting compounds across various food products (Dufossé et al. 1994). A mutation in the dwarf form of coconut 'Mu Si Khieo' (Chomcalow 1999) has yielded

V. Arunachalam (✉)
ICAR-Central Coastal Agricultural Research Institute, Ela, Goa, India
e-mail: v.arunachalam@icar.gov.in

S. V. Ramesh · M. K. Rajesh · K. S. Muralikrishna
ICAR-Central Plantation Crops Research Institute, Kasaragod, Kerala, India
e-mail: ramesh.sv@icar.gov.in

M. K. Rajesh
e-mail: rajesh.mk@icar.gov.in

K. S. Muralikrishna
e-mail: muralikrishna.ks@icar.gov.in

M. K. Rajesh et al. (eds.), *The Coconut Genome*, Compendium of Plant Genomes,
https://doi.org/10.1007/978-3-030-76649-8_11

the aromatic coconut form. The young fresh coconuts of this aromatic genotype are sweet and have the characteristic pandan-like aroma similar to that of pandan leaf (*Pandanus amaryllifolius* Roxb.). The aroma or fragrance is due to the presence of aromatic compound 2-acetyl-1-pyrroline (2AP). About five coconut genotypes were analysed for the composition of tender coconut water with special emphasis for volatile profile. Thailand Aromatic Green Dwarf genotype showed high concentrations of ethyloctonate and hexanols (Prades et al. 2012) compared to other non-aromatic coconut varieties. The compound 2AP, which is a major volatile chemical responsible for the pandan-like fragrance, was exclusively present in aromatic coconuts and conspicuously lacking in non-aromatic coconuts (Saensuk et al. 2016). 2AP was found to accumulate in aromatic coconut fruits since 5.5 months of pollination until 11 months (Luckanatinvong et al. 2018). The aroma in coconut fruit was found to accumulate when palms experience drought stress and the trait was less prominent during rainy season (Luckanatinvong et al. 2018). Since pandan-like aroma is an economically important trait for food crops it is imperative to comprehend the genetic, biochemical basis so as to aid the crop improvement programmes of coconut.

11.2 Biochemistry of Aroma

The biosynthetic pathway of the aromatic compound 2AP, also known as 1-(3,4-dihydro-2H-pyrrol-5yl) ethanone, is found downstream of biochemical reactions of polyamine catabolism mediated through the formation of ϒ-aminobutyraldehyde (ABAL) (Bradbury et al. 2008). The *Oryza sativa Os2AP*, identified to be responsible for 2AP biosynthesis, is a paralog to betaine aldehyde dehydrogenase 1 (*BAD1*) and *BAD*2 (Bradbury et al. 2008), whereas the corresponding genes of other plants are identified as orthologs of *O. sativa BADH2* (Juwattanasomran et al. 2011; Yundaeng et al. 2013, 2015). All the BADH proteins are categorised as aminoaldehyde dehydrogenases (AMADHs), belonging to the superfamily of aldehyde dehydrogenases (ADHs) (Hou and Bartels 2015). Biochemically, ADHs are a group of $NAD(P)^+$-dependent enzymes that metabolise cellular and exogenous aldehyde molecules into their corresponding carboxylic acids and participate in amino acid metabolism besides in the generation of osmoprotectants (Yoshida et al. 1998). Amino-aldeyde dehydrogenase (AMADH) enzyme plays a key role in preventing the formation of the aromatic compound 2AP, and consequently, aroma in plants. Isotopic studies have identified the role of proline as nitrogen source for the synthesis of 2AP (Yoshihashi et al. 2002). Further, the amino acids proline and arginine, through putrescine, are converted to form either amino butyraldehye (GABald) or acetyl or guanidine-butyaraldehyde. GABald later is converted into delta 1-pyrolline in the absence of BAD2. The end product later forms 2AP by addition of acetyl group in rice (Bradbury et al. 2008). GABA is formed in non-aromatic rice due to the presence of betaine aminoaldehyde dehydrogenase (BAD2) activity. Thus, the aroma characteristic feature results from dysfunctional AMADHs. It is also proposed that the main physiological role underlying the synthesis of 2AP has been the detoxification mechanism as non-functional BADHs would cause toxic cellular concentration of GABald (Saensuk et al. 2016).

Aroma is an important trait controlled by a single recessive gene *BADH* across many plant species (Somta et al. 2019). The mutations in the gene, by either small or large deletions or base substitutions on or before the active cysteine encoding site, lead either to premature truncation of gene by frame-shift mutation or change of amino acid due to change of codon by substitution. The SNPs and deletion events in exon 7 region of the fragrance gene *BAD* of *O. sativa* results in a stop codon and premature truncation of the product of *BAD2* which leads to the synthesis and accumulation of 2AP (Bradbury et al. 2005). However, in coconut, G to C base substitution leading to change of amino acid from alanine to proline at substrate binding site of position 442 was shown to impart aromatic genotype (Saensuk et al. 2016;

Vongvanrungruang et al. 2016). Similarly, in Yellow Aromatic Dwarf coconut, Dumhai et al. (2019) showed that a single nucleotide change from G to A, resulting in base substitution in the conserved cysteine (C) active site (ALDH) at exon 9, causes amino acid substitution from arginine to histidine at 299 in *CnAMADH2*. Contrarily, a long deletion of 806 bases in *BADH* led to the development of aromatic winter melon (Ruangnam et al. 2017). A dimeric deletion (TT) in the exon 10 of the *BADH*2 gene produced aromatic genotype in soybean (Arikit et al. 2011). The deletions in the gene regions in both these crops resulted in the truncated gene products due to the occurrence of premature stop codon by frame-shift mutation.

11.3 Structure Prediction of Coconut AMADHs

Protein structure 4I8P chain A of maize was utilised as a template to attempt homology modeling of coconut BADH2 (Vongvanrungruang et al. 2016). This study revealed that the change of amino acid from alanine to proline at substrate binding site of position 442 was shown to impart aromatic genotype in coconut. The homology modelling-based study indicated the role of ring side chain in proline compared to alanine with a methyl group in deactivating the enzyme BADH2 (Vongvanrungruang et al. 2016). Similarly, protein structure of 3IWJ of garden pea was used as a template to derive the structural model of *BADH2* of Aromatic Green Dwarf coconut of Thailand (Saensuk et al. 2016). The A442P mutation in this form creates a steric conformational change in the loop region leading to destabilised dimer formation of the enzyme (Saensuk et al. 2016). The same template was used in Thailand Aromatic Yellow Dwarf (Dumhai et al. 2019), and the H299R mutation was also shown to destabilise near the cytosine active site (C294) due to its close proximity it leading to abolition of enzyme activity.

11.4 Genetics and Genomics of Aroma

Presence of aroma or fragrance trait in most crops is controlled by a monogenic recessive gene action with loss-of-function of gene encoding enzyme betaine aldehyde dehydrogenase 2 (BADH2) belonging to aminoaldehyde dehydrogenase family (Somta et al. 2019). Nevertheless, the intensity of fragrance is always found to be quantitative in nature. The aromatic trait is controlled by a single gene, known as *Os2AP* in rice and *GmAMADH* in soybean (Arikit et al. 2011), *CnAMADH2* in coconut (Dumhai et al. 2019) and *BhAMADH* in winter melon (Ruangnam et al. 2017). However, functionally these diverse genes are similar encoding deactivated aminoaldehyde dehydrogenase enzyme consequently leading to the accumulation of 2AP. Interestingly, in rice, the levels of 2AP was not only correlated with the downregulation of *BADH2* and glycerladehyde-3-phosphate dehydrogenase (*GAPDH*) genes but also with the upregulation of genes such as Δ1-pyrolline-5-carboxylic acid synthesis (*P5CS*), and triose phosphate isomerase (*TPI*). Concomitant with this, environmental factors such as planting time and nutrient status also influenced the production of 2AP (Mathureet al. 2014; Yoshihashi et al. 2002). In soybean too, the BADH encoding loci responsible for the fragrance and aroma of vegetable soybean could account for only 17% and 43% of the 2AP and fragrance variations, respectively, in gene mapping studies (Juwattanasomran et al. 2011) and 61% variation for fragrance in Chamame-based mapping population analysis (Juwattansomran et al. 2012). It thus suggests the polygenic nature of this trait in these crops. The betaine aldehyde dehydrogenase (BAD) gene *BAD*2, which controls fragrance in rice through a recessive (*fgr*) gene (Bradbury et al. 2005), is located in chromosome 8 (Chen et al. 2006), even though there are minor genes located at chromosome 4 that also modify the gene action of fragrance in rice. The *CnAMADH2* gene of Aromatic Green Dwarf coconut of Thailand has been characterised and is 5.6 Kb long with 15 exons and 14 introns (Saensuk et al. 2016). Aromatic coconuts in Thailand have characteristic strong Pandan-like aroma similar to the aroma of aromatic rice linked to the biochemical 2AP. The mutation is found to be associated with the two different SNPs in the exon region of aminoaldehyde dehydrogenase (*CnAMADH2*) encoding gene in Aromatic Green Dwarf (Saensuk et al. 2016) and Yellow Aromatic Dwarf (Dumhai et al. 2019). Ten AFLP (Amplified Fragment Length Polymorphism) markers could not differentiate the aromatic from non-aromatic coconut (Nartvaranant 2019) varieties of Thailand. CAPS markers were designed to discriminate aromatic from non-aromatic coconut using *Bcc*I restriction enzyme recognising the base substitution of G-A and amplifying a 320 bp fragment in yellow dwarf aromatic coconut (Dumhai et al. 2019). A biallelic SNP marker was used to distinguish the aromatic from non-aromatic green dwarf coconut since the size of the PCR products of aromatic genotype (587 bp) differed significantly from the non-aromatic (238 bp) genotypes (Saensuk et al. 2016). Two DNA based markers, viz. CnBadh-m1 primer, specifically amplifying a 290 bp genomic region in aromatic coconut, and CnBadh-m2 primer, specifically amplifying a 535 bp genomic segment in non-aromatic coconut, discriminates the Aromatic Green Dwarf coconut of Thailand (Vongvanrungruang et al. 2016). Unlike the 2AP biosynthesis in cucumber, pollen also confers xenia effect in 2AP biosynthesis in coconut. Pollen from the non-aromatic varieties yielded coconut fruits with low to very low quantum of 2AP, whereas pollen from aromatic varieties produced nuts with high 2AP content (Pooprasert et al. 2019).

11.5 Germplasm and Genetic Improvement of Aromatic Coconuts

Dua is an aromatic coconut variety of Vietnam (Long 2005). Nam Hom or Aromatic Green Dwarf (Chomcalow 1999), Thailand Green Dwarf, and Aromatic Yellow Dwarf (Dumhai et al. 2019) are different types/names of aromatic coconut varieties of Thailand. Aromatic coconut types are also grown in Indonesia, Malaysia, the Philippines, China and other countries. The genotype, Weyne 4, is a selection from aromatic coconut with pleasant aroma in the endosperm developed at Hainan island of China (Fan et al. 2011).

11.6 Suggestions and Concluding Remarks

The genomics of aroma in Thailand Aromatic Green Dwarf and Yellow Aromatic Dwarf are understood from their respective gene sequences (Saensuk et al. 2016; Dumhai et al. 2019). However, the genomics of Dua and other aromatic coconut populations require similar DNA sequence-based analysis for *BADH* gene variations and structural significance of these mutations. The trait-linked marker database and map positions (https://tropgenedb.cirad.fr/ accessed on 20.04.2020) are available for coconut genome (Ruiz et al. 2017) which is an ever-growing resource for coconut genomic science. Aromatic gene of rice is localised in the chromosome 8 (Chen et al. 2006). The whole draft genomes of Hainan Tall (Xiao et al. 2017), Catigan Green Dwarf (Lantican et al. 2019) and Chowghat Green Dwarf (Rajesh et al. 2020) offer immense scope to understand the precise genomic location of the *BADH* gene and its structure in coconut chromosome. Though genetic dissection in coconut has revealed that aroma character is governed by single-recessive gene, investigations in monocot such as rice delineating the role of environment and other factors contributing to the quantitative nature of the trait requires further study. Also, the potential xenia effect observed in coconut further requires thorough investigation to understand the complex genetic nature of the trait. Current state of knowledge that a single-gene governs aromatic trait in coconut could offer clue to coconut breeder to harness the same in breeding programs. High-yielding aromatic coconuts with more volume of tender nut water can boost the income of coconut farmers.

References

Arikit S, Yoshihashi T, Wanchana S, Uyen TT, Huong NT, Wongpornchai S, Vanavichit A (2011) Deficiency in the amino aldehyde dehydrogenase encoded by *GmAMADH2*, the homologue of rice *Os2AP*, enhances 2-acetyl-1-pyrroline biosynthesis in soybeans (*Glycine max* L.). Plant Biotechnol J 9 (1):75–87

Bradbury LM, Fitzgerald TL, Henry RJ, Jin Q, Waters DL (2005) The gene for fragrance in rice. Plant Biotechnol J 3(3):363–370

Bradbury LM, Gillies SA, Brushett DJ, Waters DL, Henry RJ (2008) Inactivation of an aminoaldehyde dehydrogenase is responsible for fragrance in rice. Plant Mol Biol 68(4–5):439–449

Buttery RG, Ling LC, Juliano BO (1982) 2-Acetyl-1-pyrroline: an important aroma component of cooked rice. Chem Ind 23:958–959

Chen S, Wu J, Yang Y, Shi W, Xu M (2006) The *fgr* gene responsible for rice fragrance was restricted within 69 kb. Plant Sci 171(4):505–514

Chen W, Zhang G, Chen W, Zhong Q, Chen H (2018) Metabolomic profiling of matured coconut water during post-harvest storage revealed discrimination and distinct changes in metabolites. RSC Adv 8 (55):31396–31405

Chomcalow N (1999) Coconut varieties in Thailand. AU J Technol 3(1):19–30. Retrieved from https://repository.au.edu/handle/6623004553/13765

Dufossé L, Latrasse A, Spinnler HE (1994) Importance des lactones dans les arômesalimentaires: structure, distribution, propriétéssensoriellesetbiosynthèse. Sci Aliment 14:17–21

Dumhai R, Wanchana S, Saensuk C, Choowongkomon K, Mahatheeranont S, Kraithong T, Toojinda T, Vanavichit A, Arikit S (2019) Discovery of a novel *CnAMADH2* allele associated with higher levels of 2-acetyl-1-pyrroline (2AP) in yellow dwarf coconut (*Cocos nucifera* L.). Sci Hortic 243:490–497

Fan H, Feng M, Huang L, Ma Z, Tang L, Qin W, Wu D, Liu L, Zhao S (2011) A new coconut cultivar 'Wenye 4.' Acta Hortic Sinica 38(4):803–804

Fushimi T, Masuda R (2001) 2-acetyl-1-pyrroline concentration of the aromatic vegetable soybean "Dadacha-mame". In: Lumpkin TA, Shanmugasundaram S (eds) Proceedings of the 2nd international vegetable soybean conference, Washington State University, Tacoma, Washington, USA, p 39. Retrieved from https://ci.nii.ac.jp/naid/10018871922/

Hou Q, Bartels D (2015) Comparative study of the aldehyde dehydrogenase (*ALDH*) gene superfamily in the glycophyte *Arabidopsis thaliana* and *Eutrema halophytes*. Ann Bot 115(3):465–479

Juwattanasomran R, Somta P, Chankaew S, Shimizu T, Wongpornchai S, Kaga A, Srinives P (2011) A SNP in *GmBADH2* gene associates with fragrance in vegetable soybean variety "Kaori" and SNAP marker development for the fragrance. Theor Appl Genet 122 (3):533–541. https://doi.org/10.1007/s00122-010-1467-6

Juwattanasomran R, Somta P, Kaga A, Chankaew S, Shimizu T, Sorajjapinun W, Srinives P (2012) Identification of a new fragrance allele in soybean and development of its functional marker. Mol Breed 29 (1):13–21. https://doi.org/10.1007/s11032-010-9523-0

Kottur GL (1919) Classification and description of the jowars of the Bombay Karnataka. Bombay Dep Agri Bull 92:16

Lantican DV, Strickler SR, Canama AO, Gardoce RR, Mueller LA, Galvez HF (2019) *De novo* genome sequence assembly of dwarf coconut (*Cocos nucifera* L.'Catigan Green Dwarf') provides insights into genomic variation between coconut types and related palm species. G3: Genes Genomes Genet 9(8):2377–2393

Long VV (2005) Status of coconut genetic resources research in Vietnam. In: Batugal PV, Rao R, Oliver J (eds) Coconut genetic resources international plant genetic institute-regional office for Asia, pp 625–633. *Selangor* DE, Malaysia: the Pacific and Oceania (IPGRI-APO), Sardang

Luckanatinvong V, Mahatheeranont S, Siriphanich J (2018) Variation in the aromatic nature of Nam-Hom coconut depends on the presence and contents of 2-acetyl-1-pyrroline. Sci Hortic 233:277–282

Mathure SV, Jawali N, Thengane RJ, Nadaf AB (2014) Comparative quantitative analysis of headspace volatiles and their association with *BADH2* marker in non-basmati scented, basmati and non-scented rice (*Oryza sativa* L.) cultivars of India. Food Chem 142:383–391. https://doi.org/10.1016/j.foodchem.2013.07.066

Nartvaranant P (2019) Analysis of markers of the genetic variants of (Nam Hom coconut *Cocos nucifera* L.) grown in western Thailand. J Thai Interdisciplinary Res 14(1):12–20

Nguyen QT, Bandupriya HD, Foale M, Adkins SW (2016) Biology, propagation and utilization of elite coconut varieties (makapuno and aromatics). Plant Physiol Biochem 109:579–589

Pooprasert A, Imsabai W, Arikit S, Boonruangrod R (2019) Xenia effect on aroma of aromatic coconut (*Cocos nucifera* L.). Songklanakarin J Pl Sci 6(4):25–31

Prades A, Assa RRA, Dornier M, Pain JP, Boulanger R (2012) Characterization of the volatile profile of coconut water from five varieties using an optimised HS-SPME-GC analysis. J Sci Food Agric 92 (12):2471–2478

Pramnoi P, Somta P, Chankaew S, Juwattanasomran R, Srinives P (2013) A single recessive gene controls fragrance in cucumber (*Cucumis sativus* L.). J Genet 92(1):147–149. https://doi.org/10.1007/s12041-013-0228-0

Rajesh MK, Chowdappa P, Behera SK, Kasaragod S, Gangaraj KP, Kotimoole CN, Nekrakalaya B, Mohanty V, Sampgod RB, Banerjee G, Das AJ, Niral V, Karun A, Mahato AK, Gaikwad K, Singh NK, Keshava Prasad TS (2020) Assembly and annotation of the nuclear and organellar genomes of a dwarf coconut (Chowghat Green Dwarf) possessing enhanced disease resistance. OMICS: J Integrative Biol 24(12):726–742

Ruangnam S, Wanchana S, Phoka N, Saeansuk C, Mahatheeranont S, de Hoop SJ, Toojinda T, Vanavichit A, Arikit S (2017) A deletion of the gene encoding amino aldehyde dehydrogenase enhances the "pandan-like" aroma of winter melon (Benincasahispida) and is a functional marker for the development of the aroma. Theor Appl Genet 130(12):2557–2565

Ruiz M, Sempéré G, Hamelin C (2017) Using TropGeneDB: a database containing data on molecular markers, QTLs, maps, genotypes, and phenotypes for tropical crops. Plant Genomics Databases. Humana Press, New York, NY, pp 161–172

Saensuk C, Wanchana S, Choowongkomon K, Wongpornchai S, Kraithong T, Imsabai W, Chaichoompu E, Ruanjaichon V, Toojinda T, Vanavichit A, Arikit S (2016) *De novo* transcriptome assembly and identification of the gene conferring a "pandan-like" aroma in coconut (*Cocos nucifera* L.). Plant Sci 252:324–334

Somta P, Kuswanto K, Srinives P (2019) The genetics of Pandan-like fragrance, 2-acetyl-1-pyrroline, in crops. AGRIVITA, J Agric Sci 41(1). https://agrivita.ub.ac.id/index.php/agrivita/article/view/2085

Vongvanrungruang A, Mongkolsiriwatana C, Boonkaew T, Sawatdichaikul O, Srikulnath K, Peyachoknagul S (2016) Single base substitution causing the fragrant phenotype and development of a type-specific marker in aromatic coconut (*Cocos nucifera*). Genet Mol Res 15(3):gmr.15038748

Xiao Y, Xu P, Fan H, Baudouin L, Xia W, Bocs S, Xu J, Li Q, Guo A, Zhou L, Li J (2017) The genome draft of coconut (*Cocos nucifera*). Giga Sci 6(11):gix095

Yoshida A, Rzhetsky A, Hsu LC, Chang C (1998) Human aldehyde dehydrogenase gene family. Eur J Biochem 251(3):549–557

Yoshihashi T, Huong NTT, Inatomi H (2002) Precursors of 2-acetyl-1-pyrroline, a potent flavor compound of an aromatic rice variety. J Agric Food Chem 50 (7):2001–2004. https://doi.org/10.1021/jf011268s

Yundaeng C, Somta P, Tangphatsornruang S, Wongpornchai S, Srinives P (2013) Gene discovery and functional marker development for fragrance in sorghum (*Sorghum bicolor* (L.)Moench). Theor Appl Genet 126(11):2897–2906

Yundaeng C, Somta P, Tangphatsornruang S, Chankaew S, Srinives P (2015) A single base substitution in *BADH/AMADH* is responsible for fragrance in cucumber (*Cucumis sativus* L.), and development of SNAP markers for the fragrance. Theor Appl Genet 128(9):1881–1892

Resistance Gene Candidates (RGCs) in Coconut Palm: A Molecular Platform for the Genetic Improvement of Resistance to Pathogens

12

Carlos Puch-Hau, Carlos Oropeza-Salín, Santy Peraza-Echeverría, Iván Córdova-Lara, and Luis Sáenz-Carbonell

Abstract

Coconut (*Cocos nucifera* L.) encounters many problems due to the lethal diseases that severely affect its growth and development. Over the last decades, plant breeders have realized that certain genes are directly involved in disease resistance, known as resistance genes (*R* genes). Most of them encode for proteins (R proteins) with a nucleotide-binding site and a leucine-rich repeat domain (NBS-LRR). However, there are certain sequences of unknown function that code for these motifs and have been linked to known *R* genes and have been denominated as 'resistance gene candidates' (RGC). Currently, 340 RGCs have been identified in the genome of *C. nucifera,* and 231 sequences have been well characterized phylogenetically indicating the existence of specific clades of genes in *C. nucifera.* Other clades also comprise a mixture of NBS sequences from *C. nucifera, E. guineensis,* and *P. dactylifera,* indicating that similar resistance genes are still shared in different genera of the Arecales. The identification of RGCs in plants is a valuable resource for the discovery of new resistance genes. It has great potential in breeding programs using marker-assisted selection, with special emphasis on the resistance to phytoplasma, a pathogen that causes a high mortality on the plantations of this species.

12.1 Introduction

The coconut palm is considered a crop of great importance since ancient times, due to the wide variety of products obtained from it. Over the past 10 years, the international commerce for coconut derived value added products has grown exponentially, particularly in the forms of coconut water, virgin coconut oil, coconut milk and coconut sugar, etc. (Prades et al. 2016). However, coconut crops face many challenges, particularly diseases, which severely affect their productivity. An important alternative to cope with these challenges is to produce palms capable of

C. Puch-Hau
Departamento de Recursos del Mar, Centro de Investigación Y de Estudios Avanzados del Instituto Politécnico Nacional, Unidad Mérida, Mérida, Mexico
e-mail: cashi026@hotmail.com

C. Oropeza-Salín · S. Peraza-Echeverría · I. Córdova-Lara · L. Sáenz-Carbonell (✉)
Centro de Investigación Científica de Yucatán, Mérida, México
e-mail: vyca@cicy.mx

C. Oropeza-Salín
e-mail: cos@cicy.mx

S. Peraza-Echeverría
e-mail: santype@cicy.mx

I. Córdova-Lara
e-mail: cordoval@cicy.mx

M. K. Rajesh et al. (eds.), *The Coconut Genome*, Compendium of Plant Genomes,
https://doi.org/10.1007/978-3-030-76649-8_12

resisting the diseases afflicting them. Over the past century, plant breeders have realized that some genes are directly related to disease resistance, these resistance genes (*R* genes) provide an effective immunity to the host plant (Belkhadir et al. 2004). Most of these *R* genes encode proteins with NBS-LRR domain. Many *R* genes have been isolated and characterized in different plant species, but none so far in coconut palm. However, many genes containing NBS domains have been mapped in other species and have shown to be linked to known *R* genes, providing evidence of their potential role in the resistance against pathogens (Pflieger et al. 2001). These genes are referred as 'resistance gene candidates' (RGC) and could be an important tool for the identification of novel *R* genes or to develop molecular markers associated to disease-resistant cultivars, as has been done in the case of coconut (Puch-Hau et al. 2016). In this chapter, we present the latest advances in the characterization of RGCs from coconut palms and their phylogenetic relationships to other plants. The coconut palm RGCs represent a valuable molecular platform for the genetic improvement of this crop in the fight against devastating diseases.

12.2 Diseases and Pests Affecting Coconut Palm

Among the most devastating diseases that affect coconut palms are those caused by phytoplasmas. Phytoplasmas are small bacteria parasitizing phloem tissue in plants and are the main causative organisms of serious diseases such as lethal yellowing (LY) in the Americas and similar lethal yellowing diseases (LYD) in other continents (Hogenhout et al. 2008). In the Americas, LY is caused by '*Candidatatus* (*Ca.*) phytoplasma palmae' (Yankey et al. 2018). In Africa, coconut plantations of Mozambique and Nigeria are affected by '*Ca* phytoplasma palmicola,' while those in Ghana and Côte d'Ivoire are affected by '*Ca.* phytoplasma palmicola-related strain' (Harrison et al. 2014). In Asia, different phytoplasmas affect plantations of different countries. For instance, coconuts in Malaysia are affected by coconut yellow decline associated with '*Ca.* Phytoplasma malaysianum' (Nejat et al. 2012); in Kerala, India, root (wilt) disease is associated with '*Ca.* Phytoplasma oryzae' (Manimekalai et al. 2014); Weligama coconut leaf wilt disease in Sri Lanka is associated with '*Ca.* Phytoplasma oryzae' (Perera et al. 2012), and the Indonesian Kalimantan root wilt is associated with a still unidentified phytoplasma (Warokka et al. 2006). Also, a disease denominated as 'Bogia coconut syndrome' has been reported in Oceania, and it has tentatively been associated with the phytoplasma '*Ca.* Phytoplasma palmae' (Kelly et al. 2011).

Other important diseases affecting coconut plantations are foliar decay in Vanuatu caused by a virus (Rohde et al. 1990); Tinangaja disease in the island of Guam and cadang-cadang disease in the Philippines caused by viroids (Hanold and Randles 1991); lethal wilt of the palms in South America caused by protozoa (*Phytomonas staheli*) (Parthasarathy et al. 1978) and the bud rot disease in Asia and Latin America caused by the oomycete *Phytophthora palmivora* (Joseph and Radha 1975).

Additionally, important pests affecting coconut cultivation are the red palm weevil *Rhynchophorus palmarum*, which can cause direct damage to the palms but also transmits the nematode *Bursaphelenchus cocophilus* that causes the red ring disease in Latin America and Caribbean Islands (Griffith 1987), and the mite *Aceria guerreronis* Keifer, that causes very serious damage to coconut fruits and may cause premature nut fall (Moore and Howard 1996).

12.3 Plant Defense Mechanisms

Plants have several lines of defense against pathogens. The first line consists of physical barriers, such as the waxy cuticle on the leaf surface, that prevent pathogen entry. The second line of defense involves biochemical compounds, especially secondary metabolites, that act against pathogenic microorganisms inside the plant (Bednarek and Osbourn 2009). Examples of such compounds are saponins, alkaloids, and essential oils. In addition to the aforementioned non-specific defence response of hosts, pathogens have to face two specific defense mechanisms at the cellular level. The first one is carried out by cell surface receptors called pathogen recognition receptors (PRRs) that recognize specific pathogen-associated molecular patterns (PAMPs); this mechanism is known as PAMP-triggered immunity (PTI) (Boller and Felix 2009). To suppress this mechanism, pathogens produce molecules known as effectors capable of entering the host cells and suppress the PTI (Jones and Dangl 2006). Therefore, effectors help pathogen colonize plant cells and tissues (Jones and Dangl 2006). As a strategy to curb the action of the effectors, plant cells produce R proteins encoded by *R* genes that can directly or indirectly recognize effector molecules in a mechanism known as effector-triggered immunity (ETI) (Fu and Dong 2013). ETI is associated with the hypersensitive response, consisting of rapid programmed cell death at the infection site, impeding the pathogen from spreading across healthy tissue. Furthermore, ETI triggers the biosynthesis of a series of compounds (i.e., salicylic acid, azaleic acid, glycerol 3-phosphate) that are produced at the site of infection and mobilized to other tissues, initiating a process known as systemic acquired response (SAR) (Fu and Dong 2013). SAR, in turn, produces a series of metabolic changes that results in the production of pathogenic-related proteins (PR) that confer resistance to a broad-spectrum of pathogenic microorganisms (Durrant and Dong 2004).

Besides, R proteins are also known to act in recognition of effector proteins of insect pests. For instance, the *R* gene *Mi* confers resistance to the aphid *Macrosiphum euphorbiae* that affects tomato and other crops (Rossi et al. 1998). Another example is the *Bph* resistance gene which confers resistance to the brown planthopper insect (*Nilaparvata lugens* Stål) that causes severe damage to rice crops (Du et al. 2009). An interesting case is the *Vat* gene, encoded by an NBS-LRR gene, that confers cotton and cucurbits with resistance to both the aphid *Aphis gossypii* and the viruses transmitted by it (Dogimont et al. 2014).

In the case of phytoplasmas, identifying potential R proteins capable of recognizing pathogen effector proteins has just emerged. An effector protein (secreted AY-WB protein 11) associated with Aster Yellows phytoplasma has been identified. This effector protein inhibits plant defense against the insect vector (*Macrosteles quadrilineatus*), enhancing the egg deposition and hatching of the insects on leaves of *Arabidopsis thaliana* (Sugio et al. 2011). The identification of this effector protein suggests the existence of R proteins involved in resistance to phytoplasmas. The discovery of new R proteins in coconut is an important research area to expand our possibilities of managing coconut pests and diseases.

12.4 Importance of Resistance Genes and Resistance Gene Candidates in Plant Breeding

R genes represent an important platform for the genetic improvement of disease resistance in the main crops of the world. The introduction of *R* genes into susceptible cultivars confer efficient immunity against pathogens expressing the same effector. For many years, plant breeders have introgressed 'wild' *R* genes into economically important cultivars through plant breeding programs, a process that, depending on the species, could be very time-consuming. The first *R* gene was isolated in 1992; since then, 314 functional *R* genes have been identified in species of economic importance (Kourelis and van der Hoorn 2018). However, the number of *R* genes known is still limited, and in some species, such as coconut palms, no *R* genes have been reported. A total of 61% of the total *R* genes cloned encode proteins with a nucleotide-binding site leucine-rich repeats domain (NBS-LRR) (Kourelis and van der Hoorn 2018). The NBS domains contain highly conserved motifs, which are useful targets for isolating sequences homologous to *R* genes, commonly known as resistance gene candidates (RGC). The RGCs are sequences of unknown function, but many of these sequences have been mapped and shown to be linked to known

R genes, providing evidence that certain genomic regions probably code for resistance against pathogens (Pflieger et al. 2001). Moreover, several RGCs are phylogenetically closely related to known *R* genes, suggesting a similar role in disease resistance for these genes. The use of degenerate primers, designed from highly conserved motifs, for PCR amplification in species of plants for which there is no sequenced genome available, has been one of the main strategies employed for the isolation of RGCs (Peraza-Echeverria et al. 2008; Miller et al. 2008; Wan et al. 2012).

An instance of the importance of the RGCs as a tool for genetic improvement of crops is shown in the study reported by Peraza-Echeverria et al. (2008). In this study, RGCs of a wild banana exhibiting resistance to the fungal pathogen *Fusarium oxysporum* f.sp. *cubense*, the causal agent of Fusarium wilt or Panama disease, were isolated and characterized. Interestingly, one of these genes (*RGC2*) was associated with resistant lines to *F. oxysporum*. This gene was utilized for genetic transformation in an economically important banana cultivar. The transgenic lines have shown resistance against Panama disease in field trials for more than three years (Dale et al. 2017), demonstrating the relevance of considering RGCs in plant breeding programs for food security. Therefore, the identification of RGCs in plants is a practical approach for discovering new resistance genes and could be used in molecular breeding programs.

12.5 RGCs in Coconut Palm

The main focus of the genetic studies on *C. nucifera* L. has been the identification of genes with biotechnological applications, especially those involved in defense mechanisms, which have gained increased relevance in this crop (Fan et al. 2013; Huang et al. 2014; Puch-Hau et al. 2015, 2016; Nejat et al. 2015). A total of 231 RGCs (genomic NBS-type sequences) have been cloned from the coconut genome (Puch-Hau et al. 2015, 2016). Of these sequences, 195 (84%) contain an open reading frame (ORF) that could potentially be expressed as proteins and might represent *R* genes. In contrast, the RGCs that present an interrupted ORF (36: 16%) are probably pseudogenes derived from point mutations, insertions, deletions, or substitutions of nucleotides (Fig. 12.1) (Chandrasekaran and Betran 2008). Comparative genome analysis indicates that the number of *R* genes (NBS-class) in plants neither increases nor decreases in proportion to genome size (Wan et al. 2012). This number may depend on the nature of the pathogens' selective pressure or because of different life histories.

The number of these genes always varies in different plants, including plants within the same family or genus. For instance, in the grass species, the numbers are quite variable, ranging from approximately 109 *R* genes in maize (Xu et al. 2018) to 500 in rice (Luo et al. 2012) (Fig. 12.2). In the case of palms (Arecaceae), two genomes have been fully sequenced. Out of them, 144 RGCs have been reported in *Phoenix dactylifera* (Al-Mssallem et al. 2013) and 95 in the genome of *Elaeis guineensis* (Rosli et al. 2018). Particularly for *C. nucifera*, the total number of RGCs was unknown until recently, when Lantican et al. (2018) sequenced the whole genome of a dwarf variety and identified 340 RGCs.

12.6 Evolutionary History of RGCs in Coconut Palm

Studies on the evolutionary history of the RGCs in the coconut genome started recently, defining phylogenetic relationships with other plant species: Arecaceae (*P. dactylifera* and *E. guineensis*), Poaceae (*O. sativa* and *Sorghum bicolor*) and Brassicaceae (*A. thaliana, Brassica napus* and *Brassica oleracea*) families (Puch-Hau et al. 2016). The great variation in the number of RGCs in each phylogenetic tree clade (Fig. 12.3) might indicate the complex evolutionary pattern of these genes in the Arecaceae family. This high number of RGCs could, in turn, represent a fast turnover of gene copies through repeated cycles of duplication and divergence.

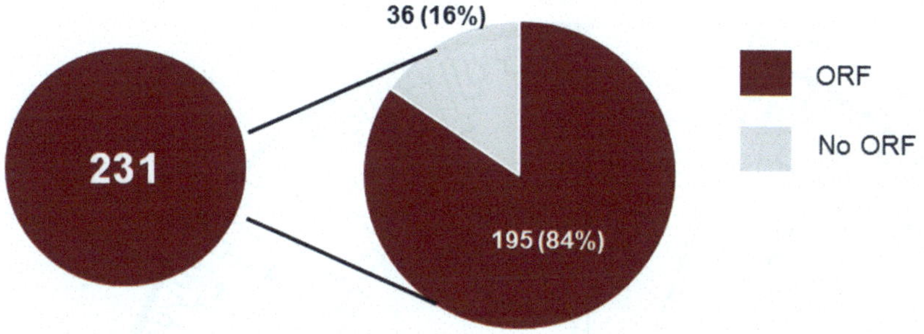

Fig. 12.1 Total number of RGCs isolated in the coconut palm by Puch-Hau et al. (2015, 2016). RGCs with ORF are indicated in red, while RGCs without ORF are gray

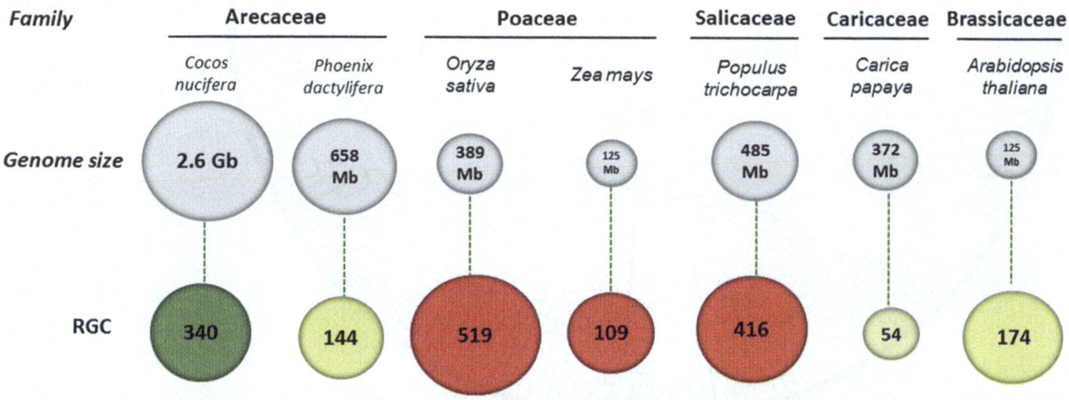

Fig. 12.2 Number of RGCs characterized in different plant genomes. Family Arecaceae: *C. nucifera* and *P. dactylifera*. Poaceae: *O. sativa* and *Z. mays*. Salicaceae: *P. trichocarpa*. Caricaceae: *C. papaya* and Brassicasceae: *A. thaliana*

Comparing NBS sequences of coconut palm with those of other Arecales indicates the existence of specific clades for *C. nucifera*. The existence of these clades is understandable due to the interaction of *C. nucifera* with specific pathogens. Alternatively, several clades include a mixture of NBS sequences from *C. nucifera*, *E. guineensis,* and *P. dactylifera,* indicating that similar resistance genes are still shared in different genera of the Arecales, highlighting a monophyletic origin of these three genera and the high conservation of some RGCs in these palms (Fig. 12.3). In this scenario, the rapid expansion or contraction of the RGCs in the coconut genome is an important adaptation strategy to the ever-changing spectrum of pathogens.

Recently, our research group built a phylogenetic tree to infer the evolutionary history of RGCs in *C. nucifera* and compared it to other plant species (monocots and dicots). In this analysis, six clades were identified (I-VI) (Fig. 12.4). The clades I-IV mainly comprised RGCs from monocots (*C. nucifera, E. guineensis, P. dactylifera, O. sativa* and *S. bicolor*). In contrast, clades V and VI mainly comprise NBS sequences of dicotyledonous plants (*A. thaliana, B. napus,* and *B. oleracea*). As expected, most of the RGCs identified in the coconut palm were closely related to NBS sequences of plants of the same class (monocots) and even more so with those of the same family (Arecaceae: *E. guineensis* and *P. dactylifera*).

On the other hand, the phylogenetic relationships identified between monocots and dicots (mainly in clades V and VI) support the proposal made by Cannon et al. (2002) by which several ancient clades of RGCs still exist and that some of them predate the split between monocots and

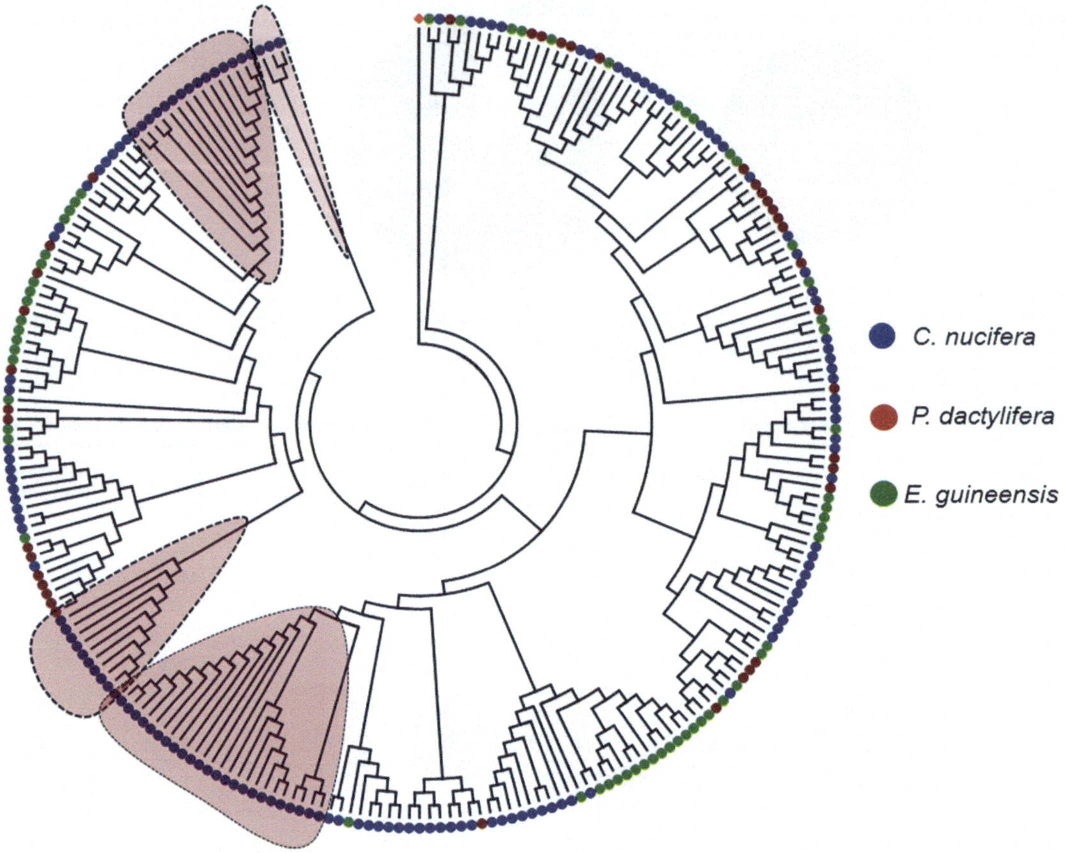

Fig. 12.3 Phylogenetic relationship of NBS sequences from *C. nucifera*, *P. dactylifera*, and *E. guineensis*. The clades indicated with dashed lines represent unique NBS sequences of coconut palm. The blue, red, and green circles represent NBS sequences of *C. nucifera*., *P. dactylifera* and *E. guineensis*, respectively

dicots. The distribution of the coconut palm RGCs across the six clades shows a high level of divergence (Fig. 12.4), possibly due to the nature of the selection pressure imposed by its pathogens. The evolutionary history of coconut palm RGCs indicates that their diversification could have occurred through potential recombination, tandem, or genomic duplication events and by their domestication history.

12.7 Perspectives

The description of RGCs in *C. nucifera* opens several avenues for future research in coconut. New *R* genes can potentially be identified from them and could accelerate developing disease-resistant plants through biotechnological techniques such as genetic transformation or gene editing. Moreover, RGCs can be used to develop molecular markers or find molecular markers associated with resistance or susceptibility to specific diseases that would help conventional breeding develop improved coconut varieties.

In our laboratory, we are working to develop molecular markers associated with resistance (or susceptibility) to the phytoplasma that causes lethal yellowing in *C. nucifera*. However, this resistance could also be associated with the plant's defense against the vector insects. It would, therefore, be interesting to expand the search to include molecular markers associated

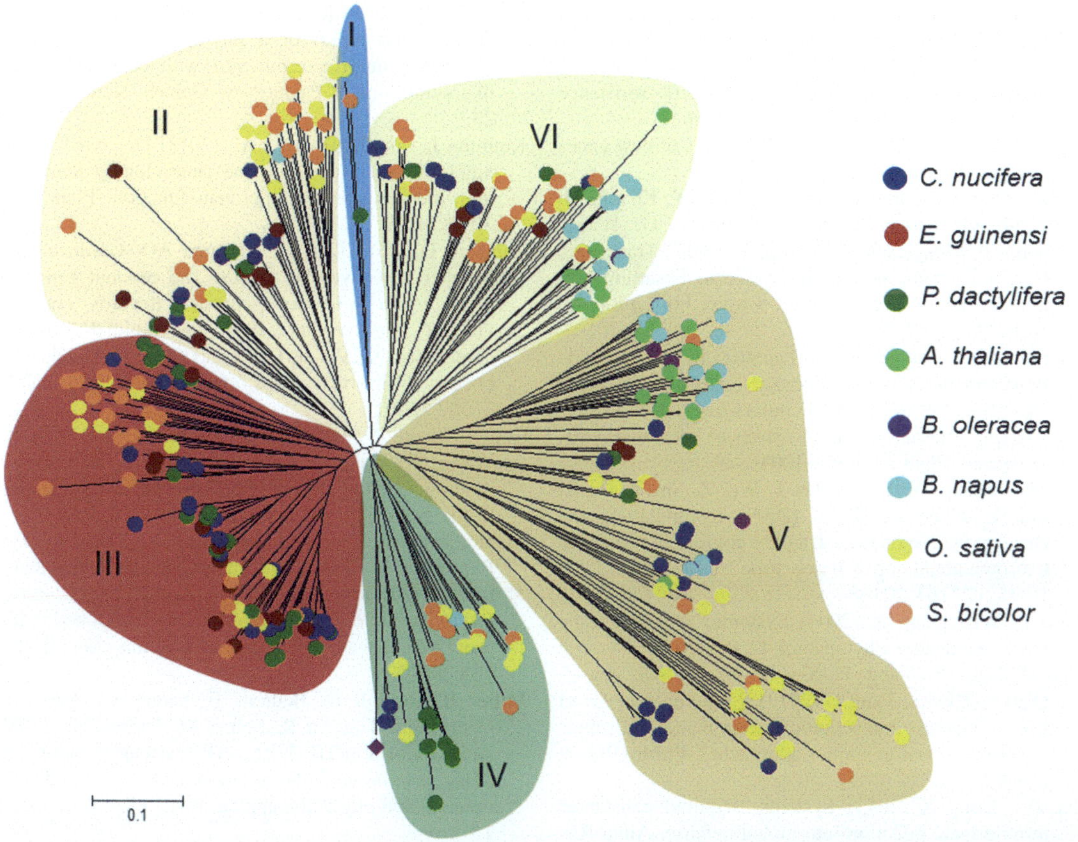

Fig. 12.4 Phylogenetic relationship among RGCs from different plant species. Six major clades were identified. Monocots are represented by *C. nucifera, E. guineensis, P. dactylifera, O. sativa,* and *S. bicolor* and dicots by *A. thaliana, B. napus,* and *B. oleracea*

with resistance against insects. Additionally, the RGCs could search for resistance to other diseases such as the red ring and the infestation of the palms with the black weevil.

The recently sequenced genome of coconut (Xiao et al. 2017; Lantican et al. 2018; Rajesh et al. 2020) will allow us to identify and characterize more RGCs and study their regulatory sequences, which will accelerate the research on the mechanisms of resistance to diseases or pests in the coconut palm and other related palm species.

Acknowledgements Part of the work was carried out in the EU H2020 R & I program project "TROPICSAFE" GA No 727459.

References

Al-Mssallem IS, Hu S, Zhang X, Lin Q, Liu W, Tan J, Yu X, Liu J, Pan L, Zhang T, Yin Y (2013) Genome sequence of the date palm *Phoenix dactylifera* L. Nat Commun 4:2274

Bednarek P, Osbourn A (2009) Plant-microbe interactions: chemical diversity in plant defense. Sci 324:746–748

Belkhadir Y, Subramaniam R, Dangl JL (2004) Plant disease resistance protein signalling: NBS-LRR proteins and their partners. Curr Opin Plant Biol 7:391–399

Boller T, Felix G (2009) A renaissance of elicitors: perception of microbe-associated molecular patterns and danger signals by pattern-recognition receptors. Annu Rev Plant Biol 60:379–406

Cannon SB, Zhu H, Baumgarten AM, Spangler R, May G, Cook DR, Young ND (2002) Diversity, distribution, and ancient taxonomic relationships within the TIR and Non-TIR NBS–LRR resistance gene subfamilies. J MolEvol 54:548–562

Chandrasekaran C, Betran E (2008) Origins of new genes and pseudogenes. Nat Educ 1:1

Dale J, Garcia-Bastidas F, James A, Paul JY, Khanna H, Smith M, Peraza-Echeverria S, Kema G, Waterhouse P, Mengersen K, Harding R (2017) Transgenic Cavendish bananas with resistance to Fusarium wilt tropical race 4. Nat Commun 8:1496. https://doi.org/10.1038/s41467-017-01670-6

Dogimont C, Chovelon V, Pauquet J, Boualem A, Bendahmane A (2014) The *Vat* locus encodes for a CC-NBS-LRR protein that confers resistance to *Aphis gossypii* infestation and *A. gossypii* mediated virus resistance. Plant J 80:993–1004

Du HB, Zhang W, Liu B, Hu J, Wei Z, Shi Z, He R, Zhu L, Chen R, He G (2009) Identification and characterization of *Bph14*, a gene conferring resistance to brown planthopper in rice. Proc Natl Acad Sci USA 106(52):22163–22168

Durrant WE, Dong X (2004) Systemic acquired resistance. Annu Rev Phytopathol 42:185–209

Fan H, Xiao Y, Yang Y, Xia W, Mason AS, Xia Z, Qiao F, Zhao S, Tang H (2013) RNA-Seq analysis of *Cocos nucifera*: transcriptome sequencing and subsequent functional genomics approaches. PLoS ONE 8: e59997

Fu ZQ, Dong X (2013) Systemic acquired resistance: turning local infection into global defense. Annu Rev Plant Biol 64:839–863. https://doi.org/10.1146/annurev-arplant-042811-105606

Griffith R (1987) Red ring disease of coconut palm. Plant Dis 71:193–196

Hanold D, Randles JW (1991) Coconut cadang-cadang disease and its viroid agent. Plant Dis 75:330–335

Harrison NA, Davis RE, Oropeza C, Helmick E, Narvaez M, Eden-Green S, Dollet M, Dickinson M (2014) '*Candidatus* Phytoplasma palmicola', a novel taxon associated with a lethal yellowing-type disease (LYD) of coconut (*Cocos nucifera* L.) in Mozambique. Int J Systemat Evol Microbiol 64:1890–1899

Hogenhout SA, Oshima K, Ammar EL-D, Kakizawa S, Kingdom HN, Namba S (2008) Phytoplasmas: bacteria that manipulate plants and insects. Mol Plant Pathol 9(4):403–423. https://doi.org/10.1111/J.1364-3703.2008.00472.X

Huang Y, Lee CP, Fu FL, Chang B, Matzke A (2014) *De novo* transcriptome sequence assembly from coconut leaves and seeds with a focus on factors involved in RNA-directed DNA methylation. G3 Genes Genom Genet 4:2147–2157

Jones JD, Dangl JL (2006) The plant immune system. Nature 444:323–329

Joseph T, Radha K (1975) Role of *Phytophthora palmivora* in bud rot of coconut. Plant Dis Rep 59:1014–1017

Kelly PL, Reeder R, Kokoa P, Arocha Y, Nixon T, Fox A (2011) First report of a phytoplasma identified in coconut palms (*Cocos nucifera*) with lethal yellowing-like symptoms in Papua New Guinea. New Dis Rep 23:9

Kourelis J, van der Hoorn RAL (2018) Defended to the nines: 25 years of resistance gene cloning identifies nine mechanisms for R protein function. Plant Cell 30:285–299

Lantican DV, Strickler SR, Canama AO, Gardoce RR, Mueller L, Galvez HF (2018) The coconut genome: providing a reference sequence towards coconut varietal improvement. Plant and animal genome conference XXVI, at Town and Country Hotel, San Diego, California. https://doi.org/10.13140/RG.2.2.21362.56001

Luo S, Zhang Y, Hu Q, Chen J, Li K, Lu C, Liu H, Wang W, Kuang H (2012) Dynamic nucleotide-binding site and leucine-rich repeat-encoding genes in the grass family. Plant Physiol 159:197–210

Manimekalai R, Soumya VP, Nair S, Thomas GV, Baranwal VK (2014) Molecular characterization identifies 16SrXI-B group Phytoplasma ('*Candidatus* Phytoplasma oryzae'-related strain) associated with root wilt disease of coconut in India. Sci Hortic 165:288–294

Miller R, Bertioli D, Baurens F, Santos C, Alves P, Martins N, Togawa R, Souza M, Pappas G (2008) Analysis of non-TIR NBS-LRR resistance analogs in *Musa acuminata* Colla: isolation, RFLP marker development, and physical mapping. BMC Plant Biol 15:1–15

Moore D, Howard FW (1996) Coconuts. In: Lindquist EE, Sabelis MW, Bruin J (eds) Eriophyid mites—their biology, natural enemies, and control. Elsevier Science BV, pp 561–569

Nejat N, Vadamalai G, Davis RE, Harrison NA, Sijam K, Dickinson M, Nor S, Abdullah A, Zhao Y (2012) '*Candidatus* Phytoplasma malaysianum', a novel taxon associated with virescence and phyllody of Madagascar periwinkle (*Catharanthus roseus*). Int J Syst Evol Microbiol 63. https://doi.org/10.1099/ijs.0.041467-0

Nejat N, Cahill DM, Vadamalai G, Ziemann M, Rookes J, Naderali N (2015) Transcriptomics-based analysis using RNA-Seq of the coconut (*Cocos nucifera*) leaf in response to yellow decline phytoplasma Infection. Mol Genet Genomics 290(5):1899–1910. https://doi.org/10.1007/s00438-015-1046-2

Parthasarathy M, van Slobbe W, Soudant YC (1978) Hartrot or fatal wilt of palms. 1. Coconuts (*Cocos nucifera*). Principes 22:3–14

Peraza-Echeverria S, Dale JL, Harding RM, Smith MK, Collet C (2008) Characterization of disease resistance gene candidates of the nucleotide binding site (NBS) type from banana and correlation of a transcriptional polymorphism with resistance to *Fusarium oxysporum* f. sp. *cubense* race 4. Mol Breed 22:565–579

Perera L, Meegahakumbura MK, Wijesekara HRT, Fernando WBS, Dickinson MJ (2012) A Phytoplasma is associated with the Weligama coconut leaf wilt disease in Sri Lanka. J Plant Pathol 94:205–209

Pflieger S, Lefebvre V, Causse YM (2001) The candidate gene approach in plant genetics: a review. Mol Breed 7:275–291

Prades A, Salum UN, Pioch D (2016) New era for the coconut sector. What prospects for research? Oilseeds Fats Crops Lipids J 23:D607. https://doi.org/10.1051/ocl/2016048

Puch-Hau C, Oropeza-Salín C, Peraza-Echeverría S, Gongora-Paredes M, Córdova-Lara I, Narvaez-Cab M, Zizumbo-Villareal D, Sáenz-Carbonell L (2015) Molecular cloning and characterization of disease-resistance gene candidates of the nucleotide binding site (NBS) type from *Cocos nucifera* L. Physiol Mol Plant Pathol 89:87–96

Puch-Hau C, Oropeza C, Góngora-Paredes M, Cordova I, Tun-Suárez J, Sáenz L (2016) New insights into the evolutionary history of resistance gene candidates in coconut palms and their expression profiles in palms affected by lethal yellowing disease. Genes Genom 39 (9):793–807

Rajesh MK, Chowdappa P, Behera SK, Kasaragod S, Gangaraj KP, Kotimoole CN, Nekrakalaya B, Mohanty V, Sampgod RB, Banerjee G, Das AJ, Niral V, Karun A, Mahato AK, Gaikwad K, Singh NK, Keshava Prasad TS (2020) Assembly and annotation of the nuclear and organellar genomes of a dwarf coconut (Chowghat Green Dwarf) possessing enhanced disease resistance. OMICS: J Integrative Biol 24(12):726–742

Rohde W, Randles J, Langridge P, Hanold D (1990) Nucleotide sequence of a circular single-stranded DNA associated with coconut foliar decay virus. Virol 176:648–651

Rosli R, Amiruddin N, Mohd AAH, Chan PL, Chan K-L, Azizi N, Moriis PE, Low E-T, Sambanthamurthi O-A, Singh R, Murphy DJ (2018) Comparative genomic and transcriptomic analysis of selected fatty acid biosynthesis genes and CNL disease resistance geness in oil palm. PLoS ONE 13(4):e0194792. https://doi.org/10.1371/journal.pone.0194792

Rossi M, Goggin FL, Milligan SB, Kaloshian I, Ullman DE, Williamson VM (1998) The nematode resistance gene *Mi* of tomato confers resistance against the potato aphid. Proc Natl Acad Sci USA 95:9750–9754

Sugio A, Kingdom HN, MacLean AM, Grieve VM, Hogenhout SA (2011) Phytoplasma protein effector SAP11 enhances insect vector reproduction by manipulating plant development and defense hormone biosynthesis. Proc Natl Acad Sci USA 108 (48):1254–1263

Wan H, Yuan W, Ye Q, Wang R, Ruan M, Li Z, Guozhi Z, Yao Z, Zhao J, Liu S, Yang Y (2012) Analysis of TIR- and non-TIR-NBS-LRR disease resistance gene analogous in paper: characterization, genetic variation, functional divergence and expression patters. BCM Genomics 13:1–15

Warokka WA, Jones P, Dickinson M (2006) Detection of phytoplasmas associated with Kalimanthan wilt disease of coconut by polymerase chain reaction. J Penelit Tanam Ind 12:154–160

Xiao Y, Xu P, Fan H, Baudouin L, Xia W, Wu Y, Xu J, Li Q, Guo A, Zhou L, Li J, Peng M, Ma Z, Armero A, Issali AE, Liu N, Peng M, Yang Y (2017) The genome draft of coconut (*Cocos nucifera*). Giga Sci 6:1–11

Xu Y, Liu F, Zhu S, Li X (2018) The maize NBS-LRR gene ZmNBS25 enhances disease resistance in rice and arabidopsis. Front Plant Sci 9:1033. https://doi.org/10.3389/fpls.2018.01033

Yankey EN, Bila J, Arocha-Rosete Y, Oropeza C, Pilet F (2018) Phytoplasma diseases of palms. In: Rao IGP et al (eds) Phytoplasmas: plant pathogenic bacteria, Springer Nature Singapore Pte Ltd, pp 267–285. https://doi.org/10.1007/978-981-13-0119-3_9

Rajesh MK, Chowdappa P, Behera SK, Kasaragod S, Gangaraj KP, Kotimoole CN, Nekrakalaya B, Mohanty V, Sampgod RB, Banerjee G, Das AJ, Niral V, Karun A, Mahato AK, Gaikwad K, Singh NK, Keshava Prasad TS (2020) Assembly and annotation of the nuclear and organellar genomes of a dwarf coconut (Chowghat Green Dwarf) possessing enhanced disease resistance. OMICS: J Integrative Biol 24(12):726–742

13 Epigenetic Considerations on Altered Phenotypes of the Coconut Endosperm

Jorge Gil C. Angeles, Jickerson P. Lado, Evangeline D. Pascual, Antonio C. Laurena, and Rita P. Laude

Abstract

Epigenetics, which affects the phenotypes of various plants during development, is brought about by several environmental influences. In this chapter, we discuss DNA methylation in plants and cite some instances of phenotypic variations associated with such epigenetic changes, with special emphasis on the coconut. We describe how methylomic variation affects *in vitro* culture and endosperm development. We further cite how such methylomic changes can alter the phenotype of oil palm (*Elaeis guineensis* Jacq.) endosperm, a closely related species of coconut, which led to its decreased viability and loss of oil yield. These epigenetic changes are most likely caused by the alteration in the levels of DNA methylation in the affected somaclonal plants leading to phenotypic plasticity. Naturally occurring variants in the soft coconut endosperms such as the Makapuno and Lono are mass-cultivated through *in vitro* means, such as via tissue culture.

J. G. C. Angeles (✉) · A. C. Laurena
Philippine Genome Center—Program for Agriculture, Livestock, Forestry and Fisheries, Office of the Vice Chancellor for Research and Extension, University of the Philippines Los Baños, Los Baños, Laguna, Philippines
e-mail: gcangeles@up.edu.ph; jgcangeles.pgc.uplb@gmail.com

A. C. Laurena
e-mail: aclaurena@yahoo.com

J. P. Lado · E. D. Pascual · R. P. Laude
Genetics and Molecular Biology Division, Institute of Biological Sciences, College of Arts and Sciences, University of the Philippines Los Baños, Los Baños, Laguna, Philippines
e-mail: jplado@up.edu.ph

E. D. Pascual
e-mail: edpascual2@up.edu.ph

R. P. Laude
e-mail: ritalaude50@gmail.com

13.1 Epigenetics

13.1.1 Introduction

Conrad Waddington (1942) first used the term 'epigenetics' to describe the study of the causal interaction between genes and their products, which resulted in phenotypic changes. Epigenetics has evolved not only to elucidate the phenotypic changes due to altered DNA methylation status but also encompasses such chemical modifications on RNAs (epi-transcriptomics), the activity of non-coding RNAs, chemical modifications of histone-specific amino acids (histone code modifications), and silencing of one strand of the DNA to affect parental or genome imprinting (Ibeagha-Awemu and Zhao 2015; Yong et al. 2016; Triantaphyllopoulos et al. 2016; Springer and Schmitz 2017). Thus, the current

M. K. Rajesh et al. (eds.), *The Coconut Genome*, Compendium of Plant Genomes,
https://doi.org/10.1007/978-3-030-76649-8_13

understanding of epigenetics encompasses mechanisms underlying changes in chromosome-bound proteins and also chemical changes to the DNA nucleotides, ultimately affecting the expression of gene(s) that are not dependent on changes to the DNA sequence and that these are either meiotically or mitotically inherited (Wu and Morris 2001; Deans and Maggert 2015).

DNA methylation is one of the well-studied epigenetic changes and is the fundamental basis of gene and transposon silencing resulting in gene and genome imprinting (Langevin and Kelsey 2013; Ibeagha-Awemu and Zhao 2015; Yong et al. 2016; Triantaphyllopoulos et al. 2016). In eukaryotes, the addition of a methyl group at the fifth position of cytosine signifies an important epigenetically associated change in gene expression. Modification to 5-methylcytosine (5mC) directly hinders the process of transcription activation by preventing the binding of the transcription factor(s) to their target sites. This inactivation has been attributed either to steric hindrance, increased nucleosome compaction, or the recruitment of repressive protein complexes that specifically bind to such methylated sites. Epigenetic modulations can also directly or indirectly recruit other enzymes which detect, modify, or erase such chemical modifications; leading to alternative splicing of the resultant transcript (Zhang and Pradhan 2014; Liyanage et al. 2014; Ong-Abdullah et al. 2015); or, silence transposons or activate gene bodies (Langevin and Kelsey 2013; Ibeagha-Awemu and Zhao 2015).

13.1.2 Epigenetics in Plants

In plants, DNA methylation is found as symmetrical cytosine methylation in CG and CHG sites and also in asymmetrical CHH regions (Springer and Schmitz 2017) (where H refers to either A or C or T), with CG sites most preferred followed by CHG regions then, lastly, CHH. In *Arabidopsis,* methylation occurs at CG (24%), CHH (6–7%) and CHG (1.7%) sites (Cokus et al. 2008; Law and Jacobsen 2010; King et al. 2010; Khan et al. 2015; Springer and Schmitz 2017). *De novo* methylation at CH, CHH and CHG sites is accomplished through the Argonaute-dependent RNA-directed DNA methylation (RdDM) system. The CG methylated regions are maintained following replication by METHYLTRANSFERASE 1 (METI) enzyme, which recognizes the hemimethylated strand to methylate the opposite, daughter strand. The *Arabidopsis* MET1 (AtMET1) methyltransferase enzyme is essential for developing the embryo and viable seed (Xiao et al. 2006).

The maintenance of DNA methylation at CHG methylated strands is performed by a KRYPTONITE (KYP) or various members of the SUPPRESSOR OF VARIEGATION HOMOLOGY family of enzymes (including SUVH4, SUVH5, and SUVH6); CHG methyltransferase; and CHROMOMETHYLASE3 (CMT3) involving a characteristic self-reinforcing loop (Pooggin 2013; Springer and Schmitz 2017). CHROMOMETHYLTRANSFERASES (CMTs) are distinctly found in plants and are involved in the methylation of non-CG sites and in maintaining CHG methylation. They also maintain RdDM in the asymmetric CHH sites (Henikoff and Comai 1998; Bender 2004). The methylation of these CHH sites could also occur through CMT2, which characteristically recognizes DNA sequence regions with H3K9me2 patterns and have long transposable elements usually residing in euchromatic DNA regions such as at CAA and CAT sites in tomato, CAA and CTA in *Arabidopsis* and maize, and CTA in rice (Gouil and Baulcombe 2016). Interestingly, methylation of these CHH regions is directed by siRNAs which influence *de novo* methyltransferase called DOMAINS REARRANGED METHYLTRANSFERASE 2 (DRM2) to perform the methylation (Springer and Schmitz 2017).

Mutations in the rice *MET1* cause a global hypomethylation, which results in non-germination of seeds, but these seeds could be rescued through tissue culture. However, such rescued plants were found to be significantly retarded (Hu et al. 2014). In maize, mutations in *ZMET2* and *ZMET5*, which are paralogues of the CG methyltransferase, are associated with seed lethality after fertilization and during the growth and development of the seed but not during gametogenesis (Gouil and Baulcombe 2016).

Huang et al. (2014) studied the relative levels of RdDM-associated gene transcripts obtained from various coconut tissues, including the endosperm, embryo, and the leaf. They found that transcripts encoding for MET1 and CMT3 were most abundant in the endosperm compared to these tissues. Further, relative quantification of RdDM genes (*MET1*, *DRM*, *NRPD1*, and *NRPE1*) across three endosperm developmental stages (gelatinous, thin layer solid, and thick layer solid regions) divulged that their expression increased during the endosperm development and ultimately reducing when the endosperm attains a thick solid layer. It suggests the interplay of these RdDM factors and small RNAs in the developing endosperm (Huang et al. 2014).

13.1.3 Plant Phenotypes Associated with Epigenetics

Other epigenetic-related mechanisms involving plant endosperms include (i) changes in methylation status such as in the mantled phenotype of *Elaeis guineensis* Jacq., (ii) paramutation exemplified by the maize *b1* locus altering anthocyanin pigment production resulting in purple-colored endosperms, and (iii) long non-coding RNA (lncRNA) found in various genome-imprinted lncRNAs in maize and rice endosperms (Luo et al. 2011; Zhang et al. 2011).

The mantled phenotype of *Elaeis guineensis* Jacq. is associated with the methylation of a long interspersed nuclear element (LINE) located between exons 5 and 6 of the *DEFENSIN1* gene (*EgDEF1*). This methylation results in the truncation of the DEF1 transcript that led to the shift from the normal to the mantled oil palm endosperm (Ong-Abdullah et al. 2015).

In corn, endosperm color is partly influenced by the *b1* gene, which encodes a transcription factor affecting the biosynthetic pathway of purple anthocyanin pigments (Chandler 2007). The locus has two possible alleles, viz., B-1 and B′. Both the B-1 and B′ alleles of *b1* are identical in sequence but have different DNAseI hypersensitivity and methylation states (Stam et al. 2002). The B-1 epiallele is responsible for darkly pigmented endosperms, while the B′ is for weakly pigmented kernels (Chandler 2007), with the B-1 epiallele being more methylated than the B' allele (Stam et al. 2002). In a *b1* heterozygous state, the B-1 epiallele (dark-colored endosperm) is paramutated to its light-colored variant due to the B' epiallele. This conversion occurs by the hypomethylation of the B-1 epiallele in the seven tandem repeats located $\sim$100 kb upstream from the transcription start site (TSS) of this *b1* locus via the RNA-dependent RNA polymerase (RdRP) MEDIATOR OF PARAMUTATION1 (MOP1), resulting in the B' epiallele (Stam et al. 2002; Chandler 2007).

Genome imprinting is an epigenetic phenomenon causing gene expression changes dependent on which parental genome (maternal or paternal) is affected (Reik and Walter 2001). In occasions where the maternal set of genes is silenced, the paternal set of genes is activated (paternally expressed genes). Conversely, if the paternal set of genes is silenced, the maternal set of genes is turned on (maternally expressed genes). In angiosperms, the phenomenon of genome imprinting is found to be pervasive in the early embryo and the endosperm, but not in more developed tissues (Bauer and Fischer 2011). Berger (2003) considered plant endosperms as a functional analog to the placenta found in mammals since both support and nurture the growing embryo or fetus (Jiang and Köhler 2012; Pires and Grossniklaus 2014).

Some plant genes are also differentially controlled by methylating their DNA based on which parent such genes are derived from. This genome or parental imprinting results in differential gene expression based solely on which parental side was silenced due to the differential methylation. In plant endosperms, some examples include *FERTILIZATION-INDEPENDENT SEED2* (*FIS2*) and *MEDEA* (*MEA*), which are both involved in cell proliferation. Mutations in these genes result in aborted seeds. Both these genes are maternally expressed since only the maternal allele is active. For *FIS2*, demethylation of the maternal side by the DEMETER (DME) DNA glycosylase allows the expression of the maternal genome, while in MEA, the methylation-associated silencing of the gene is reinforced by the FIS-PRC2 via trimethylation of H3K27 (H3K27me3) (Luo et al. 2000; Jullien et al. 2006, 2008; Raissig et al. 2011; Yuan et al. 2012; Jiao and Liu 2015; Moreno-Romero et al. 2016).

Certain amino acids on different histones that package DNA may be chemically modified in various processes such as methylation, acetylation, sumoylation, and ubiquitination, leading to altered transcription. For example, the addition of three methyl groups on the lysine residue at the 27th position of histone 3 (H3K27me3) is a widely studied histone mark associated with the suppression of gene expression. Trimethylation at this histone position (H3K27me3) leads to the recognition by the Polycomb Repression Complex 2 (PRC2) repressing transcription through the maintenance of a heterochromatin state (Schuettengruber and Cavalli 2009; Tammen et al. 2013).

13.2 Epigenetically Known Effects of Plant Tissue Culture Process and Its Constituents

Several highly valued soft endosperm types of the coconut, viz., Makapuno and Lono phenotypes (Fig. 13.1), have been successfully propagated via tissue culture (de Guzman and del Rosario 1964; Schwabe et al. 1970; de Guzman and Manuel 1977; Carpio et al. 2005a, b). Makapuno is borne among the Laguna Tall variety (Torres 1937; Zuniga 1953; Ramirez and Mendoza 1998; Islam et al. 2013). This coconut endosperm variant could occur at 25% if the nuts are from Makapuno-bearing parent plants (Cedo et al. 1984; Islam et al. 2013), whereas the probability of natural occurrence of Lono is less than 25% (Philippine Coconut Authority–Albay Research Center, personal communication). However, *in vitro* techniques such as embryo rescue have enhanced the chances of their occurrence in Makapuno and Lono to about 85% and 50%, respectively (Carpio et al. 2005a, b). Menon and Pandalai (1958) and Niral et al. (2010) have recognized the efforts of the Philippine Coconut Authority—Albay Research Center, which initiated tissue culture efforts for the Lono phenotype.

Plant tissue culture is a fundamental tool for studying the biological, physiological, and biotechnological attributes (Berdasco et al. 2008; de la Peña et al. 2012; Moricova et al. 2013) and is extensively used for mass production, improvement, and conservation (Valledor et al. 2007; Huang et al. 2012). Though the *in vitro* culture method has been very useful, novel morphological, physiological, and biochemical changes can arise affecting their biological and economic qualities (such as enhanced stress tolerance and quality improvement) (Huang et al. 2012). These somaclonal variants are derived from single donors via *in vitro* clonal propagation of plants using cell and tissue culture techniques (Larkin and Scowcroft 1981; Russo et al. 1996), which could also result from epigenetic changes such as changes in DNA methylation levels (Rival et al. 2008). Global methylation level changes in tissue-cultured regenerants and their progenies leading to phenotypic changes had been documented in various plants (Table 13.1).

Fig. 13.1 Endosperm phenotypes of coconut (*Cocos nucifera* L.) variants derived from the Laguna Tall Variety: (**a**) Normal, (**b**) Makapuno, and (**c**) Lono

Table 13.1 Plant tissue culture-associated epigenetic changes and corresponding altered phenotypes

Tissue culture	Organism	Epigenetic changes and corresponding phenotypes	References
Tissue culture process	*Arabidopsis*	Methylation of promoters of single-copy genes represses gene expression and contributes to the regulation of calli and suspension calli undifferentiation	Berdasco et al. (2008)
		Promoter demethylations result in higher WUSCHEL gene expression leading to early shoot primordia initiation and higher regeneration of shoots in root explant-derived calli	Li et al. (2011)
	Barley	More methylation state variations in tissue-cultured plants	Bednarek et al. (2007)
	Cacao (*Theobroma cacao*)	Increase in global methylation levels associated with somatic embryo differentiation. These methylation levels decrease in cultures going through additional somatic embryogenesis and are associated with reduced embryogenic potential	de la Peña et al. (2012), Pila Quinga et al. (2017)
	Carrot	Hypomethylation occurs during the initial stages of the culture process; however, as cells shift to the torpedo embryo and plantlet formation stages, global methylation is dramatically increased	LoSchiavo et al. (1989)
	Clivia miniata	Cytosine hypermethylation results in yellow leaves while light-green leaves are due to cytosine hypomethylation	Wang et al. (2016)
	Oil palm (*Elaeis guineensis* Jacq.)	Mantled phenotype, producing abnormal fruits, is the result of hypomethylation of the mantled gene. The *KARMA* region is hypermethylated in the non-mantled variants, while it is hypomethylated in the mantled phenotypes.	Jaligot et al. (2002, 2011), Alwee et al. (2006), Ong-Abdullah et al. (2015)
	Poplar	Methylation levels increased between internodes of its explants and calli, but these methylation levels decrease in tissue culture regenerated internodes	Vining et al. (2013)

(continued)

Table 13.1 (continued)

Tissue culture		Organism	Epigenetic changes and corresponding phenotypes	References
		Rice	Methylation of *mPING* in cultured calli increased CHG methylation	Ngezahayo et al. (2009)
		Rye (*Secale cereal* L. cv. "Ailes")	Regenerants of immature embryo-derived calli were more globally hypomethylated than the donor plants	González et al. (2013)
		Siberian ginseng (*Eleutherococcus senticus*)	Embryonic calli have significantly lower methylation levels than non-embryogenic calli	Chakrabarty et al. (2003)
		Tomato	Changes in the levels of methylation in tomato calli harvested after the second regeneration cycle	Smulders et al. (1995)
Number of tissue culture cycles	Number of the tissue culture cycles done	*Malus xiaojinensis*	Cytosine hypomethylation observed in leaves derived from the micropropagation of the tetraploid, apomictic *M. xiaojinensis*	Huang et al. (2012)
Length of time in culture media	Long-term suspension in cultures	*Arabidopsis*	Hypomethylation of *Athila* and *LTR*-copia retroelements and the *AtMu* DNA transposon; hypermethylation in the *LTR-Gypsy* retrotransposon	Tanurdzic et al. (2008)
			Hypermethylation of the euchromatic DNA eliminated the capability of the cells to regenerate shoots readily	Tanurdzic et al. (2008)
	Incubation in MS media (3.5 months)	Maize	Hypermethylation of *Pericarp color 1 (P1-wr)* results in alterations in maize pericarp and cob-glume	Rhee et al. (2010)
	Duration of the cultures maintained at the embryoid multiplication stage	Oil palm (*Elaeis guineensis* Jacq.)	The emergence of the mantled phenotype	Corley et al. (1986), Rival et al. (2013)
		Hops	Repeated regeneration enhances 5-mC levels	Peredo et al. (2006), González et al. (2013)

The extent of DNA methylation is also affected by the tissue culture media and/or its supplementary reagents employed (Table 13.2). The strength and the support components of the tissue culture media could also influence the methylation levels of tissue-cultured plants.

13.3 Epigenetics in Related and Model Palm Species

Variation in global methylation levels is found across diverse plant tissues as illustrated by the mantled endosperm, an apparent feminization of

Table 13.2 Epigenetically associated plant phenotypes due to the tissue culture media and/or its supplementary reagents

Tissue culture		Organism	Epigenetic changes and some phenotypes	References
MS media		Potato	Decreased methylation of cytosines at CCGG sites which yielded variations in leaf morphology and degree of node branching in tissue-cultured potato plants	Joyce and Cassells (2002)
Reagents	2,4-D (2,4-dichlorophenoxyacetic acid)	*Allium sativum*	Increases sister chromatid recombination in root tips	Doležel et al. (1987)
		Carrot	Increased ethylene production resulting in accumulation of SAM causing increased SAM: SAH ratio. It leads to increased DNA methylation during torpedo stage and plantlet development	Galaud Jean-Philippe et al. (1993a, b), von Aderkas and Bonga (2000), Causevic et al. (2005)
	Auxins (NAA)		Increased DNA methylation in root explant proliferating cultures	LoSchiavo et al. (1989)
	Cytokinin (kinetin)		Decreased DNA methylation in root explants	Arnholdt-Schmitt et al. (1991)
	2,4-dichlorophenoxyacetic acid (2,4-D)	*Begonia*	High levels resulted in huge variation of leaf shape and length	De Klerk (1990), Smulders and de Klerk (2011)
		Kalanchoe blossfeldiana	Improper levels of the reagents led to abnormal formation of bud primordia	Varga et al. (1988)
	Increasing levels of the auxin alpha-naphthalene acetic acid (NAA)	Oil palm (*Elaeis guineensis* Jacq.)	Decreased occurrence of mantled ramets	Eeuwens et al. (2002)
	Kinetin	Oil palm (*Elaeis guineensis* Jacq.)	Increased incidence of the mantled phenotype	

the male flower parts, of the oil palm (*Elaeis guineensis* Jacq.), a close relative of the coconut (Jaligot et al. 2000, 2002; Eeuwens et al. 2002; Rival et al. 2009; Ong-Abdullah et al. 2015). In oil palm clones, obtained through somatic embryogenesis, the reduced levels of global methylation and overall methylation rates are associated with the mantled phenotype resulting in altered development of the nuts, thereby decreasing oil production. This phenomenon is further influenced by the severity of the observed mantled phenotype (Corley et al. 1986; Ong-Abdullah et al. 2015). The mantled phenotype only affects 5% of the regenerated oil palms and nodular compact calli, is unpredictable, and has a varied disparity in severity even if the regenerants arise from the same clonal progeny. The emergence of such mantled nuts decreases the palm's productivity since the greater proportion of nut branches is affected, causing huge losses to the oil palm industry (Matthes et al. 2001; Springer and Schmitz 2017). Interestingly, the mantled phenomenon observed in oil palm is a suitable model to investigate the role of epigenetics in the growth and development of palm species or even higher plants (Rival et al. 2009).

13.4 Development of the Coconut Endosperm and Associated Fatty Acid Profile

The coconut fruit, botanically a 'drupe', comprises of three distinct layers; (a) exocarp, smooth external layer in various colors that ranges from red-brown to green; (b) mesocarp, a middle layer having a white and hard fibrous layer during the initial stages which become fibrous later, and (c) endocarp, a hard shell which covers the endosperm. The brown seed coat layer, called testa, lies between endosperm and endocarp, covering the former (Child 1974).

In general, plant endosperm development follows four sequential phases: coenocytic or syncytial, cellularization, differentiation and death (Berger 1999). Coconut is characterized with nuclear type of endosperm; hence, it produces many free nuclei from the repetitive division of its primary endosperm with no wall formation during the syncytial stage. This stage usually begins at the six month after pollination and ends at the 12th month during the maturation phase.

The normal coconut variety forms two types of endosperm; the solid endosperm characterized as having a white, hard, firm, and compact meat near the shell of the nut and (b) the liquid endosperm commonly known as coconut water. In the more matured coconut seed, this endosperm could be 100-fold heavier than their corresponding embryos (Huang et al. 2014). The solid coconut endosperm generally comprises 46% oil, predominantly the medium-chain C12:0 fatty acid, viz., lauric acid and α-tocopherols (Naresh Kumar and Balakrishna 2009; Tangwatcharin and Khopaibool 2012; Arlee et al. 2013).

The solid coconut endosperm has three main regions. These are (a) the inner layer in the vicinity of the coconut water, which is approximately made of 16% fat, (b) the middle layer, which has a higher fat content (46%), and (c) the outer layer which is nearest to the testa and has the highest fat content (62%), predominantly lauric acid (C12:0) (Balachandran et al. 1985). The coconut fruit takes 12–13 months to mature and contains 70–75% oil (Child 1974; Oo and Stumpf 1979).

The developmental stages of the coconut palm and various phases of its endosperm were described by Tammes (1955) and Child (1974). The coconut palm starts flowering at five years after planting, and it flowers at approximately monthly intervals. Fruits begin their maturity immediately after fertilization. During fruit development, the embryo sac becomes bigger, while a large vacuole is formed in the central region (Juliano 1926).

During the initial six months of fruit development, the nut volume increases while the cavity continues to be filled with nut water or liquid endosperm. Subsequently, the formation of a solid layer of endosperm starts from the opposite end of the stalk, slowly extending into the interior region. The rapid development of solid endosperm happens from the seventh through ninth month from a thin jelly-like layer turning into a firm, solid kernel comprising 84% of oil by the ninth month. This process continues until the fruit forms a hard white flesh of endosperm due to the intracellular deposition of oil by the 13th month.

The various developmental stages of endosperm divulge variations in the component fatty acids. Free fatty acid levels rapidly decline at a rate of 2% per month between the seventh through ninth months after pollination (MAP) and become less than 1% from the tenth month stage onward as the kernel oil content proportionally increases at a rate of 13% per month (Child 1974). Generally, the process of fatty acid anabolism begins during the period of fourth to seventh month when the solid endosperm starts to form, the rate of fatty acid biosynthesis increases until the 12th month (Tammes 1955; Kartha and Narayanan 1956; Nathanael 1959; Balleza and Sierra 1972; Child 1974; Padua-Resurreccion and Banzon 1979). Regarding the fatty acid composition, it was found that the short-chain fatty acids (SCFAs of C6:0, C8:0, and C10:0) tend to increase in parallel with the development of the solid and liquid coconut endosperm. In contrast, the longer-chain fatty acids (LCFAs of C18:3 and C22:0) are formed throughout the different ontogenic stages of the liquid endosperm. However, LCFAs are conspicuously absent in the kernel. LCFAs (C16:0 and C18:1) decline to 9% and 6%, respectively, in a developed 12-month-old endosperm. Also, the content of unsaturated fatty acids [palmitoleic (C16:1) and linoleic (C18:2)] decreases to a negligible level as the nut matures (Padua-Resurreccion and Banzon 1979; Balachandran et al. 1985). The kernel of the endosperm, which comprises roughly 29% of linoleic acid (C18:1) at 6–7 months, is characterized with over 35% of lauric acid (C12:0) content by the end of 12 months and beyond (Child 1974). Further, it was deduced that endosperm synthesizes MCFAs of C10:0-C14:0 since the sixth month culminating in the production of endosperm having 50% of C12:0 (lauric acid) and C14:0 (myristic acid) (Child 1974).

13.5 Some Phenotypic Variations of the Coconut Endosperm

Aside from the normal endosperm, coconut has highly sought-after variants including the Makapuno and the Lono (Fig. 13.1b, c). The latter is studied for its observed higher oil content.

13.5.1 Makapuno or Makapuno Tall (MAKT)

Gonzales (1914) published the first scientific account on the occurrence of this coconut fruit variant with 'Maka' and 'puno' translating to 'almost' and 'full', respectively. The white meat is over-proliferating and fluffy and almost fills up the whole cavity, and has little to no water in its central space. The fruit's remaining core is nearly filled up by a viscous, white, translucent jelly (Fig. 13.1b). Such descriptions are deviant to the normal coconut endosperm consisting of a hard, dense, and crisp solid endosperm and a liquid endosperm occupied by coconut water. Except for the endosperm and inability of the nuts to germinate, Makapuno palms possess all the morphological aspects of normal coconut palms (Copeland 1931; Ramirez and Mendoza 1998).

Even within the Makapuno endosperm phenotype, there are three known types: Type A, normal, thick endosperm but with slightly viscous liquid endosperm; Type B, characterized with thicker endosperm having a soft inner layer but the liquid endosperm is more viscous and translucent; Type C, having the thickest endosperm with soft inner layers and virtually a non-existent liquid endosperm that is replaced by an oily semi-solid endosperm (Adriano and Manahan 1931). Currently, there is no explanation on how these different Makapuno endosperm types arise.

The Makapuno endosperm types possess an economic advantage as these are more expensive (4–5 times) than the normal coconut. They are utilized in desserts, delicacies, and other food and industrial applications. Embryo culture technology developed by de Guzman and peers (de Guzman and del Rosario 1964; de Guzman and Manuel 1977) was successfully utilized to propagate these nuts. Further improvements in the embryo culture techniques by Philippine Coconut Authority—Albay Research Center (Rillo 1997) resulted in the commercial production of Makapuno seedlings in the Philippines (Rillo et al. 1999). Currently, several Philippine laboratories are engaged in the mass propagation of these embryo-cultured Makapuno seedlings.

This over-proliferating endosperm type is a global phenomenon. Several countries have their local names for the 'Makapuno' or coconut curd: Kopyor (Indonesia), Dua Sap (Vietnam), Dong Kathy (Cambodia), Dahi Nariyal (Myanmar), Maphrao Kathi (Thailand), Thairu Thengai (India), Dikiri Pol (Sri Lanka), and Niu Garuk (Papua New Guinea). These variants all have a natural thick solid endosperm, less to no water, are non-germinating in nature, and are economically profitable. Makapuno is borne naturally among the tall varieties. On the other hand, Kopyor is naturally found in dwarf, tall, and hybrid types.

The studies of Torres (1937) and Zuñiga (1953) provided early evidence on the genetic basis or mode of inheritance of the Makapuno endosperm. The results strongly suggested a single Mendelian recessive gene (*m*) possibly controlling the trait derived from the heterozygous (*Mm*) Makapuno-bearing palm. Cedo et al. (1984) observed that when self- or cross-pollinated embryo-cultured Makapuno phenotypes were used, an all-Makapuno yield was obtained, while if an embryo-cultured Makapuno phenotype was crossed with a coconut palm bearing normal endosperms, all of the produced fruits will have the normal coconut endosperm (Cedo et al. 1984). Thus, it was inferred that the Makapuno trait in embryo culture-derived plantlets could be governed by homozygous recessive alleles (*mm*). In contrast, out-crossing with a normal palm ensures the restoration of normal endosperm (*Mm*) in the offspring (Cedo et al. 1984). Similar researches on Kopyor showed the recessive *k* gene as a possible factor for this phenotype (Maskromo et al. 2017). On the other hand, genetic analysis has further postulated that a single recessive gene governs the Makapuno character. Hence, the genetically heterozygous nature of Makapuno-bearing palms could be the underlying basis for the phenotype.

Biochemical features of the Makapuno endosperm have been studied by several researchers (Mendoza et al. 1982; Samonte et al. 1987). Ramirez and Mendoza (1998) found that the carbohydrate and protein content of Makapuno nuts are higher than the normal coconuts though it is dependent on the developmental stages. Biochemically, the increase in galactomannan content of the Makapuno endosperm causes the formation of viscous, jelly-like endosperm (Mujer et al. 1983). These galactomannans are composed of a ratio of 1:3 galactose:mannose residues (Balasubramaniam 1976; Samonte et al. 1987). Galactomannan catabolism is chiefly controlled sequentially by three enzymes: ALPHA-D-GALACTOSIDASE (AGAL), BETA-MANNANASE, and MANNOSIDASE. The levels of transcript and/or enzyme activities of AGAL and MANNOSIDASE are typically low in Makapuno than in normal coconut endosperm (Mujer et al. 1984a, b; Samonte et al. 1989; Diaz 2002). However, no differences in the activity of BETA-MANNANASE were found between these two phenotypes (Diaz 2002; Samonte et al. 1989). The deficiency of AGAL activity could probably block the degradation of galactomannan, resulting in the characteristic Makapuno endosperms. On the contrary, the enhanced activity of the AGAL enzyme in normal endosperm yields a solid endosperm (Mujer et al. 1984a; Samonte et al. 1989). Molecular analysis of genetic component encoding *AGAL* gene revealed several deletions, insertions, substitutions between the normal and Makapuno phenotypes (de la Cruz and Bugayong 2016). Later, Phoeurk et al. (2018) characterized three novel mutations in the *AGAL* gene of Makapuno. These sequence changes affect protein conformation and compromise enzyme stability (Phoeurk et al. 2018). Further, transcriptomic data suggest an early expression of galactomannan biosynthesis genes and a massive down-regulation of *AGAL* in Makapuno than the normal nuts (Lado et al. 2019).

Another possible explanation for the Makapuno phenomenon could be associated with the xenia effect (Cedo et al. 1984), which is the emergence of phenotypic characteristics that are controlled by a single recessive gene and is influenced by the genotype of donor pollen. This xenia effect has been recorded in maize, dates, and coconut (Satyabalan 1953; Cedo et al. 1984; Watanabe et al. 2006; Faraq et al. 2012; Urechean and Bonea 2017; Maskromo et al. 2017). In the coconut, this xenia effect influences nut size, copra weight, and hybrid vigor (Satyabalan 1953).

13.5.2 Lono or Philippine Lono Tall (PLNT)

There are very limited studies about the Lono coconut endosperm phenotype. The Lono-bearing palms are morphologically similar to other coconut tall varieties except for their soft but high oil-containing endosperms. The soft endosperm of the Lono may be eaten fresh, and like the Makapuno, its endosperm could not be made into copra. It is an economically important endosperm variety owing to its higher oil content. This Lono coconut variant is closely related to the Philippine Ordinary and Laguna Tall varieties and is observed only among the tall cultivars (Devakumar et al. 2010; Niral et al. 2013).

The oil content of the Lono nut is over 64% of the endosperm. The fatty acid composition reveals that over 93% of Lono oil samples comprise saturated fatty acids (SFAs), whereas 6–7% of unsaturated fatty acids was reported. Among these SFAs, MCFAs and LCFAs comprise 60% and 33%, respectively. The Lono phenotype has a higher average percentage of oil yield (21%) and has a slightly high percentage abundance of C12:0, C14:0, C16:0, C18:0, and C18:1 fatty acids compared to their normal counterpart (Zambiazi et al. 2007; Naresh Kumar and Balakrishna 2009; Kumar 2011; Kostik et al. 2013). These fatty acids are stored in specialized cellular organelles called oleosomes or oil bodies (Huang 1992). Initial reports on the Lono oil bodies found that these form larger bodies, rather than being more numerous, compared to those from the Philippine Laguna Tall normal variety (Aala et al. 2016).

There is a wide probability (0–25%) of obtaining a Lono nut from a Laguna Tall coconut palm (Rillo et al. 1999). Thus, it is comparatively a challenging task to obtain Lono nuts than Makapuno. However, similar to the Makapuno, the Lono nuts could not be segregated from other nuts after splitting, nor do these nuts germinate; hence, the embryo rescue technique is the only viable option (Carpio et al. 2005a, b).

There is no single hypothesis explaining the incidence of this Lono phenotype or its relatively higher oil content. However, it is postulated that this phenotype arose from a spontaneous coconut mutation (Satyabalan 1953; Jerard et al. 2013). The study of Osio et al. (2019) suggested a molecular basis for the Lono endosperm variant based on transcriptomic data. De novo sequencing, assembly, and annotation of the 6–7 MAP transcriptome of Lono endosperm revealed early differential expression of genes involved in oil biosynthesis (Osio et al. 2019). Notable oil-biosynthetic and storage genes include *DELTA (8)-DESATURASE2, ELONGATION OF FATTY ACIDS3, ACYL-ACP THIOESTERASE (FATA and FATB)*. These genes may have influenced the early high oil content in Lono endosperm.

Since both these Makapuno and Lono coconut phenotypes are primarily propagated through tissue culture-based methods and that such techniques are highly associated with epigenetics, it is postulated that such epigenetic mechanisms could also be at play in the Makapuno and Lono phenotypes (Angeles et al. 2018).

13.6 Conclusions and Perspectives

Naturally occurring variations of the normal coconut endosperm exist, such as the Makapuno and Lono. These altered phenotypes are non-germinating; hence, tissue culture methods have been standardized for mass propagation of these phenotypes. Such efforts enabled the sustainable supply of these variants as the probability of their natural occurrence is very low. Though we do not discount the known genetic controls associated with these phenotypes, for other variations that could not be fully explained based on genetic phenomenon, epigenetics could uncover the molecular basis of these novel and unexplained phenotypes (Angeles et al. 2018). In this chapter, we have discussed epigenetics, particularly the effects of altered DNA methylation associated with plants. We have also described various plant phenotypes that arise due to such altered methylation states either naturally occurring or caused by the tissue culture media composition/process with special reference to the coconut endosperm. Further investigations on various molecular targets and pathways and whether epigenetics affect these traits could explain the occurence of the Makapuno and Lono phenotypes.

References

Aala Jr WF, Custodio CJC, Lado JP, Berdos MLG, Cueto CA, Diaz MGQ (2016) Oil body enlargement and not number is responsible for increased oil content in solid endosperm of Philippine Lono tall coconut (*Cocos nucifera* L.). In: 25th International Conference and 43rd PSBMB Annual Convention. Manila, Philippines, p 152

Adriano FT, Manahan M (1931) The nutritive value of green, ripe and sport coconut (buko, niyog, and Makapuno). Phil Agric 20:195–198

Alwee SS, Van der Linden CG, Van der Schoot J, De Folter S, Angenent GC, Cheah SC, Smulders MJ (2006) Characterization of oil palm MADS box genes in relation to the mantled flower abnormality. Plant Cell Tissue Organ Cult 85:331–344. https://doi.org/10.1007/s11240-006-9084-4

Angeles JGC, Lado JP, Pascual ED, Cueto CA, Laurena AC, Laude RP (2018) Towards the understanding of important coconut endosperm phenotypes: Is there an epigenetic control? Agronomy 8:225. https://doi.org/10.3390/agronomy8100225

Arlee R, Suanphairoch S, Pakdeechanuan P (2013) Differences in chemical components and antioxidant-related substances in virgin coconut oil from coconut hybrids and their parents. Int Food Res J 20:2103–2109

Arnholdt-Schmitt B, Holzapfel B, Schillinger A, Neumann KH (1991) Variable methylation and differential replication of genomic DNA in cultured carrot root expiants during growth induction as influenced by hormonal treatments. Theor Appl Genet 82:283–288. https://doi.org/10.1007/BF02190613

Balachandran C, Arumughan C, Mathew AG (1985) Distribution of major chemical constituents and fatty acids in different regions of coconut endosperm. JAOCS 62:1583–1586. https://doi.org/10.1007/BF02541692

Balasubramaniam K (1976) Polysaccharides of the kernel of maturing and matured coconuts. J Food Sci 41:1370–1373. https://doi.org/10.1111/j.1365-2621.1976.tb01174.x

Balleza CF, Sierra ZN (1972) Proximate analysis of the coconut endosperm in progressive stages of development. Phil J Coco Studies 16:37–43

Bauer MJ, Fischer RL (2011) Genome demethylation and imprinting in the endosperm. Curr Opin Plant Biol 14:162–167. https://doi.org/10.1016/j.pbi.2011.02.006

Bednarek PT, Orłowska R, Koebner RM, Zimny J (2007) Quantification of the tissue-culture induced variation in barley (*Hordeum vulgare* L.). BMC Plant Biol 7:10. https://doi.org/10.1186/1471-2229-7-10

Bender J (2004) DNA methylation and epigenetics. Annu Rev Plant Biol 55:41–68. https://doi.org/10.1146/annurev.arplant.55.031903.141641

Berdasco M, Alcázar R, García-Ortiz MV, Ballestar E, Fernández AF, Roldán-Arjona T, Tiburcio AF, Altabella T, Buisine N, Quesneville H, Baudry A (2008) Promoter DNA hypermethylation and gene repression in undifferentiated *Arabidopsis* cells. PLoS ONE 3: e3306. https://doi.org/10.1371/journal.pone.0003306

Berger F (1999) Endosperm development. Curr Opin Plant Biol 2:28–32. https://doi.org/10.1016/S1369-5266(99)80006-5

Berger F (2003) Endosperm: The crossroad of seed development. Curr Opin Plant Biol 6:42–50. https://doi.org/10.1016/S1369526602000043

Carpio CB, Manohar E, Rillo EP, Cueto CA, Orense O, Areza-Ubaldo MB, Alfiler AR (2005a) Status of coconut genetic resources research in the Philippines. In: Coconut genetic resources. International Plant Genetic Resources Institute—Regional Office for Asia, the Pacific and Oceania (IPGRI-APO), Malaysia, pp 630–647

Carpio CB, Santos GA, Emmanuel EE, Novarianto H (2005b) Research on coconut genetic resources in Southeast and East Asia. In: Coconut genetic resources. International Plant Genetic Resources Institute—Regional Office for Asia, The Pacific and Oceania (IPGRI-APO), Malaysia, pp 533–545

Causevic A, Delaunay A, Ounnar S, Righezza M, Delmotte F, Brignolas F, Hagège D, Maury S (2005) DNA methylating and demethylating treatments modify phenotype and cell wall differentiation state in sugar beet cell lines. Plant Physiol Biochem 43:681–691. https://doi.org/10.1016/j.plaphy.2005.05.011

Cedo MLO, de Guzman EV, Rimando TJ (1984) Controlled pollination of embryo-culture “Makapuno” coconut. Philipp Agric 67:100–104

Chakrabarty D, Yu KW, Paek KY (2003) Detection of DNA methylation changes during somatic embryogenesis of Siberian ginseng (*Eleuterococcus senticosus*). Plant Sci 165:61–68. https://doi.org/10.1016/S0168-9452(03)00127-4

Chandler VL (2007) Paramutation: From maize to mice. Cell 128:641–645. https://doi.org/10.1016/j.cell.2007.02.007

Child R (1974) Coconuts, 2nd edn. Longman Group Ltd, London

Cokus SJ, Feng S, Zhang X, Chen Z, Merriman B, Haudenschild CD, Pradhan S, Nelson SF, Pellegrini M, Jacobsen SE (2008) Shotgun bisulphite sequencing of the *Arabidopsis* genome reveals DNA methylation patterning. Nature 452:215

Copeland EB (1931) The Coconut. Macmillian and Company, Ltd

Corley RHV, Lee CH, Law LH, Wong CY (1986) Abnormal flower development in oil palm clones. Planter (Kuala Lumpur) 62:233–240

de Guzman EV, del Rosario DA (1964) The growth and development of *Cocos nucifera* L. ‘Makapuno’ embryo in-vitro. Philipp Agric 48:82–94

de Guzman EV, Manuel GC (1977) Improved root growth in embryo and seedlings culture of coconut ‘Makapuno’ by the incorporation of charcoal in the growth medium. PJCS 11:35–39

de Klerk GJ (1990) How to measure somaclonal variation. Acta Botanica Neerlandica 39:129–144. https://doi.org/10.1111/j.1438-8677.1990.tb01481.x

de la Cruz RY, Bugayong VJD (2016) Mutations in the alpha-D-galactosidase gene suggest molecular basis of the mutant “Makapuno” coconut (*Cocos nucifera* L.) phenomenon. Philipp Agric Sci 99:321–325

de la Peña C, Nic-Can G, Ojeda G, Herrera-Herrera JL, López-Torres A, Wrobel K, Robert-Díaz ML (2012) *KNOX*1 is expressed and epigenetically regulated

during *in vitro* conditions in *Agave* spp. BMC Plant Biol 12:203. https://doi.org/10.1186/1471-2229-12-203
Deans C, Maggert KA (2015) What do you mean, "Epigenetic"? Genetics 199:887. https://doi.org/10.1534/genetics.114.173492
Devakumar K, Jerard BA, Dhanapal R, Damodaran V, Niral V, Naganeeswaran S (2010) Genetic relatedness of coconut populations from Andaman and Nicobar Islands traced to the Indian sub-continent using microsatellite markers. Kerala, India, pp 67–75
Diaz MGQ (2002) Differential gene expression of the enzymes of galactomannan degradation in the developing endosperms of normal and "Makapuno" coconut (*Cocos nucifera* Linn.). PhD Dissertation, University of the Philippines Los Baños
Doležel J, Lucretti S, Novák FJ (1987) The influence of 2,4-dichlorophenoxyacetic acid on cell cycle kinetics and sister-chromatid exchange frequency in garlic (*Allium sativum* L.) meristem cells. Biologia Plantarum 29:253. https://doi.org/10.1007/BF02892785
Eeuwens CJ, Lord S, Donough CR, Rao V, Vallejo G, Nelson S (2002) Effects of tissue culture conditions during embryoid multiplication on the incidence of "mantled" flowering in clonally propagated oil palm. Plant Cell Tissue Organ Cult 70:311–323. https://doi.org/10.1023/A:1016543921508
Faraq K, Elsabagh A, Elashry H (2012) Fruit characteristics of "Zaghloul" date palm in relation to metaxenic influence of used pollinator. Am Euras J Agric Environ Sci 12:842–855
Jean-Philippe G, Thomas G, Nicole B (1993) Effect of anti-DNA methylation drugs on growth, level of methylated DNA, peroxidase activity and ethylene production of *Bryonia dioica* internodes. Physiol Plant 87:528–534. https://doi.org/10.1111/j.1399-3054.1993.tb02503.x
Jean-Philippe G, Thomas G, Nicole B (1993) Inhibition of internode growth due to mechanical stress in *Bryonia dioica*: Relationship between changes in DNA methylation and ethylene metabolism. Physiol Plant 87:25–30. https://doi.org/10.1111/j.1399-3054.1993.tb08786.x
Gonzales BM (1914) The macapuno coconut. Phil Agric Forester 3:31–32
González AI, Sáiz A, Acedo A, Ruiz ML, Polanco C (2013) Analysis of genomic DNA methylation patterns in regenerated and control plants of rye (*Secale cereale* L.). Plant Growth Regul 70:227–236. https://doi.org/10.1007/s10725-013-9794-7
Gouil Q, Baulcombe DC (2016) DNA methylation signatures of the plant chromomethyltransferases. PLOS Genet 12:e1006526. https://doi.org/10.1371/journal.pgen.1006526
Henikoff S, Comai L (1998) A DNA methyltransferase homolog with a chromodomain exists in multiple polymorphic forms in *Arabidopsis*. Genetics 149:307
Hu L, Li N, Xu C, Zhong S, Lin X, Yang J, Zhou T, Yuliang A, Wu Y, Chen YR, Cao X (2014) Mutation of a major CG methylase in rice causes genome-wide hypomethylation, dysregulated genome expression, and seedling lethality. PNAS 111:10642. https://doi.org/10.1073/pnas.1410761111
Huang AHC (1992) Oil bodies and oleosins in seeds. Annu Rev Plant Physiol Plant Mol Biol 43:177–200. https://doi.org/10.1146/annurev.pp.43.060192.001141
Huang H, Han SS, Wang Y, Zhang XZ, Han ZH (2012) Variations in leaf morphology and DNA methylation following *in vitro* culture of *Malus xiaojinensis*. Plant Cell Tissue Organ Cult 111:153–161. https://doi.org/10.1007/s11240-012-0179-9
Huang Y-Y, Lee C-P, Fu JL, Chang BC, Matzke AJ, Matzke M (2014) *De novo* transcriptome sequence assembly from coconut leaves and seeds with a focus on factors involved in RNA-directed DNA methylation. G3 4:2147. https://doi.org/10.1534/g3.114.013409
Ibeagha-Awemu EM, Zhao X (2015) Epigenetic marks: regulators of livestock phenotypes and conceivable sources of missing variation in livestock improvement programs. Front Genet 6:302. https://doi.org/10.3389/fgene.2015.00302
Islam MN, Azad AK, Namuco LO, Borromeo TH, Lourdes M, Cedo O, Aguilar EA (2013) Morphometric characterization and diversity analysis of a Makapuno coconut population in U.P. Los Baños. Pakistan J Agric Res 26:254–264
Jaligot E, Adler S, Debladis É, Beulé T, Richaud F, Ilbert P, Finnegan EJ, Rival A (2011) Epigenetic imbalance and the floral developmental abnormality of the *in vitro*-regenerated oil palm *Elaeis guineensis*. Ann Bot 108:1453–1462. https://doi.org/10.1093/aob/mcq266
Jaligot E, Beulé T, Rival A (2002) Methylation-sensitive RFLPs: Characterisation of two oil palm markers showing somaclonal variation-associated polymorphism. Theor Appl Genet 104:1263–1269. https://doi.org/10.1007/s00122-002-0906-4
Jaligot E, Rival A, Beulé T, Dussert S, Verdeil JL (2000) Somaclonal variation in oil palm (*Elaeis guineensis* Jacq.): The DNA methylation hypothesis. Plant Cell Rep 19:684–690. https://doi.org/10.1007/s002999900177
Jerard BA, Damodaran V, Niral V, Samsudeen K, Rajesh MK, Sankaran M (2013) Conservation and utilization of soft endosperm coconut accession from Andaman Islands. J Plant Crops 41:14–21. https://doi.org/10.1007/s002999900177
Jiang H, Köhler C (2012) Evolution, function, and regulation of genomic imprinting in plant seed development. J Exp Bot 63:4713–4722. https://doi.org/10.1093/jxb/ers145
Jiao L, Liu X (2015) Structural basis of histone H3K27 trimethylation by an active polycomb repressive complex 2. Science 350:aac4383. https://doi.org/10.1126/science.aac4383
Joyce SM, Cassells AC (2002) Variation in potato microplant morphology *in vitro* and DNA methylation. Plant Cell Tissue Organ Cult 70:125–137. https://doi.org/10.1023/A:1016312303320

Juliano JB (1926) Origin, development and nature of the stony layer of the coconut (*Cocos nucifera*). Phil J Sci 30:187–200

Jullien PE, Kinoshita T, Ohad N, Berger F (2006) Maintenance of DNA methylation during the *Arabidopsis* life cycle is essential for parental imprinting. Plant Cell 18:1360. https://doi.org/10.1105/tpc.106.041178

Jullien PE, Mosquna A, Ingouff M, Sakata T, Ohad N, Berger F (2008) Retinoblastoma and its binding partner MSI1 control imprinting in *Arabidopsis*. PLOS Biol 6:e194. https://doi.org/10.1371/journal.pbio.0060194

Kartha ARS, Narayanan R (1956) Development of oil in the ripening coconut. Ind J Agric Sci 26:319–327

Khan AR, Shah SM, Irshad M (2015) Immediate and transgenerational regulation of plant stress response through DNA methylation. J Agr Sci 7(4). https://doi.org/10.5539/jas.v7n4p144

King GJ, Amoah S, Kurup S (2010) Exploring and exploiting epigenetic variation in crops. Genome 53:856–868. https://doi.org/10.1139/G10-059

Kostik V, Memeti S, Bauer B (2013) Fatty acid composition of edible oils and fats. JHED 4:112–116. https://doi.org/10.1016/j.foodchem.2017.05.072

Kumar SN (2011) Variability in coconut (*Cocos nucifera* L.) germplasm and hybrids for fatty acid profile of oil. J Agric Food Chem 59:13050–13058. https://doi.org/10.1021/jf203182d

Lado JP, Arellano BZ, Osio CAL, Tendero BJT, de la Vina CB, Torio MAO, Laude RP, Diaz MGQ (2019) Early differential expression of galactomannan biosynthesis genes in 'Makapuno' coconut (*Cocos nucifera* L.) revealed by the *de novo* assembly and analysis of endosperm transcriptome. Philipp Agricul Sci 102:6–24

Langevin SM, Kelsey KT (2013) The fate is not always written in the genes: Epigenomics in epidemiologic studies. Environ Mol Mutagen 54:533–541. https://doi.org/10.1002/em.21762

Larkin PJ, Scowcroft WR (1981) Somaclonal variation - a novel source of variability from cell cultures for plant improvement. Theor Appl Genet 60:197–214. https://doi.org/10.1007/BF02342540

Law JA, Jacobsen SE (2010) Establishing, maintaining and modifying DNA methylation patterns in plants and animals. Nature Rev Genet 11:204–220. https://doi.org/10.1038/nrg2719

Li W, Liu H, Cheng ZJ, Su YH, Han HN, Zhang Y, Zhang XS (2011) DNA methylation and histone modifications regulate de novo shoot regeneration in *Arabidopsis* by modulating *WUSCHEL* expression and auxin signaling. PLoS Genet 7:e1002243. https://doi.org/10.1371/journal.pgen.1002243

Liyanage VRB, Jarmasz JS, MurugeshanN DMR, Rastegar M, Davie JR (2014) DNA modifications: Function and applications in normal and disease states. Biology (Basel) 3:670–723. https://doi.org/10.3390/biology3040670

LoSchiavo F, Pitto L, Giuliano G, Torti G, Nuti-Ronchi V, Marazziti D, Vergara R, Orselli S, Terzi M (1989) DNA methylation of embryogenic carrot cell cultures and its variations as caused by mutation, differentiation, hormones and hypomethylating drugs. Theor Appl Genet 77:325–331. https://doi.org/10.1007/BF00305823

Luo M, Bilodeau P, Dennis ES, Peacock WJ, Chaudhury A (2000) Expression and parent-of-origin effects for *FIS2*, *MEA*, and *FIE* in the endosperm and embryo of developing *Arabidopsi*s seeds. Proc Natl Acad Sci USA 97:10637. https://doi.org/10.1073/pnas.170292997

Luo M, Taylor JM, Spriggs A, Zhang H, Wu X, Russell S, Singh M, Koltunow A (2011) A genome-wide survey of imprinted genes in rice seeds reveals imprinting primarily occurs in the endosperm. PLoS Genet 7: e1002125. https://doi.org/10.1371/journal.pgen.1002125

Maskromo I, Larekeng S, Novarianto H, Sudarsono S (2017) Xenia negatively affecting kopyor nut yield in Kalianda Tall kopyor and Pati Dwarf kopyor coconuts. EJFA 28:644–652. https://doi.org/10.9755/ejfa.2015-07-552

Matthes M, Singh R, Cheah SC, Karp A (2001) Variation in oil palm (*Elaeis guineensis* Jacq.) tissue culture-derived regenerants revealed by AFLPs with methylation-sensitive enzymes. Theor Appl Genet 102:971–979. https://doi.org/10.1007/s001220000491

Mendoza EMT, Mujer CV, Rodriguez FM, Ramirez DA (1982) Ontogeny of some chemical constituents in developing Makapuno and normal coconut endosperms. Kalikasan Philipp J Biol 11:293–301

Menon KP, Pandalai KM (1958) The coconut palm, a monograph. Indian Central Coconut Committee, Kerala, South India

Moreno-Romero J, Jiang H, Santos-González J, Köhler C (2016) Parental epigenetic asymmetry of PRC2-mediated histone modifications in the *Arabidopsis* endosperm. EMBO J35:1298–1311. https://doi.org/10.15252/embj.201593534

Moricova P, Ondrej V, Navratilova B, Luhova L (2013) Changes of DNA methylation and hydroxymethylation in plant protoplast cultures. Acta Biochim Pol 60:33–36

Mujer CV, Arambulo AS, Mendoza EMT, Ramirez DA (1983) The viscous components of the mutant (Makapuno) endosperm I. Isolation and characterization. Kalikasan Philipp J Biol 12:42–50

Mujer CV, Ramirez DA, Mendoza EMT (1984) α-D-galactosidase deficiency in coconut endosperm: Its possible pleiotropic effects in Makapuno. Phytochemistry 23:893–894. https://doi.org/10.1016/S0031-9422(00)85051-7

Mujer CV, Ramirez DA, Mendoza EMT (1984) Coconut α-D-galactosidase isoenzymes: Isolation, purification and characterization. Phytochemistry 23:1251–1254. https://doi.org/10.1016/S0031-9422(00)80435-5

Naresh Kumar S, Balakrishna A (2009) Seasonal variations in fatty acid composition of oil in developing coconut. J Food Qual 32:158–176. https://doi.org/10.1111/j.1745-4557.2009.00243.x

Nathanael WRN (1959) Changes in the coconut endosperm during development. Ceylon Coconut Quart 10:35–36

Ngezahayo F, Xu C, Wang H, Jiang L, Pang J, Liu B (2009) Tissue culture-induced transpositional activity of *mPing* is correlated with cytosine methylation in rice. BMC Plant Biol 9:91. https://doi.org/10.1186/1471-2229-9-91

Niral V, Samsudeen K, Nair RV (2010) Genetic resources of coconut. Kerala, India, pp 22–28

Niral V, Devakumar K, Umamaheswari TS, Naganeeswaran S, Nair RV, Jerard BA (2013) Morphological and molecular characterization of a large fruited unique coconut accession from Vaibhavwadi, Maharashtra, India. Indian J Genet 73:220–224

Ong-Abdullah M, Ordway JM, Jiang N, Ooi SE, Kok SY, Sarpan N, Azimi N, Hashim AT, Ishak Z, Rosli SK, Malike FA (2015) Loss of *Karma* transposon methylation underlies the mantled somaclonal variant of oil palm. Nature 525:533

Oo KC, Stumpf PK (1979) Fatty acid biosynthesis in the developing endosperm of *Cocos nucifera*. Lipids 14:132–143. https://doi.org/10.1007/BF02533862

Osio CAL, Aquino GMB, Aala WF Jr, Lado JP, Tendero BJT, Cueto CA, Diaz MGQ, Laude RP (2019) Transcriptomic analysis of 'Philippine Lono Tall' coconut (*Cocos nucifera* L.) endosperm reveals differential expression of genes involved in oil biosynthesis. Philipp Agricul Scientist 102:101–118

Padua-Resurreccion AN, Banzon JA (1979) Fatty acid composition of the oil from progressively maturing bunches of coconut. Phil J Coco Stud 4:1–16

Peredo EL, Ángeles Revilla M, Arroyo-García R (2006) Assessment of genetic and epigenetic variation in hop plants regenerated from sequential subcultures of organogenic calli. J Plant Physiol 163:1071–1079. https://doi.org/10.1016/j.jplph.2005.09.010

Phoeurk C, Somana J, Sornwatana T, Udompaisarn S, Traewachiwiphak S, Sirichaiyakul P, Phongsak T, Arthan D (2018) Three novel mutations in α-galactosidase gene involving in galactomannan degradation in endosperm of curd coconut. Phytochemistry 156:33–42. https://doi.org/10.1016/j.phytochem.2018.08.015

Pila Quinga LA, Pacheco de Freitas Fraga H, do Nascimento Vieira L, Guerra MP (2017) Epigenetics of long-term somatic embryogenesis in *Theobroma cacao* L.: DNA methylation and recovery of embryogenic potential. Plant Cell Tissue Organ Culture (PCTOC) 131: 295–305. https://doi.org/10.1007/s11240-017-1284-6

Pires ND, Grossniklaus U (2014) Different yet similar: Evolution of imprinting in flowering plants and mammals. F1000Prime Rep 6:63. https://doi.org/10.12703/P6-63

Pooggin MM (2013) How can plant DNA viruses evade siRNA-directed DNA methylation and silencing? IJMS 14:15233–15259. https://doi.org/10.3390/ijms140815233

Raissig MT, Baroux C, Grossniklaus U (2011) Regulation and flexibility of genomic imprinting during seed development. Plant Cell 23:16–26. https://doi.org/10.1105/tpc.110.081018

Ramirez DA, Mendoza EMT (1998) The Makapuno Coconut. The National Academy of Science and Technology (Philippines)

Reik W, Walter J (2001) Genomic imprinting: Parental influence on the genome. Nature Rev Genet 2:21–32. https://doi.org/10.1038/35047554

Rhee Y, Sekhon RS, Chopra S, Kaeppler S (2010) Tissue culture-induced novel epialleles of a *Myb* transcription factor encoded by *pericarp color1* in maize. Genetics 186:843. https://doi.org/10.1534/genetics.110.117929

Rillo EP (1997) Coconut embryo culture. CICY, Merida, Mexico

Rillo EP (1999) Coconut embryo culture. In: Oropeza C, Verdeil JL, Ashburner GR et al (eds) Current advances in coconut biotechnology. Springer, Netherlands, Dordrecht, pp 279–288

Rival A, Ilbert P, Labeyrie A, Torres E, Doulbeau S, Personne A, Dussert S, Beulé T, Durand-Gasselin T, Tregear JW, Jaligot E (2013) Variations in genomic DNA methylation during the long-term *in vitro* proliferation of oil palm embryogenic suspension cultures. Plant Cell Rep 32:359–368. https://doi.org/10.1007/s00299-012-1369-y

Rival A, Jaligot E, Beulé T, Tregear JW, Finnegan J (2009) The oil palm "mantled" somaclonal variation: A model for epigenetic studies in higher plants. Acta Horticulturae Int Soc Horticultural Sci (ISHS) 829:177–182

Rival A, Jaligot E, Beulé T, Finnegan EJ (2008) Isolation and expression analysis of genes encoding MET, CMT, and DRM methyltransferases in oil palm (*Elaeis guineensis* Jacq.) in relation to the 'mantled' somaclonal variation. J Exp Bot 59:3271–3281. https://doi.org/10.1093/jxb/ern178

Russo VEA, Martienssen RA, Riggs AD (eds) (1996) Epigenetic mechanisms of gene regulation. Cold Spring Harbor Laboratory Press, New York

Samonte JL, Mendoza EMT, Ilag LL, Norberto B, Ramirez DA (1989) Galactomannan degrading enzymes in maturing normal and makapuno and germinating normal coconut endosperm. Phytochemistry 28:2269–2273. https://doi.org/10.1016/S0031-9422(00)97966-4

Samonte JL, Ramirez DA, Mendoza EMT (1987) Galactomannan in developing normal and Makapuno coconut endosperm. Bull Philipp Biochem Soc 7:15–19

Satyabalan K (1953) A note on the occurrence of "buttery" kernel in coconut. Indian Coconut J 6:152–154

Schuettengruber B, Cavalli G (2009) Recruitment of polycomb group complexes and their role in the dynamic regulation of cell fate choice. Development 136:3531. https://doi.org/10.1242/dev.033902

Schwabe WW, Blake J, Fulford RM (1970) Application of tissue culture techniques for vegetative propagation.

Plant Production and Protection Division, Food and Agriculture Organization

Smulders MJ, Rus-Kortekaas W, Vosman B (1995) Tissue culture-induced DNA methylation polymorphisms in repetitive DNA of tomato calli and regenerated plants. Theor Appl Genet 91:1257–1264. https://doi.org/10.1007/BF00220938

Smulders MJM, de Klerk GJ (2011) Epigenetics in plant tissue culture. Plant Growth Regul 63:137–146. https://doi.org/10.1007/s10725-010-9531-4

Springer NM, Schmitz RJ (2017) Exploiting induced and natural epigenetic variation for crop improvement. Nature Rev Genet 18:563

Stam M, Belele C, Dorweiler JE, Chandler VL (2002) Differential chromatin structure within a tandem array 100 kb upstream of the maize *b1* locus is associated with paramutation. Genes Dev 16:1906–1918. https://doi.org/10.1101/gad.1006702

Tammen SA, Friso S, Choi SW (2013) Epigenetics: The link between nature and nurture. Mol Aspects Med 34:753–764. https://doi.org/10.1016/j.mam.2012.07.018

Tammes P (1955) Review of coconut selection in India. Euphytica 4:17–24. https://doi.org/10.1007/BF00042411

Tangwatcharin P, Khopaibool P (2012) Activity of virgin coconut oil, lauric acid or monolaurin in combination with lactic acid against *Staphylococcus aureus*. Southeast Asian J Trop Med Public Health 43:969–985

Tanurdzic M, Vaughn MW, Jiang H, Lee TJ, Slotkin RK, Sosinski B, Thompson WF, Doerge RW, Martienssen RA (2008) Epigenomic consequences of immortalized plant cell suspension culture. PLoS Biol 6:e302. https://doi.org/10.1371/journal.pbio.0060302

Torres J (1937) Some notes on Makapuno coconut and its inheritance. Philipp J Agric 8:27–37

Triantaphyllopoulos KA, Ikonomopoulos I, Bannister AJ (2016) Epigenetics and inheritance of phenotype variation in livestock. Epigenetics Chromatin 9:31. https://doi.org/10.1186/s13072-016-0081-5

Urechean V, Bonea D (2017) Coexistence in cultivation of genetically modified maize (MON810) with conventional maize. Rom Agric Res 34:51–58

Valledor L, Hasbún R, Meijón M, Rodríguez JL, Santamaría E, Viejo M, Berdasco M, Feito I, Fraga MF, Canal MJ, Rodríguez R (2007) Involvement of DNA methylation in tree development and micropropagation. Plant Cell Tiss Org Culture 91:75–86. https://doi.org/10.1007/s11240-007-9262-z

Varga A, Thoma LH, Bruinsma J (1988) Effects of auxins and cytokinins on epigenetic instability of callus-propagated *Kalanchoe blossfeldiana* Poelln. Plant Cell Tiss Org Culture 15:223–231. https://doi.org/10.1007/BF00033646

Vining K, Pomraning KR, Wilhelm LJ et al (2013) Methylome reorganization during *in vitro* dedifferentiation and regeneration of *Populus trichocarpa*. BMC Plant Biol 13:92. https://doi.org/10.1186/1471-2229-13-92

von Aderkas P, Bonga JM (2000) Influencing micropropagation and somatic embryogenesis in mature trees by manipulation of phase change, stress and culture environment. Tree Physiol 20:921–928

Waddington CH (1942) The epigenotype. Endeavour 1:18–20

Wang Q-M, Wang L, Zhou Y et al (2016) Leaf patterning of *Clivia miniata* var. *variegata* is associated with differential DNA methylation. Plant Cell Rep 35:167–184. https://doi.org/10.1007/s00299-015-1877-7

Watanabe S, Kamada H, Ezura H (2006) Effect of a special screened greenhouse covered by fine mesh on maize outcrossing. Plant Biotechnol 23:309–316. https://doi.org/10.5511/plantbiotechnology.23.309

Wu CT, Morris JR (2001) Genes, genetics, and epigenetics: A correspondence. Science 293:1103. https://doi.org/10.1126/science.293.5532.1103

Xiao W, Custard KD, Brown RC, Lemmon BE, Harada JJ, Goldberg RB, Fischer RL (2006) DNA methylation is critical for *Arabidopsis* embryogenesis and seed viability. Plant Cell 18:805. https://doi.org/10.1105/tpc.105.038836

Yong WS, Hsu FM, Chen PY (2016) Profiling genome-wide DNA methylation. Epigenetics Chromatin 9:26. https://doi.org/10.1186/s13072-016-0075-3

Yuan W, Wu T, Fu H, Dai C, Wu H, Liu N, Li X, Xu M, Zhang Z, Niu T, Han Z (2012) Dense chromatin activates Polycomb Repressive Complex 2 to regulate H3 lysine 27 methylation. Science 337:971. https://doi.org/10.1126/science.1225237

Zambiazi R, Przybylski R, Zambiazi M, Mendonca C (2007) Fatty acid composition of vegetable oils and fats. Boletim Centro De Pesquisa De Processamento De Alimentos 25:111–120

Zhang G, Pradhan S (2014) Mammalian epigenetic mechanisms. IUBMB Life 66:240–256. https://doi.org/10.1002/iub.1264

Zhang M, Zhao H, Xie S, Chen J, Xu Y, Wang K, Zhao H, Guan H, Hu X, Jiao Y, Song W (2011) Extensive, clustered parental imprinting of protein-coding and non-coding RNAs in developing maize endosperm. Proc Natl Acad Sci USA 108:20042. https://doi.org/10.1073/pnas.1112186108

Zuñiga LC (1953) The probable inheritance of the Makapuno character of the coconut. Philipp Agric 36:402–413

Genome Editing: Prospects and Challenges

14

K. A. Lineesha and Ginny Antony

Abstract

Coconut palm, a ubiquitous landscape feature of the tropics, is perennial in nature and therefore poses inherent constraints to the breeding programs. To date, the breeding strategies for trait improvement in coconut have relied on conventional approaches like selection and hybridization. Apart from this, coconut is mostly cross-pollinated, and lack of methods for mass propagation of elite varieties is a major bottleneck in the breeding efforts. The feasibility of the use of gene editing technologies for breeding desirable traits in a few tree species has opened the door for possibilities in similar plants, and this chapter provides an overview of genome editing systems available with an emphasis on the more popular clustered regularly interspaced short palindromic repeats (CRISPR)-based systems, possibilities, and challenges regarding coconut improvement.

14.1 Introduction

Coconut (*Cocos nucifera* L.), a member of the family Arecaceae, is considered the "tree of life" by millions of people living in the coastal areas of the tropics. This palm is cropped in 93 countries and traditionally grown for its coir and oil, and the latter is used in cooking by the local population and largely by the cosmetic industry. Recently, value-added products like virgin coconut oil, palm sugar, and coconut-based drinks have become popular and have rejuvenated the coconut-based industries. Coconut growers are usually smallholder farmers who often cultivate landraces and are unable to keep up the production with the growing demands due to a myriad of challenges. These include low productivity, aging plantations, non-availability of high yielding planting materials, non-uniformity in the productivity of palms, lethal diseases, drought, and emerging pests to name a few. Coconut being a perennial has its challenges in the breeding programs owing to its biology. Breeding for improved varieties for yield and resistance to pests and diseases is costly and cumbersome due to its long juvenile phase preceding the fruiting phase and requirement of a large cropping area for screening. Therefore, the improvements made through conventional breeding are few and slow. In conventional breeding programs, the desirable genetic variations in plants that occur naturally or through induction via physical, chemical, or biological means serve as the source for new traits. However, the generation of spontaneous or induced mutations is rare and random. Genetic improvements are also made by crossing among the different genotypes of coconut (*Cocos nucifera* L.). Worldwide, many high-yielding hybrids

K. A. Lineesha · G. Antony (✉)
Central University of Kerala, Kasaragod, Kerala, India
e-mail: ginnyantony@cukerala.ac.in

M. K. Rajesh et al. (eds.), *The Coconut Genome*, Compendium of Plant Genomes,
https://doi.org/10.1007/978-3-030-76649-8_14

have been developed through the inter-varietal crossing, but very few are grown commercially due to lack of awareness, non-availability of good hybrid seedlings and the high-input requirements.

Developments in the field of molecular biology, and the targeted manipulation of DNA sequence have paved the way for the identification and characterization of genes in a large number of plant species. But this strategy requires the regeneration of plants from single cells through micropropagation methods and one of the serious bottlenecks for DNA manipulation in coconut is that it is highly *in vitro* recalcitrant. The recent publishing of the draft genome of coconut has provided genomic insights that will greatly facilitate molecular-assisted breeding and holds promise for a better understanding of the genetic characteristics of this palm (Xiao et al. 2017). The coconut genome is estimated to be 2.42 GB in size with an estimated 28,039 protein-coding genes. The developments in the field of genome editing offer hope for unlocking the potential of the coconut genome.

14.2 Molecular Tools for Gene Editing

14.2.1 Programmable Sequence-Specific Nucleases

14.2.1.1 Zinc Finger Nucleases

More than a decade ago, the first-generation genome editing tool, zinc finger nucleases or ZFN, was introduced which theoretically permits modification in any sequence of a genome (Fig. 14.1a). The fundamental structure of a ZFN comprises mainly two components: (a) zinc finger DNA-binding modules and (b) non-specific cleavage domain of *Fok*1 (Nemudryi et al. 2014). The *Fok*1 belongs to the category of Type IIs restriction enzyme which can recognize the 5'GGATG 3' sequence and cleaves the DNA 9 bp downstream and 13 bp upstream on the sense and antisense strand, respectively (Kim et al. 1996). It has been already reported that the plant genomes are rich in zinc finger modules, each of which holds ~30 Cys2-His2 amino acids in ββα confirmation. Individual zinc finger DNA module can recognize 3 bp sequence on a DNA. Therefore, each zinc finger array with ~3–6 DNA module can target 9–18 nucleotides on the target site (Xiong et al. 2015). The proper binding of a ZFN to the desired target site is only possible when the α helix of the surface amino acids in ZFN domain effectively interacts with the major groove of the DNA (Chen and Gao 2014). Each ZFN array targets either parallel or antiparallel strands of the DNA. However, it should maintain ~5–6 bp spacer distance while binding, which would enable the dimerization of *Fok*I and its action. So far, ZFNs-based gene editing has been applied on many crops such as maize, sorghum, tobacco, rice, etc. Low-frequency mutation (0.0–0.3%) per allele was reported in poplar tree while targeting the *LEAFY* and *AGAMOUS* orthologues (Lu et al. 2016). Furthermore, the tedious protein engineering, increased off-targets when using more than four domains, less efficiency, unavailability of ZFN domains in the target site, and the cell toxicity limit the use of ZFN as a popular or widely utilized gene-editing tool (Gaj et al. 2013; Gupta and Musunuru 2014).

14.2.1.2 Transcription Activator-Like Effector Nuclease (TALENS)

Transcription activator-like effector nuclease (TALENS) is the second kind of "molecular scissors" which is derived from the Type III effectors of *Xanthomonas* bacteria, a major threat to crop plants (Fig. 14.1b) (Yang and White 2004). Once inside the nucleus, these effectors hijack the plant transcriptional machinery altering the gene expression of certain host targets, which eventually benefits the pathogen survival and multiplication (Yang and White 2004). The vital structure of TALE effectors contains an N-terminal Type III secretion signal domain, DNA-binding domain (DBD), and a C-terminal activation domain. The DBD is a tandem array of 34 amino acid repeats, in which each repeat could recognize a single nucleotide (Mak et al. 2014). (NI-A, NG-T, NN-G or A, HD-C, NH, or NK-G)

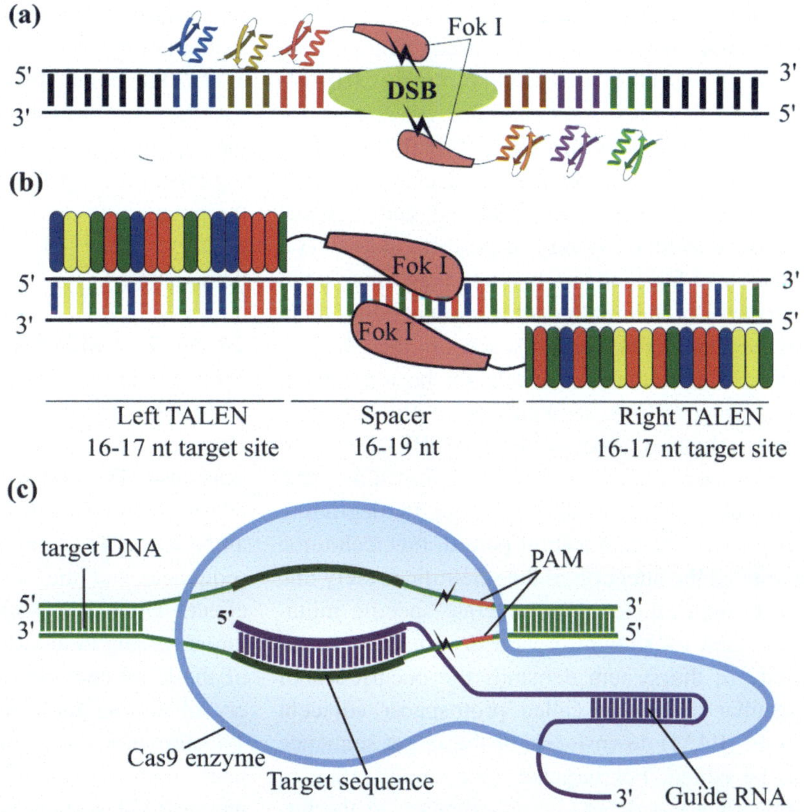

Fig. 14.1 Pictorial representation of three different gene-editing platforms, **a** zinc finger nucleases, **b** transcription activator-like effector nucleases, **c** clustered regularly interspersed short palindromic sequence and CRISPR-associated proteins

(Boch et al. 2009). The repeat variable di-residue (RVD), which is situated at the 12th and 13th position of each repeat of the DNA-binding domain, determines the unique binding of TAL effector (Boch et al. 2014). TALEs recognize the host targets, which are preceded by thymine (T), which is essential for their activity. Deciphering the code of TALE-binding phenomena led to the invention of a versatile site-specific nuclease, called TALENs. These are the fusion of TALE effector repeats with the catalytic module of the *Fok*1. After delivery into a cell, individual TALENs would assemble on the forward and reverse strand of the DNA as a dimer and maintain a ~12–25 bp distance in between for specific binding and endonuclease action (Wright et al. 2014). TALENs possess some advantages over ZFN. It is easier to engineer a TALEN than a ZFN. The repeat array of the TALEN can be increased to a desirable length, whereas in ZFN, it is limited up to 9–18 bp length. Moreover, TALENs do not induce cell toxicity when delivered into a cell. TALENs generate few off-target effects compared to ZFN. The construction of the TALEN vectors is tedious due to the repetitive nature of the TALEs and the high cost of protein engineering. The most popular method of TALEN assembly is the Golden Gate method (GG method), which is step-wise cloning using Type-II s restriction enzymes and ligases (Cermak et al. 2015). TALEN technology was popular due to its versatility and enabled targeted modifications in several crops such as rice, maize, *Brachypodium*, tobacco, wheat, tomato, potato, *Arabidopsis,* and sugarcane with very high efficiencies (Christian et al. 2013; Jaganathan et al. 2018; Li et al. 2012; Lor et al. 2014; Shan et al. 2013; Zhang et al. 2013). TALEs have also been used to create fusion proteins containing domains that direct epigenetic modifications or site-specific recombination that are heritable (Day 2014).

14.2.1.3 CRISPR-Cas

Clustered Regularly Interspaced Palindromic Repeats and their associated protein (CRISPR-Cas) is the most versatile tool developed for targeted gene modifications. (Fig. 14.1c). This RNA-guided endonuclease mechanism was identified as part of the bacterial and archaeal adaptive immune system. It protects the bacteria from invading viruses by first capturing small segments of the viral genome and encoding them in the host genome called spacers in the CRISPR array. If the same virus attacks the bacteria again, it produces RNA segments from the CRISPR array which is used by the CRISPR-associated endonuclease (Cas) to identify the invading virus and subsequently disable the virus by degrading its genome. Within a short period, this technique garnered the attention of the scientific society due to its high efficiency to generate specific mutations and its straightforward design. The specificity of the system demands the occurrence of peculiar sequences called protospacer adjacent motif (PAM) downstream of the target sequence to be edited. For targeted gene modification, a guide RNA (sgRNA) complementary to the target site in the DNA is introduced along with Cas endonuclease. The most widely accepted and the first introduced, Cas 9, endonuclease was isolated from *Streptococcus pyrogenes*, which recognize the 5'NGG3' as a PAM (Abdallah et al. 2015). Hence, the active complex formed between the RNA hybrid and the Cas9 can screen along the DNA sequence until it contacts the PAM. The Cas 9 nuclease domains such as HNH and the RuvC facilitate a double-stranded break in the DNA (Gudeta 2019). Subsequently, the endogenous repair pathway of the plants such as NHEJ and HDR mend the nick in the DNA which results in InDels or gene replacement in the target region (Zhang and Sheen 2014; Zhang et al. 2017). The attractive feature of this system is the multiplexing by using two or more sgRNAs, thereby targeting multiple loci simultaneously (Gao et al. 2015; Lowder et al. 2015; Ma et al. 2016). Ideal sgRNAs have a length of 20-22 bp, GC content (40–60%), fewer off-targets, and the requirement of PAM sequence. Several software resources are freely available that assist in the identification of guide RNAs that target the desired loci (Labun et al. 2019; Lei et al. 2014; Minkenberg et al. 2018). CRISPR, as an efficient gene-editing tool, offers researchers the opportunity for the rapid engineering of organisms that perform novel functions. It has the potential to revolutionize and shape synthetic biology, some of which have already been demonstrated.

14.2.1.4 CRISPR Variants

To date, the CRISPR Cas system has gained wide popularity due to its simplicity and potency. However, the system possesses certain limitations also. The first one being PAM dependency, editing with CRISPR Cas 9, is only achievable if an NGG or NAG motif is present near the target sequence, and the second one is the off-target effect, which is a major drawback of the technique (Zhang et al. 2017). Thirdly, the large size of the Cas9 enzyme impedes the efficient delivery of editing tools in therapeutic applications. To eliminate these problems, The SpCas9 has been modified or engineered to recognize alternative PAM sequences (Kleinstiver et al. 2015). This has resulted in Cas9 variants which could cleave the DNA by identifying the relaxed or different PAM sequence.

The native Cas9 enzyme holds two different catalytic sites such as the RuvC domain and HNH domain which facilitates the cleavage of non-complementary and complementary strands, respectively (Townson 2017). To mitigate the off targeting by SpCas9, a variant of Cas9 was created by introducing the alanine substitution in the key residues of RuvC domain (D10A) (Fig. 14.2a) and HNH (N863A or H840A) (Fig. 14.2b) (Trevino and Feng 2014). The DNA break is further repaired by an innate plant repair mechanism, which results in precise editing like Cas9 with low off-targets. The paired nickases have been employed to generate the staggered cut after maintaining proper spacing and orientation on the dsDNA, unlike the wild nickase (Gopalappa et al. 2018; Mikami et al. 2016).

The mutation in both nuclease domains of wild SpCas9 generated a specific variant called dead Cas9 (SpdCas9). The mutation-induced in the catalytic residues abolished the endonuclease activity but retained the DNA binding property of the wild Cas9. The dCas9 has been widely employed in gene editing along with sgRNA and *Fok*I domain. In addition, dead Cas9 has widened its scope as a tool for cell imaging (Dreissig et al. 2017), epigenetic modulation (Moradpour and Abdulah 2019), and transcriptional regulation by CRISPRi and CRISPRa (Jensen 2018; Piatek et al. 2015; Tiwari et al. 2012).

Apart from redesigned or modified Cas9, orthologues of Cas9 were identified as part of the adaptive immune mechanism from several bacteria and harnessed as a gene-editing tool with novel PAM specificities. Such orthologues are smaller in size compared to the Cas9. The Cas9 from *Staphylococcus aureus* (*S. aureus*) (Fig. 14.2c) has been extensively used in rice, tobacco, and *Arabidopsis thaliana* (Kaya et al. 2016; Steinert et al. 2015). This enzyme induces a double-stranded nick at 21–24nt length target sequence on the availability of relaxed PAM sequences such as (5′NNGAGT-3′, 5′ NNGGGT3′, and 5′NNGAAT3′), SaCas9 precisely cut the DNA four nucleotide upstream of the PAM sequence and induce large deletion in case of 5′NNGGGT3′ (Endo et al. 2016; Murovec et al. 2017). *Streptococcus thermophilus*-derived Cas9 (*St1*Cas9) (Fig. 14.2d) identifies the 5′NNAGAA3′ and 5′NNGGAA3′ as PAM motif and cuts the DNA 2–3 bp upstream of the PAM sequence (Murovec et al. 2017). St3Cas9 derived from *Streptococcus thermophilus* recognizes the 5′-NGGNG-3′ PAM (Cong et al. 2013). Cas9 orthologue in *Neisseria meningitidis* called

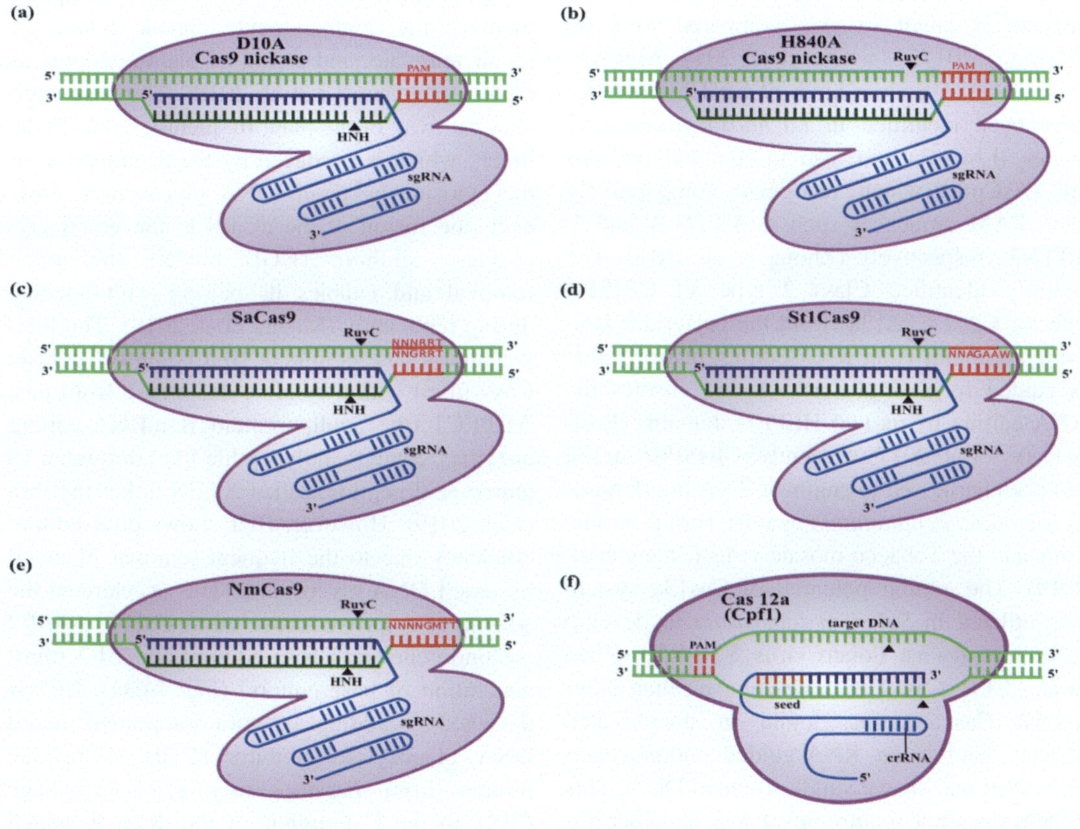

Fig. 14.2 Schematic representation of variants of CRISPR Cas 9 system: **a** D10A Cas9 nickase, **b** H840A Cas9 nickase, **c** *Sa* cas9 (*S.* aureus Cas9), **d** *St1* cas9 (*S. thermophilus* Cas9), **e** *Nm* Cas9 (*N.meningiditis* Cas9), **f** CRISPR-Cpf1

NmCas9 (Fig. 14.2e) targets the sequence upstream of 5′ NNNNGMTT 3′. SpCas9-NG is a highly efficient version of SpCas9 that can recognize the relaxed NG PAM sequences rather than NGG (Ren et al. 2019). xcas9 is a variant of SpCas9 which possesses broad PAM compatibilities such as NG, GAA, and GAT and also shows potent editing efficiency with low off-targeting effects (Wang et al. 2019).

CRISPR-Cas systems are classified into six types and two classes. The putative type V CRISPR effector Cpf1 (Cas12a) (Fig. 14.2f) belongs to CRISPR class II systems and was first identified in *Francisella novicida* (FnCpf1) targeting an A/T-rich PAM 5′TTTV 3′ (Mahfouz 2017). The system has its own RNase activity and does not require the tracrRNA along with the crRNA for the editing. The Cpf1 system induces a double-stranded staggered cut in the sequence which is distant from the PAM motif, and this enzyme is small in size compared to Cas9 (Mahfouz 2017; Malzahn et al. 2017; Murovec et al. 2017). The other kinds of Cpf1 orthologues have been identified in *Lachnospiraceae bacterium* (LbCpf1) and also in *Acidaminococcus* sp. (AsCpf1) which recognize comparatively short PAM sequences such as 5′TTN 3′ and 5′ TTTN3′, respectively (Zhong et al. 2018). The recently identified Class 2 type VI CRISPR effector C2c2 (Cas13a) from the bacterium *Leptotrichia shahii* is an RNA-editing platform that is guided by a single crRNA and ensures the RNA editing by its two HEPEN domains (East-Seletsky et al. 2017). Recently, CRISPR/Cas13a has been harnessed to engineer RNA interference in *Nicotiana benthamiana* against Turnip mosaic virus and the Tobacco mosaic virus (Zhang et al. 2019). The editing potential of Cas13a system was utilized in *Solanum tuberosum* to develop resistance against Potato virus Y (PVY) (Zhan et al. 2019). Cas14 is more compact than other known Cas systems, found in uncultivated archaea, and is an RNA-guided endonuclease that target and cleave single-stranded DNA. This system does not require any PAM sequence for the cleavage of the target sequence (Harrington et al. 2018).

14.3 Base Editing

High throughput genome sequencing and genome-wide association studies of different traits in many plants have identified that single nucleotide polymorphisms in target alleles confer many specific traits. Precise gene editing provides an opportunity to convert unfavorable alleles into favorable ones without the hassle of conventional breeding approaches that takes several years. The base editors accelerate the conversion of the C: T-G: A in a programmable manner which is an alternative tool to HDR-mediated gene replacement (Komor et al. 2016; Li et al. 2017). The catalytically inefficient version of the CRISPR Cas such as Cas9 nickase and dead Cas9 fused with cytosine deaminase or adenosine deaminase is the fundamental feature of a base editor (Komor et al. 2016). As compared to other platforms, it avoids the requirement of the double-stranded break, homology repair template, and action of plant endogenous repair pathways (Li et al. 2018). The base excision pathway of the plant frequently removes the uracil, which is produced by the deamination of the cytosine by uracil DNA glycosylase; however, the fusion of the dCas9 to the uracil glycosylase inhibitor (UGI) hinders the uracil removal and enables its pairing with adenine during replication (Mishra et al. 2019). The first-generation base editors or BE1 comprises a dead Cas9 fused with a cytidine deaminase from rats, AOPEC1 (the apolipoprotein B mRNA-editing enzymes, catalytic polypeptide-like) through a 16 amino acid peptide, called XTEN linker (Mishra et al. 2019). However, BE1 shows poor editing efficiency due to the frequent removal of uracil by uracil DNA glycosylase. This accelerated the advent of improved base editors such as BE2 (second-generation base editors) and BE3 (third generation of base editors) (Fig. 14.3a). BE2 is designed by adding an extra component, uracil DNA glycosylase inhibitor (UGI), 83-residue protein from *Bacillus subtilis* bacteriophage PBS1 to the C terminus of the dCas 9, which could surmount the limitations of BE1. Most of the agronomic traits are targeted and improved in crop plants with the aid of BE3 editors which employ the fusion of Cas9 nickase with rat cytidine deaminases (rAPOBEC1) at N-terminus via XTEN linker and UGI protein in C terminus by a four amino acid linker. An orthologue of, rAPOBEC1 from sea lamprey, PmCDA1 was linked with Cas9 nickase to generate a base editor termed as Target-AID (activation-induced cytidine deaminase) (Fig. 14.3b) (Komor et al. 2016; Nishida et al. 2016). The purity of the desired product can be achieved by reducing the InDel formation by combining the Mu bacteriophage-derived Gam with the BE3, SaBE3, BE4, and SaBE4 editors, (Fig. 14.3c) which employ catalytically inactive *S. pyrogens* cas9 (SpCas9) and *S. aureus* (SaCa9) together with two UGI components (Komor et al. 2017). The Gam protein promotes non homologous end joining (Bhattacharyya et al. 2018). The application of the Spcas9-based editors is limited to an extent due to the unavailability of the PAM sequence after the base to be edited and the increased off-targeting has been reported for cytosine base editor due to the presence of more than one cytosine in the catalytic window. This issue is resolved by the Cas9 variants which demand canonical PAM sequences (Ecol 2017).

Targeted aid-mediated mutagenesis (TAM) (Fig. 14.3d) was invented to induce genetic diversification of human disease-related genes by linking the dead Cas9 (dcas9) with activation-induced cytidine deaminase (AID) (Ma et al. 2016). In adenosine base editor (ABE) (Fig. 14.3f), spCas9 nickase is fused with adenosine deaminase, and the deamination of adenine converts it to inosine which is recognized as guanine and results in AT to GC conversion (Li et al. 2018). The advent of more efficient base editors such as CRISPR-X (Fig. 14.3e) (Hess et al. 2016), Cpf1-based cytidine deaminase base editor dLbCpf1-BE0 (Li et al. 2018), adenine base editors (Gaudelli et al. 2017; Hua et al. 2019), and the RNA base editors (ADAR) (Fig. 14.3g) (Montiel-Gonzalez et al. 2019) could expand the scope of editing and expedite the strategies for the breeding and crop improvement in the near future. The base editors were tested in rice at two different sites in the gene *OsSBEIIb* encoding starch branching enzyme IIb and the editing efficiency ranged from 10 to 19% and resulted in high amylose rice (Xu et al. 2021). Similar editing through base editors has been reported in other plants like tomato, potato, cotton, and maize for trait improvement (Li et al. 2018; Qin et al. 2020; Shimatani et al. 2019; Veillet et al. 2020). CBE system can also generate mutations at multiple sites simultaneously and enable crop improvement by targeting multiple traits in one attempt. Although the scope and range of the base editing tool are enormous, several challenges limit the universal application like high off-target effects and wide editing window.

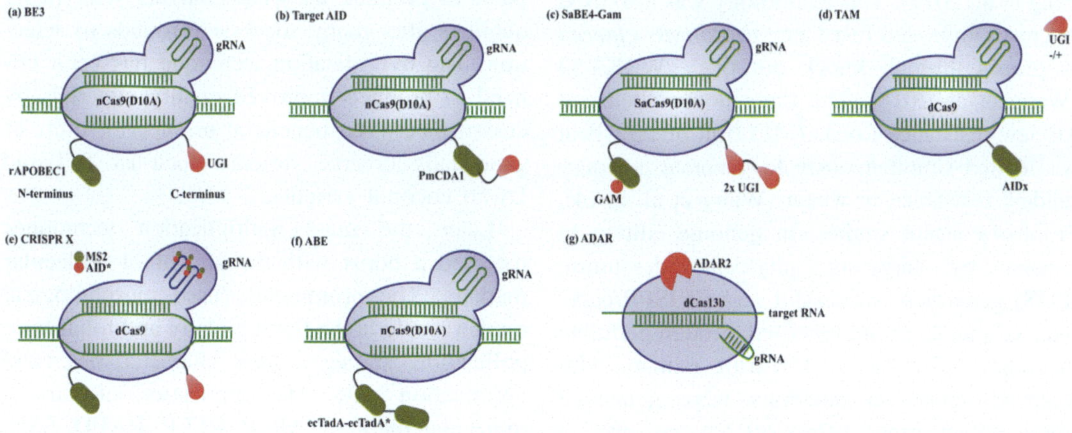

Fig. 14.3 Schematic of various base editing tools, **a** BE3, **b** target AID, **c** SaBE4 Gam, **d** TAM, **e** CRISPR X, **f** ABE, **g** ADAR

14.4 CRISPR-Based Editing in Tree Crops

CRISPR-based genome editing has been successfully deployed in several tree species (Bewg et al. 2018). The ease of use and the species independent nature of the CRISPR-based method are the reasons behind the growing popularity of this technique over other editing methods. Unlike other genome editing technologies, CRISPR-based system does not require time-consuming and labor-intensive protein engineering. Also, this technology can be used for targeting multiple loci simultaneously and therefore provides an opportunity for improving multiple traits in a single approach. The application of this technology in several model systems has helped our understanding of the system and improvement of the technique by generating knowledge on the parameters and external factors that affect editing efficiency and off-target effects. Based on this knowledge, the CRISPR system has been effectively deployed in many non-model crop systems to induce tailored alterations in the plant genome to understand its biology or for the improvement of plant varieties.

To date, CRISPR-induced mutations are reported in poplar, citrus, apple, grape, coffee, kiwi, and grapefruit (Bewg et al. 2018). CRISPR-based editing of susceptibility gene *LOB*1 promoter resulted in citrus plants resistant to *Xanthomonas citri* subsp. *citri* (Jia et al. 2017; Peng et al. 2017). This technology was also used to create improved resistance to *Botrytis cinerea* in grapes through knock out of VvWRKY52 (Wang et al. 2018). The three homo alleles of Mildew resistance Locus (MLO) in bread wheat was edited simultaneously to generate powdery mildew resistance in wheat (Wang et al. 2014). Proof-of-concept studies on genome editing in banana by targeting phytoene desaturase (PDS) gene were successful and CRISPR/cas9-induced mutations are inherited through germline in grapes and citrus. In kiwifruit, mutants with large chromosomal deletions were generated using polycistronic tRNA-sgRNA cassette in which paired sgRNAs were used to target distant loci (Wang et al. 2018). *CENTRORADIALIS* like genes (*CEN*) in kiwi are identified as repressors of flowering and biallelic mutations in AcCEN4, or AcCEN 1 led to early flowering and mutations affecting both the genes led to compact growth habit (Varkonyi-Gasic et al. 2019). Poplar being most amenable to genetic transformation has seen a multitude of studies using CRISPR-based knockout as functional genomics tool (Azeez and Busov 2020; Ding et al. 2020; Fan et al. 2015; Wang et al. 2020).

14.5 Limitations of Breeding in Coconut

Genetic improvement in coconut is cumbersome as the breeding cycle takes at least 15 years and requires a large plantation for screening. Added to that palm produces very few nuts per year and throughout the coconut growing tracts of the world, coconut plantations are old, past their prime producing age, and slowly declining. There is an urgent need for re-plantation with high yielding palms that flower early. Currently, conventional breeding strategies to mass-produce planting materials like mass selection and hybridization are used in major coconut growing regions. Mass selection produces the high-quality plant materials in which a breeder depends on the high-quality fruit traits of the mother palm such as size, weight, shape, the yield of copra per plant to produce the elite cultivars with better qualities after many successive rounds of selection. The hybridization technique has been utilized to produce improved quality materials by crossing between beneficial or superior traits of genetically diverse varieties such as Tall and Dwarf coconut varieties.

Later, the mass multiplication techniques received a boost with the advent of molecular markers. The prominently used morphological markers include the floral biology and foliar trait, pollination strategy, fruit characteristics, and germination time. The application of various molecular markers, AFLP, RFLP, RAPD, SSR, and SCoT helped improve the identification of genetic variation in coconut germplasm and thereby improve the mapping for different agronomic traits (Dasanayaka et al. 2009; Lebrun et al. 1998; Rajesh et al. 2014; Perera et al. 2008; Rajesh et al. 2015; Teulat et al. 2000).

The role of tissue culture techniques to improve the quality of elite cultivars from the coconut explants was inevitable. The different parts of the coconut such as embryo, ovary, pollen, inflorescence, endosperm, and plumule have been *in vitro* cultured for the elite cultivar production. However, none of the attempts was fruitful. Several studies were conducted to improve the somatic embryogenesis of plumule explant by adjusting the medium composition and manipulation of exotic factors like hormones, etc. However, the number of plantlets derived from each somatic embryo was very less and the regeneration potential of the somatic embryo into complete plantlets is also poor.

14.6 Prospects of Genome Editing in Coconut

Although genome engineering through a CRISPR-based system holds great promise for genetic improvements in coconut, the major limitation is its complex genome coupled with its recalcitrance to *in vitro* regeneration. To date, successful genome editing has not been reported in any palm species. In some plants with low transformation efficiency like maize, overexpression of somatic embryogenesis regulators like baby boom (Bbm) and Wuschel2 (Wus2) has overcome this limitation and improved the transformation efficiency (Lowe et al. 2016). Although this strategy has not been attempted in coconut, improving the regeneration process through this method is worth attempting. Another strategy is to identify key miRNA regulators in the process of somatic embryogenesis in coconut and fine-tune their levels in callus to improve the somatic embryo turn over and regeneration (Sabana et al. 2020). Recently, carbon nanoparticles were used as a delivery agent of siRNA into plants for efficient gene knock down (Demirer et al. 2020). Another interesting development is the *de novo* meristem induction in dicot plants to generate genome-edited plants without the need for tissue culture (Maher et al. 2020). Multiple combinations of developmental regulators and vectors of viral origin were used to edit plants and induce shoot development (Maher et al. 2020). *De novo* meristem induction and gene editing were observed when soil-grown plants expressing Cas9 was treated with developmental regulators and sgRNA. This approach is quick and eliminates the need for traditional tissue culture techniques; however, in coconut, development of Cas9 expressing coconut is itself challenging.

14.7 Ethical Concerns of CRISPR Cas9 Technology

Although the CRISPR Cas9 system provides better and precise gene modifications, there are some drawbacks inherent to the technique. One of the concerns is that the variations created are not uniform in each case with the same guide RNA. This can lead to a collection of mutant alleles with different genetic consequences. Another concern is the development of mosaicism or genetic chimeras in primary transgenic plants (Jang et al. 2016). Multiple strategies to reduce the production of chimeric mutants in poplars have been attempted with some success (Zhou et al. 2015; Elorriaga et al. 2018). A third major concern is the probability of off-target mutations, and several strategies have been introduced to reduce off-target mutations and the most effective method so far is to use ligand-dependent ribozymes called aptazymes. This strategy enables temporal and spatial precision in the production of guide RNAs reducing off-targeting effects (Tang et al. 2017; Hajiahmadi et al. 2019).

The regulation of genome-edited plants is the point of contention and debate between two schools of thought. The group that considers gene-edited plants as less risk and requires no regulation and those who consider it as high risk and akin to GMOs and require stringent regulation (Sharma et al. 2019; Waltz 2016). In Europe, all organisms produced through genome editing are subject to the legal framework applicable to GMOs. In the USA, all bio-engineered foods must be disclosed by food manufacturers and importers (Sharma et al. 2019; Waltz 2016).

14.8 Conclusion

Conventional breeding strategies are the main mode for crop improvement in coconut. Conventional breeding via selective hybridization, progeny evaluation, selection, and mass propagation is time-consuming due to the perennial biology of the palm, requires large space and human resources, and is limited by available germplasm. Even with all these efforts, the outcome of the breeding program is unpredictable. Therefore, targeted interventions or precision breeding through genetic engineering or gene editing can bring about a quantum leap in the breeding programs of coconut. The challenge is to develop methods to improve somatic embryo turn over in coconut or develop *in planta* transformation method that enables the development of genome-edited plants.

References

Abdallah NA, Prakash CS, McHughen AG (2015) Genome editing for crop improvement: challenges and opportunities. GM Crops Food 6(4):183–205

Azeez A, Busov V (2020) CRISPR/Cas9 mediated single and biallelic knockout of poplar STERILE APETALA (PopSAP) leads to complete reproductive sterility. Plant Biotechnol J 19(1):23–25. https://onlinelibrary.wiley.com/doi/abs/10.1111/pbi.13451

Bewg WP, Ci D, Tsai CJ (2018) Genome editing in trees: from multiple repair pathways to long-term stability. Frontiers Plant Sci 9 23. https://www.frontiersin.org/article/10.3389/fpls.2018.01732/full. (October 4, 2020)

Bhattacharyya S, Soniat MM, Walker D, Jang S, Finkelstein IJ, Harshey RM (2018 Dec 11) Phage Mu Gam protein promotes NHEJ in concert with *Escherichia coli* ligase. Proc Natl Acad Sci USA 115(50):E11614–E11622. https://doi.org/10.1073/pnas.1816606115

Boch J, Scholze H, Schornack S, Landgraf A, Hahn S, Kay S, Lahaye T, Nickstadt A, Bonas U (2009) Breaking the code of DNA binding specificity of TAL-Type III effectors. Science 326(5959):1509–1512

Boch J, Bonas U, Lahaye T (2014) TAL effectors–pathogen strategies and plant resistance engineering. New Phytol 204(4):823–832

Cermak T, Starker CG, Voytas DF (2015) Efficient design and assembly of custom TALENs using the Golden Gate platform. Methods Mol Biol 1239:133–159. https://doi.org/10.1007/978-1-4939-1862-1_7. PMID: 25408404.

Chen K, Gao C (2014) Targeted genome modification technologies and their applications in crop improvements. Plant Cell Rep 33(4):575–583

Christian M, Qi Y, Zhang Y, Voytas DF (2013) Targeted mutagenesis of *Arabidopsis thaliana* using engineered TAL effector nucleases. G3: Genes Genomes Genet 3 (10):1697–1705

Cong L, Ran FA, Cox D, Lin S, Barretto R, Habib N, Hsu PD, Wu X, Jiang W, Marraffini LA, Zhang F (2013) Multiplex genome engineering using CRISPR/Cas systems. Sci 339(6121):819–823

Dasanayaka PN, Nandadasa HG, Everard JM, Karunanayaka EH (2009) Analysis of coconut (*Cocos nucifera* l.) diversity using microsatellite markers with emphasis on management and utilisation of genetic resources. J Nat Sci Found Sri Lanka 37(2):99–109

Day J (2014) New approaches to manipulating the epigenome. Dialogues Clin Neurosci 16(3):345–357

Demirer GS, Zhang H, Goh NS, Pinals RL, Chang R, Landry MP (2020) Carbon nanocarriers deliver siRNA to intact plant cells for efficient gene knockdown. Sci Adv 6(26):eaaz0495. http://advances.sciencemag.org/

Ding L, Chen Y, Ma Y, Wang H, Wei J (2020) Effective reduction in chimeric mutants of poplar trees produced by CRISPR/Cas9 through a second round of shoot regeneration. Plant Biotechnol Rep 14(5):549–558. https://link.springer.com/article/10.1007/s11816-020-00629-2

Dreissig S, Schiml S, Schindele P, Weiss O, Rutten T, Schubert V, Gladilin E, Mette MF, Puchta H, Houben A (2017) Live-cell CRISPR imaging in plants reveals dynamic telomere movements. Plant J 91 (4):565–573

East-Seletsky A, O'Connell MR, Burstein D, Knott GJ, Doudna JA (2017) RNA targeting by functionally orthogonal type VI-A CRISPR-Cas enzymes. Mol Cell 66(3):373-383.e3. https://doi.org/10.1016/j.molcel.2017.04.008

Ecol M (2017) HHS Public Access 25(5):1032–1057

Elorriaga E, Klocko AL, Ma C, Strauss SH (2018) Variation in mutation spectra among CRISPR/Cas9 mutagenized poplars. Frontiers Plant Sci 9:594

Endo A, Masafumi M, Kaya H, Toki S (2016) Efficient targeted mutagenesis of rice and tobacco genomes using Cpf1 from *Francisella novicida*. Sci Rep 6 (1):1–9. https://doi.org/10.1038/srep38169

Fan D, Liu T, Li C, Jiao B, Li S, Hou Y, Luo K (2015) Efficient CRISPR/Cas9-mediated targeted mutagenesis in Populus in the first Generation. Sci Rep 5(1):1–7. /pmc/articles/PMC4507398/report abstract (October 4, 2020)

Gaj T, Gersbach CA, Barbas CF III (2013) ZFN, TALEN, and CRISPR/Cas-based methods for genome Engineering. Trends Biotechnol 31(7):397–405. https://doi.org/10.1016/j.tibtech.2013.04.004

Gao J, Wang G, Ma S, Xie X, Wu X, Zhang X, Wu Y, Zhao P, Xia Q (2015) CRISPR/Cas9-mediated targeted mutagenesis in *Nicotiana tabacum*. Plant Mol Biol 87(1–2):99–110

Gaudelli NM, Komor AC, Rees HA, Packer MS, Badran AH, Bryson DI, Liu DR (2017) Programmable base editing of T to G C in genomic DNA without DNA cleavage. Nature 551(7681):464–471

Gopalappa R, Suresh B, Ramakrishna S, Kim H (2018) Paired D10A Cas9 Nickases are sometimes more efficient than individual nucleases for gene disruption. Nucleic Acids Res 46(12):e71

Gudeta D (2019) Genome editing: tools and application in plants. Open Access J Microbiol Biotechnol 4 (1):000135

Gupta RM, Musunuru K (2014) Expanding the genetic editing tool kit: ZFNs, TALENs, and CRISPR-Cas9. J Clin Investigation 124(10):4154–4161

Hajiahmadi Z, Movahedi A, Wei H, Li D, Orooji Y, Ruan H, Zhuge Q (2019) Strategies to increase on-target and reduce off-target effects of the CRISPR/Cas9 system in plants. Int J Mol Sci 20(15):3719

Harrington LB, Burstein D, Chen JS, Paez-Espino D, Ma E, Witte IP, Cofsky JC, Kyrpides NC, Banfield JF, Doudna JA (2018) Programmed DNA destruction by miniature CRISPR-Cas14 Enzymes. Sci 362(6416):839–842. https://pubmed.ncbi.nlm.nih.gov/30337455/

Hess GT, Frésard L, Han K, Lee CH, Li A, Cimprich KA, Montgomery SB, Bassik MC (2016) Directed evolution using dCas9-targeted somatic hypermutation in mammalian Cells. Nature Methods 13(12):1036–1042. http://www.ncbi.nlm.nih.gov/pubmed/27798611

Hua K, Tao X, Zhu JK (2019) Expanding the base editing scope in rice by using Cas9 variants. Plant Biotechnol J 17(2):499–504

Jaganathan D, Ramasamy K, Sellamuthu G, Jayabalan S, Venkataraman G (2018) CRISPR for crop improvement: an update review. Frontiers Plant Sci 9:985

Jang G, Lee S, Um TY, Chang SH, Lee HY, Chung PJ, Kim JK, Do Choi Y (2016) Genetic chimerism of CRISPR/Cas9-mediated rice mutants. Plant Biotechnol Rep 10(6):425–435

Jensen MK (2018) Design principles for nuclease-deficient CRISPR-based transcriptional regulators. FEMS Yeast Res 18(4):foy039

Jia H, Zhang Y, Orbović V, Xu J, White FF, Jones JB, Wang N (2017) Genome editing of the disease susceptibility gene *CsLOB*1 in citrus confers resistance to citrus canker. Plant Biotechnol J 15(7):817–823. https://onlinelibrary.wiley.com/doi/full/10.1111/pbi.12677

Kaya H, Mikami M, Endo A, Endo M, Toki S (2016) Highly specific targeted mutagenesis in plants using *Staphylococcus aureus* Cas9. Sci Rep 6(1):1–9

Kim YG, Cha J, Chandrasegaran S (1996) Hybrid restriction enzymes: zinc finger fusions to *Fok* I cleavage domain. Proc Natl Aca Sci USA 93(3):1156–1160

Kleinstiver BP, Prew MS, Tsai SQ, Topkar VV, Nguyen NT, Zheng Z, Gonzales AP, Li Z, Peterson RT, Yeh JR, Aryee MJ (2015) Engineered CRISPR-Cas9 nucleases with altered PAM specificities. Nature 523(7561):481–485. https://www.nature.com/articles/nature14592

Komor AC, Kim YB, Packer MS, Zuris JA, Liu DR (2016) Programmable editing of a target base in genomic DNA without double-stranded DNA cleavage. Nature 533(7603):420–424. https://doi.org/10.1038/nature17946

Komor AC, Zhao KT, Packer MS, Gaudelli NM, Waterbury AL, Koblan LW, Kim YB, Badran AH, Liu DR (2017) Improved base excision repair inhibition and bacteriophage Mu Gam protein yields C: G-to-T: A base editors with higher efficiency and product purity. Sci Adv 3(8):1–9

Labun K, Montague TG, Krause M, Torres Cleuren YN, Tjeldnes H, Valen E (2019) CHOPCHOP v3: expanding the CRISPR web toolbox beyond genome editing. Nucleic Acids Res 47(W1): W171–W174. https://academic.oup.com/nar/article/47/W1/W171/5491735

Lebrun P, N'cho YP, Seguin M, Grivet L, Baudouin L (1998) Genetic diversity in Coconut (*Cocos nucifera* L.) revealed by restriction fragment length polymorphism (RFLP) markers. Euphytica 101(1):103–108

Lei Y, Lu L, Liu HY, Li S, Xing F, Chen LL (2014) CRISPR-P: a web tool for synthetic single-guide RNA design of CRISPR-system in plants. Mol Plant 7 (9):1494–1496

Li C, Zong Y, Wang Y, Jin S, Zhang D, Song Q, Zhang R, Gao C (2018) Expanded base editing in rice and wheat using a Cas9-adenosine deaminase fusion. Genome Biol 19(1):1–9

Li J, Sun Y, Du J, Zhao Y, Xia L (2017) Generation of targeted point mutations in rice by a modified CRISPR/Cas9 system. Mol Plant 10(3):526–529

Li T, Liu B, Spalding MH, Weeks DP, Yang B (2012) High-efficiency TALEN-based gene editing produces disease-resistant rice. Nature Biotechnol 30(5):390–392

Lor VS, Starker CG, Voytas DF, Weiss D, Olszewski NE (2014) Targeted mutagenesis of the tomato PROCERA Gene using TALENs. Plant Physiol 166:1288–1291. http://www.ncbi.nlm.nih.gov/pubmed/25217528

Lowder LG, Zhang D, Baltes NJ, Paul JW, Tang X, Zheng X, Voytas DF, Hsieh TF, Zhang Y, Qi Y (2015) A CRISPR/Cas9 toolbox for multiplexed plant genome editing and transcriptional regulation. Plant Physiol 169(2):971–985

Lowe K, Wu E, Wang N, Hoerster G, Hastings C, Cho MJ, Scelonge C, Lenderts B, Chamberlin M, Cushatt J, Wang L (2016) Morphogenic regulators Baby boom and Wuschel improve monocot transformation. Plant Cell 28(9):1998–2015. www.plantcell.org/cgi/doi/10.1105/tpc.16.00124

Lu H, Klocko AL, Dow M, Ma C, Amarasinghe V, Strauss SH (2016) Low frequency of zinc-Finger nuclease-induced mutagenesis in Populus. Mol Breed 36(9):1–13. https://link.springer.com/article/10.1007/s11032-016-0546-z

Ma Y, Zhang J, Yin W, Zhang Z, Song Y, Chang X (2016) Targeted AID-mediated mutagenesis (TAM) enables efficient genomic diversification in mammalian cells. Nat Methods 13(12):1029–1035

Maher MF, Nasti RA, Vollbrecht M, Starker CG, Clark MD, Voytas DF (2020) Plant gene editing through denovo induction of meristems. Nature Biotechnol 38(1):84–89. https://pubmed.ncbi.nlm.nih.gov/31844292/

Mahfouz MM (2017) Genome editing: the efficient tool CRISPR-Cpf1. Nature Plants 3:1–2. https://doi.org/10.1038/nplants.2017.28

Mak AN, Philip B, Adam JB, Barry LS (2014) NIH Public Access 23(1):93–99

Malzahn A, Levi L, Yiping Q (2017) Plant genome editing with TALEN and CRISPR. Cell Biosci 7(1):1–18

Mikami M, Seiichi T, Masaki E (2016) Precision targeted mutagenesis via Cas9 paired nickases in rice. Plant Cell Physiol 57(5):1058–1068

Minkenberg B, Jianwei Z, Kabin X, Yinong Y (2018) CRISPR-PLANT v2: an online resource for highly specific guide RNA spacers based on improved off-target analysis. Plant Biotechnol J 17(1):5–8. http://doi.wiley.com/10.1111/pbi.13025

Mishra R, Raj KJ, Kaijun Z (2019) Base editing in crops: current advances, limitations, and future implications. Plant Biotechnol J: 1–12

Montiel-Gonzalez MF, Quiroz JF, Rosenthal JJ (2019) Current strategies for site-directed RNA editing using ADARs. Methods (San Diego, Calif.) 156: 16–24. http://www.ncbi.nlm.nih.gov/pubmed/30502398

Moradpour M, Siti NAA (2019) CRISPR/dCas9 platforms in plants: strategies and applications beyond genome editing. Plant Biotechnol J 18(1):32–44

Murovec J, Žan P, Bing Y (2017) New variants of CRISPR RNA-guided genome editing enzymes. Plant Biotechnol J 15(8):917–926

Nemudryi AA, Valetdinova KR, Medvedev SP, Zakian SM (2014) TALEN and CRISPR/Cas genome editing systems: tools of discovery. Acta Naturae 6 (22):19–40

Nishida K, Arazoe T, Yachie N, Banno S, Kakimoto M, Tabata M, Mochizuki M, Miyabe A, Araki M, Hara KY, Shimatani Z (2016) Targeted nucleotide editing using hybrid prokaryotic and vertebrate adaptive immune systems. Science 16:353(6305)

Peng A, Chen S, Lei T, Xu L, He Y, Wu L, Yao L, Zou X (2017) Engineering canker-resistant plants through CRISPR/Cas9-targeted editing of the susceptibility gene *CsLOB*1 promoter in Citrus. Plant Biotechnol J 15(12):1509–1519. https://onlinelibrary.wiley.com/doi/full/10.1111/pbi.12733

Perera PI, Perera L, Hocher V, Verdeil JL, Yakandawala DM, Weerakoon LK (2008) Use of SSR markers to determine the anther-derived homozygous lines in coconut. Plant Cell Rep 27(11):1697–1703. http://www.ncbi.nlm.nih.gov/pubmed/18712524

Piatek A, Ali Z, Baazim H, Li L, Abulfaraj A, Al-Shareef S, Aouida M, Mahfouz MM (2015) RNA-guided transcriptional regulation in planta via synthetic dCas9-based transcription factors. Plant Biotechnol J 13(4):578–589

Qin L, Li J, Wang Q, Xu Z, Sun L, Alariqi M, Manghwar H, Wang G, Li B, Ding X, Rui H (2020) High-efficient and precise base editing of C•G to T•A in the allotetraploid cotton (*Gossypium hirsutum*) genome using a modified CRISPR/Cas9 system. Plant Biotechnol J 18(1):45–56. https://onlinelibrary.wiley.com/doi/abs/10.1111/pbi.13168

Rajesh MK, Jerard BA, Preethi P, Thomas RJ, Karun A (2014) Application of RAPD markers in hybrid verification in coconut. Crop Breed Appl Biotechnol 14(1):36–41

Rajesh MK, Sabana AA, Rachana KE, Rahman S, Jerard BA, Karun A (2015) Genetic relationship and diversity among coconut (*Cocos nucifera* L.) accessions revealed through SCoT analysis. 3 Biotech 5 (6):999–1006

Ren B, Liu L, Li S, Kuang Y, Wang J, Zhang D, Zhou X, Lin H, Zhou H (2019) Cas9-NG greatly expands the targeting scope of the genome-editing toolkit by recognizing NG and other atypical PAMs in rice. Mol Plant 12(7):1015–1026

Sabana AA, Rajesh MK, Antony G (2020) Dynamic changes in the expression pattern of miRNAs and associated target genes during coconut somatic embryogenesis. Planta 251(4):1–18. https://link.springer.com/article/10.1007/s00425-020-03368-4

Shan Q, Wang Y, Chen K, Liang Z, Li J, Zhang Y, Zhang K, Liu J, Voytas DF, Zheng X, Zhang Y (2013) Rapid and efficient gene modification in rice and brachypodium using TALENs. Molecular Plant 6 (4):1365–1368. http://www.ncbi.nlm.nih.gov/pubmed/23288864

Sharma S, Seonghee L, Heqiang H (2019) Current status of research, regulations, and future challenges for CRISPR gene-editing in crop improvement. Edis 4:4

Shimatani Z, Ariizumi T, Fujikura U, Kondo A, Ezura H, Nishida K (2019) Targeted base editing with CRISPR-deaminase in tomato. In: Methods in molecular biology, Humana Press Inc, pp 297–307. https://pubmed.ncbi.nlm.nih.gov/30610645/, (October 4, 2020)

Steinert J, Simon S, Friedrich F, Holger P (2015) Highly efficient heritable plant genome engineering using Cas9 orthologues from *Streptococcus thermophilus* and *Staphylococcus aureus*". Plant J 84(6):1295–1305

Tang W, Johnny HH, David RL (2017) Aptazyme-embedded guide RNAs enable ligand-responsive genome editing and transcriptional activation. Nature Commun 8:1–8. https://doi.org/10.1038/ncomms15939

Teulat B, Aldam C, Trehin R, Lebrun P, Barker JH, Arnold GM, Karp A, Baudouin L, Rognon F (2000) An analysis of genetic diversity in coconut (*Cocos nucifera*) populations from across the geographic range using sequence-tagged microsatellites (SSRs) and AFLPs. Theor Appl Genet 100(5):764–771

Tiwari SB, Belachew A, Ma SF, Young M, Ade J, Shen Y, Marion CM, Holtan HE, Bailey A, Stone JK, Edwards L (2012) The EDLL Motif: a potent plant transcriptional activation domain from AP2/ERF transcription factors. Plant J 70(5):855–865

Townson J (2017) Recent developments in genome editing for potential use in plants. Biosci Horizons: Int J Student Res 10

Trevino AE, Feng Z (2014) Genome editing using Cas9 nickases. In: Methods in enzymology. Academic Press Inc, pp 161–174

Varkonyi-Gasic E, Wang T, Voogd C, Jeon S, Drummond RS, Gleave AP, Allan AC (2019) Mutagenesis of kiwifruit CENTRORADIALIS-like genes transforms a climbing woody perennial with long juvenility and axillary flowering into a compact plant with rapid terminal flowering. Plant Biotechnol J 17(5):869–880. https://onlinelibrary.wiley.com/doi/abs/10.1111/pbi.13021

Veillet F, Kermarrec MP, Chauvin L, Chauvin JE, Nogué F (2020) CRISPR-induced indels and base editing using the *Staphylococcus aureus* Cas9 in potato. PLOS ONE 15(8):e0235942. https://doi.org/10.1371/journal.pone.0235942

Waltz E (2016) CRISPR-edited crops free to enter market, skip regulation. Nature Biotechnol 34(6):582–582. https://doi.org/10.1038/nbt0616-582

Wang J, Huaitong W, Yingnan C, Tongming Y (2020) Efficient CRISPR/Cas9-mediated gene editing in an interspecific hybrid poplar with a highly heterozygous genome. Frontiers Plant Sci 11: 996. https://www.frontiersin.org/article/10.3389/fpls.2020.00996/full

Wang J, Meng X, Hu X, Sun T, Li J, Wang K, Yu H (2019) XCas9 expands the scope of genome editing with reduced efficiency in rice. Plant Biotechnol J 17 (4):709–711. https://onlinelibrary.wiley.com/doi/abs/10.1111/pbi.13053

Wang X, Tu M, Wang D, Liu J, Li Y, Li Z, Wang Y, Wang X (2018) CRISPR/Cas9-mediated efficient targeted mutagenesis in grape in the first generation. Plant Biotechnol J 16(4):844–855. http://doi.wiley.com/10.1111/pbi.12832

Wang Y, Cheng X, Shan Q, Zhang Y, Liu J, Gao C, Qiu JL (2014) Simultaneous editing of three homoalleles in hexaploid bread wheat confers heritable resistance to powdery mildew. Nature Biotechnol 32(9):947–951. https://www.nature.com/articles/nbt.2969

Wang Z, Wang S, Li D, Zhang Q, Li L, Zhong C, Liu Y, Huang H (2018) Optimized paired-SgRNA/Cas9 cloning and expression cassette triggers high-efficiency multiplex genome editing in kiwifruit. Plant Biotechnol J 16(8):1424–1433. http://doi.wiley.com/10.1111/pbi.12884

Wright DA, Ting L, Bing Y, Martin HS (2014) TALEN-mediated genome editing: prospects and perspectives. Biochemical J 462(1):15–24

Xiao Y, Xu P, Fan H, Baudouin L, Xia W, Bocs S, Xu J, Li Q, Guo A, Zhou L, Li J (2017) The genome draft of Coconut (*Cocos nucifera*). Giga Sci 6(11):1–11. https://academic.oup.com/gigascience/article/6/11/gix095/4345653

Xiong JS, Jing D, Yi L (2015) Genome-editing technologies and their potential application in horticultural crop breeding. Hortic Res 2:1–10

Xu Y, Lin Q, Li X, Wang F, Chen Z, Wang J, Li W, Fan F, Tao Y, Jiang Y, Wei X (2021) Fine-tuning the amylose content of rice by precise base editing of the Wx gene. Plant Biotechnol J 19(1):11–13. https://onlinelibrary.wiley.com/doi/10.1111/pbi.13433

Yang B, Frank FW (2004) Diverse members of the AvrBs3/PthA family of Type III effectors are major virulence determinants in bacterial blight disease of rice. 17(11):1192–1200

Zhan X, Zhang F, Zhong Z, Chen R, Wang Y, Chang L, Bock R, Nie B, Zhang J (2019) Generation of virus-resistant potato plants by RNA genome targeting. Plant Biotechnol J 17(9):1814–1822. http://doi.wiley.com/10.1111/pbi.13102

Zhang H, Zhang J, Lang Z, Botella JR, Zhu JK (2017) Genome editing—principles and applications for functional genomics research and crop Improvement. Critical Rev Plant Sci 36(4):291–309. https://doi.org/10.1080/07352689.2017.1402989

Zhang T, Zhao Y, Ye J, Cao X, Xu C, Chen B, An H, Jiao Y, Zhang F, Yang X, Zhou G (2019) Establishing CRISPR/Cas13a immune system conferring RNA virus resistance in both dicot and monocot plants. Plant Biotechnol J 17(7):1185–1187. https://onlinelibrary.wiley.com/doi/abs/10.1111/pbi.13095

Zhang Y, Zhang F, Li X, Baller JA, Qi Y, Starker CG, Bogdanove AJ, Voytas DF (2013) Transcription activator-like effector nucleases enable efficient plant genome engineering. Plant Physiol 161(1):20–27

Zhong Z, Zhang Y, You Q, Tang X, Ren Q, Liu S, Yang L, Wang Y, Liu X, Liu B, Zhang T (2018) Plant genome editing using *FnCpf*1 and *LbCpf*1 nucleases at redefined and altered PAM sites. Mol Plant 11 (7):999–1002

Zhou X, Jacobs TB, Xue LJ, Harding SA, Tsai CJ (2015) Exploiting SNPs for biallelic CRISPR mutations in the outcrossing woody perennial *Populus* reveals 4-Coumarate: CoA ligase specificity and redundancy. New Phytol 208(2):298–301

Coconut: The Tree of Life-Endless Possibilities

15

M. K. Rajesh, S. V. Ramesh, and Chittaranjan Kole

Abstract

In 2017, coconut joined the elite list of sequenced plant genomes. The genome sequence is of immense value to coconut researchers working in various fields, ranging from plant breeding, biotechnology to physiology and biochemistry. The recent accomplishments of genome sequences of two green dwarfs, via independent sequencing efforts in the Philippines and India, will provide an improved coconut genome assembly and annotation. The dense map produced recently through a genotyping-by-sequencing (GBS) approach has aided the consolidation of the genome assembly of 'Hainan Tall' accession, the first to be sequenced. These genome sequencing efforts will immensely benefit plant breeders, who will now be able to utilize dense sets of molecular markers which are evenly distributed across the coconut genome and linked to traits of agronomic importance. More importantly, these will enable the development of genomic selection (GS) models and their implementation, the selection procedure based solely on the molecular profiling of genotypes, rather than the conventional phenotypic evaluation of traits, which is quite time-consuming and cumbersome in crop like coconut for obvious reasons. This concluding chapter provides a perspective view of both the processes and the outcomes of the genome sequencing endeavors in coconut as a foundation and roadmap for future discoveries which would serve coconut researchers for the years to come.

It is as important to take a comprehensive retrospective view on the achievements made along the years to enable putting forth the way forward in a proper perspective. With all parts finding a use, either directly for human consumption and as construction material, or indirectly as a raw material for numerous industries, the coconut palm has aptly earned the sobriquet the 'tree of life.' The amazing diversity existing in coconut populations has captured the attention of geneticists and plant breeders for more than half a century now. Genetic characterization studies have unearthed the rich diversity of coconut populations from around the globe. Unfortunately, stagnancy in the global production of coconuts, price instability and negative growth rate in productivity, witnessed in the last decade, are worrisome facts. Hitherto, coconut breeders

M. K. Rajesh (✉) · S. V. Ramesh
ICAR-Central Plantation Crops Research Institute, Kasaragod, Kerala, India
e-mail: rajesh.mk@icar.gov.in

S. V. Ramesh
e-mail: ramesh.sv@icar.gov.in

C. Kole
Raja Ramanna Fellow, Government of India, ICAR-National Institute for Plant Biotechnology, Pusa, New Delhi, India
e-mail: ckoleorg@gmail.com

M. K. Rajesh et al. (eds.), *The Coconut Genome*, Compendium of Plant Genomes,
https://doi.org/10.1007/978-3-030-76649-8_15

had only limited choice of strategies, adopting conventional breeding, for evaluation of traits of agronomic importance such as copra yield, resistance to phytoplasmal diseases and tolerance to water stress. An enhancement of production and productivity, via sustainable coconut cultivation, would definitely need employment of effective and novel tools. It is in this scenario that genomics-based tools come into rescue.

There has been an exponential increase in omics-related publications in coconut in PubMed (https://pubmed.ncbi.nlm.nih.gov/?term=coconut) since 2014. Undoubtedly, seminal work on host–pathogen interactions, embryogenesis, nut quality parameters and responses to water stress, undertaken previously, have provided the impetus toward the need for whole genome sequencing of coconut. Applications of genomic science in coconut are greatly hindered due to the complexities of heterozygosity owing to outbreeding behavior, history of evolution including domestication, perennial nature of the palm, etc. These inherent constraints are further compounded by the high percentage of repeat regions/transposons in the coconut genome. Presently, the availability of genome sequences of coconut accessions can be leveraged toward the development of useful molecular markers and dense linkage maps to anchor and orient the genome scaffolds. Initial insights into the evolution of genes involved in adaptation to salinity stress, tolerance to biotic and abiotic stresses and oil biosynthesis have been obtained. The rich repertoire of repeat regions and transposon elements has also been revealed.

The genome sequences of two other important members of Arecaceae, viz. date palm (*Phoenix dactylifera*) and oil palm (*Elaeis guineensis* and *E. oleifera*) have been uncovered in 2011 and 2013, respectively, before the publication of the coconut genome sequence in 2017. Comparative phylogenomic analysis has revealed an extensive homology among the palm genomes, in addition to two whole genome duplication (WGD) events in palms during the course of evolution. The reference genomes of coconut published will definitely constitute a solid base for future resequencing efforts, involving diverse coconut accessions, which could reveal the immense genetic variations existing in these accessions, and corroborate the results of numerous molecular marker analyses undertaken previously and also to identify and characterize novel molecular markers. Besides, it is critical to further improve the size of the pseudo-molecule physical map, constructed recently, to unravel the complete potential of genome-assisted biology for coconut. Hopefully, these initiatives will definitely provide the much needed fillip to a shift toward genomics-assisted breeding, culminating in the decrease of time required to develop new coconut varieties catering to preferences of all stakeholders. Concurrently, there is a need to develop mapping populations which could be utilized for high-resolution mapping and identification of genomic regions linked to key agronomic traits.

The availability of resequencing data will also shed light on important traits, and genetic loci associated with these traits, which would in turn reveal the complex genetic architecture underlying the foundation of the cultivated coconut genome, and more importantly, facilitate the development and implementation of genomic selection (GS) models in coconut. These studies will advance our comprehension of traits of interest in coconut like tolerance to biotic and abiotic stresses, climate resilience, oil biosynthesis and somatic embryogenesis.

The recent paradigm shift in genomics, from a single reference genome to a pan-genome, has been eliciting wide interest among researchers. The pan-genome concept has been proposed to symbolize the entire repertoire of genes possessed by a species. Pan-genomes are available for many model crop plants and have aided the identification of structural variations (SVs) which are associated with traits of agronomic importance. Characterization of the coconut pan-genome, with the availability of multiple, high-quality reference genomes, will definitely shed light on the enormous landscape of coconut diversity. Nevertheless, data acquired from pan-genomics studies entail integration with quantitative trait loci (QTL)/genome-wide association studies (GWAS), and also resequencing studies for the identification of useful alleles/genes would benefit the coconut breeding and research community and could be employed in effective coconut breeding programs.

The advent of large-scale omics-driven sequencing efforts has unleashed an unprecedented quantum of data warranting an user-friendly pipeline to store, analyze and share data in a swift and less laborious manner. Further, integration of enormous quantum of data from multiple omics platforms (genomics, transcriptomics, proteomics, metabolomics, ionomics, etc.) have become a major challenge for plantation biologists. Similar to plant-centric genome databases, namely Phytozome (https://phytozome.jgi.doe.gov) and Ensembl Plants (http://plants.ensembl.org), a dedicated resource for palm genomics is required to be established. To achieve an effortless visualization of genomic data from palm sequencing projects, taking cues from the resources such as The Genome Database for Rosaceae (GDR; https://www.rosaceae.org/), The Sol Genomics Network (SGN; https://solgenomics.net/), The Brassica Database (BARD; http://brassicadb.org/brad/), etc., palm genome database covering the genomic information of all the palms is mandatory.

To conclude, a collaboration, which brings together dedicated coconut researchers from around the globe, would be needed to unravel the enormous possibilities of multi-omics-enabled technologies in coconut.

The manufacturer's authorised representative in the EU is Springer
Nature Customer Service Centre GmbH, Europaplatz 3, 69115 Heidelberg,
Germany. If you have any concerns regarding our products, please
contact ProductSafety@springernature.com

Printed and bound by CPI Group (UK) Ltd, Croydon, CR0 4YY
30/09/2026
02225691-0015